NEUROACANTHOCYTOSIS SYNDROMES

NEUROACANTHOCYTOSIS SYNDROMES

Edited by

ADRIAN DANEK

Neurologische Klinik, Ludwig-Maximilians Universität München, Germany

Publication Sponsors:

Jung-Stiftung, Hamburg, Germany

Novartis AG, Nürnberg, Germany

The Advocacy for Neurocanthocytosis Patients, London, U.K.

John Grooms – Working with Disabled People, London, U.K.

John Grooms
Working with Disabled People

Springer

A C.I.P. Catalogue record for this book is available from the Library of Congress.

ISBN 1-4020-2897-0 (HB)
ISBN 1-4020-2898-9 (e-Book)

Published by Springer,
P.O. Box 17, 3300 AA Dordrecht, The Netherlands.

Sold and distributed in North, Central and South America
by Springer,
101 Philip Drive, Norwell, MA 02061, U.S.A.

In all other countries, sold and distributed
by Springer,
P.O. Box 322, 3300 AH Dordrecht, The Netherlands.

Printed on acid-free paper

springeronline.com

Printed in the Netherlands.

Corresponding Authors

Alice Andreu, MSc
Imperial College Genetic Therapies Centre
Department of Chemistry, Flowers Building, Armstrong Rd
Imperial College of Science, Technology and Medicine
GB London SW7 2AZ, United Kingdom
e-mail: alice.andreu@ic.ac.uk
telephone: +44-0207-594-3156

G.J.C.G.M. Bosman, PhD
Department of Biochemistry
Nijmegen Center for Molecular Life Sciences UMC-Nijmegen (160)
P.O. Box 9101
NL-6500 HB Nijmegen, The Netherlands
e-mail: G.Bosman@ncmls.kun.nl
telephone: +31-24-3615390
fax: +31-24-3616413

Adrian Danek, Prof. Dr. med.
Neurologische Klinik und Poliklinik
Ludwig-Maximilians-Universität
Marchioninistr. 15
D-81366 München, Germany
e-mail: danek@lmu.de
telephone: +49-89-7095-4821
fax: +49-89-7095-4801

Geoff Daniels, PhD, FRCPath
Bristol Institute for Transfusion Sciences
Southmead Road
GB Bristol BS10 5ND, United Kingdom
e-mail: geoff.daniels@nbs.nhs.uk
telephone: +44-117-991-2116
fax: +44-117-959-1660

Lucia de Franceschi, MD
Department of Clinical and Experimental Medicine
Section of Internal Medicine
University of Verona, Policlinico GB Rossi
Piazzale L. Scuro, 10
I-37134 Verona, Italy
e-mail: lucia.defranceschi@univr.it
telephone: +39-045-8074918
fax: +39-045-580111

Justo García de Yébenes, Prof. Dr. med.
Servicio de Neurología, Fundación Jiménez Díaz
Avda. de Reyes Católicos, 2
SP-Madrid 28040, Spain
e-mail: jgyebenes@fjd.es
telephone: +34-91-5449008
fax: +34-91-5449008

Matthias Dose, Prof. Dr. med.
Bezirkskrankenhaus Taufkirchen
Bräuhausstr. 5
D-84416 Taufkirchen, Germany
e-mail: m.dose@bkh-taufkirchen.de
telephone: +49-08084-934-212
fax: +49-08084-934-400

Maria Teresa Dotti, Prof. Dr. med.
Institute of Neurological Sciences, University of Siena
Policlinico Le Scotte
Viale Bracci
I-53100 Siena, Italy
e-mail: dotti@unisi.it
telephone: +39-0577-585763
fax: +39-0577-40327

Hans H. Jung, Priv.-Doz. Dr. med.
Department of Neurology
University Hospital Zürich
Frauenklinikstr. 26
CH-8091 Zürich, Switzerland
e-mail: hans.jung@nos.usz.ch
telephone: +41-1-255-55-45
fax: +41-1-255-45-07

Richard J. Hardie, MD, FRCP
Department of Neurology, Atkinson Morley's Hospital
Copse Hill
GB London SW20 0NE, United Kingdom
e-mail: rhardie@sghms.ac.uk
telephone: +44-20-8725-4765
fax: +44-20-8944-9927

Robert A. Hegele, MD, FRCP, FACP
Blackburn Cardiovascular Genetics Laboratory, Robarts Research Institute
100 Perth Drive
London, Ontario, Canada N6A 5K8
e-mail: hegele@robarts.ca
telephone: +1-519-663-3461
fax: +1-519-663-3037

Michael Hengartner, PhD
Institute of Molecular Biology
Winterthurerstr. 190
CH-8057 Zürich, Switzerland
e-mail: michael.hengartner@molbio.unizh.ch
telephone: +41-1-635-3140
fax: +41-1-635-6861

Thomas Klopstock, Priv.-Doz. Dr. med.
Neurologische Klinik und Poliklinik
Ludwig-Maximilians-Universität
Marchioninistr. 15
D-81366 München, Germany
e-mail: klopstock@nefo.med.uni-muenchen.de
telephone: +49-89-7095-0
fax: +49-89-7095-3677

Jan Kobal, MD, MSc
Division of Neurology, University Medical Centre
Zaloška 7
SL-1000 Ljubljana, Slovenia
e-mail: jan.kobal@kclj.si
telephone: +386-1-522-23-11
fax: +386-1-522-22-08

Christoph M. Kosinski, Priv.-Doz. Dr. med.
Neurologische Klinik Universitätsklinikum Aachen
Pauwelsstr. 30
D-52057 Aachen, Germany
e-mail: Christoph.Kosinski@post.rwth-aachen.de
telephone: +49-241-8089827
fax: +49-241-8082444

Hartmut Meierkord, Priv.-Doz. Dr. med.
Universitätsklinikum Charité
Humboldt-Universität zu Berlin
Neurologische Klinik und Poliklinik
Schumannstrae 20/21
D-10117 Berlin, Germany
e-mail: hartmut.meierkord@charite.de
telephone: +49-30-560-105

Mariarosa A.B. Melone, Prof. Dr. med.
Department of Neurological Sciences
Second University of Naples, School of Medicine
Edificio 10 Policlinico "Federico II"
Via Sergio Pansini, 5
I-80131 Naples, Italy
e-mail: marina.melone@unina2.it
telephone: +39-081-566-6810
fax: +39-081-566-6805

Saidi A. Mohiddin MD, MRCP
Cardiovascular Branch, NHLBI
National Institutes of Health
Building 10, Room 7B15
10 Center Drive
Bethesda, MD, USA. 20892
e-mail: mohiddis@nhlbi.nih.gov
telephone: +1-301-496-5678
fax: +1-301-402-0888

Anthony P. Monaco, MD, PhD
The Wellcome Trust Centre for Human Genetics
Roosevelt Drive, Headington
GB Oxford OX3 7BN United Kingdom
e-mail: anthony.monaco@well.ox.ac.uk
telephone: +44-1865-287502
fax: +44-1865-287650

Colvin M. Redman, PhD
Lindsley F Kimball Research Institute of the New York Blood Center
310 East 67 Street
New York, NY 10021, USA
e-mail: credman@nybc.org
telephone: +1-212-570-3059
fax: +1-212-879-0243

Ralf Reilmann, Dr. med.
Klinik und Poliklinik für Neurologie
Universitätsklinikum Münster
Albert-Schweitzer-Str. 33
D-48129 Münster, Germany
e-mail: R.Reilmann@uni-muenster.de
telephone: +49-251-83-45303
fax: +49-251-83-47716

Akira Sano, MD, PhD
Department of Psychiatry
Kagoshima University Graduate School of Medical and Dental Sciences
8-35-1 Sakuragaoka
Kagoshima 890-8520, Japan
e-mail: sano@m3.kufm.kagoshima-u.ac.jp
telephone: +81-99-275-5346
fax: +81-89-265-7089

Benedikt G.H. Schoser, Priv.-Doz. Dr. med.
Friedrich-Baur-Institut, Neurologische Klinik
Ludwig-Maximilians-Universität
Ziemssenstr. 1a
D-80801 München, Germany
e-mail: bschoser@fbs.med.uni-muenchen.de
telephone: +49-89-5160-7443
fax: +49-89-5160-7402

Alexander Storch, Prof. Dr. med.
Technical University of Dresden
Department of Neurology
Fetscherstrasse 74
D-01307 Dresden, Germany
e-mail: alexander.storch@neuro.tu-dresden.de
telephone: +49-351-458-2532
fax: +49-351-458-4352

François Tison, MD, PhD
Service de Neurologie, Hôpital Haut Lévèque
CHU de Bordeaux
Avenue de Magellan
F-33600 Pessac, France
e-mail: Francois.tison@chu-bordeaux.fr
telephone: +33-5-56-55-64-20
fax: +33-5-56-55-68-15

Maarten van den Buuse, PhD
Behavioural Neuroscience Laboratory, Mental Health Research Institute
155 Oak Street, Parkville
Melbourne, Victoria 3052, Australia
e-mail: m.vandenbuuse@papyrus.mhri.edu.au
telephone: +61-3-9388-1633
fax: +61-3-9387-5061

Jens Volkmann, Priv.-Doz. Dr. med.
Department of Neurology
Christian-Albrechts-Universität
D-24105 Kiel, Germany
e-mail: j.volkmann@neurologie.uni-kiel.de
telephone: +49-431-597-2631
fax: +49-431-597-2712

Ruth H. Walker, MB, ChB, PhD
Department of Neurology
Veterans Affairs Medical Center
130 W. Kingsbridge Road
Bronx, NY 10468, USA
e-mail: ruth.walker@mssm.edu
telephone: +1-718-584-9000 x5915
fax: +1-718-741-4708

PREFACE

Mark Hallett
Human Motor Control Section, National Institute of Neurological Disorders and Stroke, NIH, Bethesda, MD, USA

Why have a book on the neuroacanthocytosis syndromes? Most physicians likely think that they would never see a patient with any of these rare disorders. But this is, of course, the first reason for having and reading such a book. If a physician does not know about a disorder, then even if a patient with it does appear, the diagnosis will be missed. And, why should the book be called *Neuroacanthocytosis Syndromes*? Isn't there only one disorder of neuroacanthocytosis? As the reader of this book will find out, if they don't know already, there are a number of neurologic disorders with acanthocytes, and neuroacanthocytosis proper has at least two distinct forms with separate defined genetic basis (chorea-acanthocytosis and the McLeod syndrome).

Neuroacanthocytosis was first described in the United States. Estes, Morley, Levine and Emerson reported a family from New England in 1967, and more details were given in an article by Levine, Estes and Looney in 1968. Critchley, Clark and Wikler presented a family from Kentucky at the American Neurological Association in 1967 and published their full paper in 1968. Now the disorder has been found all over the world.

Neuroacanthocytosis (proper) clinically can be characterized by chorea particularly in the orofacial region, tics including vocalizations and lip and tongue biting, dysarthria, dysphagia, seizures, peripheral neuropathy, myopathy with raised plasma CK and behavioral and cognitive dysfunction. If the whole syndrome is present, it should be relatively easy to diagnose, but when only partial it can masquerade as many other conditions.

In this book, the neurology of the neuroacanthocytosis syndromes is expanded in detail with a good overview of the differential diagnosis. Readers of the book will be able to find out about the pathophysiology of the clinical manifestations, cell biology of acanthocytes, the genes that cause the disorder and their biology, possible animal models, relationship to Huntington's disease, and approaches to therapy including surgery and gene therapy.

Many thanks to Dr. A. Danek for putting this all together and the team of experts that he has gotten to contribute to this useful work.

CONTENTS

ORGAN INVOLVEMENT

TREATMENT

SUMMARY

APPENDIX

CHAPTER 1

NEUROACANTHOCYTOSIS SYNDROMES: WHAT LINKS RED BLOOD CELLS AND NEURONS?

ADRIAN DANEK
Neurologische Klinik und Poliklinik, Ludwig-Maximilians-Universität, München, Germany

Abstract. Neuroacanthocytosis is an umbrella term for neurological conditions that occur together with misshapen acanthocytic red cells. Although the name is used largely synonymously with Levine-Critchley syndrome, it is hard to decide what exactly the three Levine-Critchley families, which originated in the US and the UK, suffered from. None of them has yet been re-investigated with the modern methods that allow increasing distinctions within neuroacanthocytosis. How one distinct subtype of neuroacanthocytosis was recognized through an uncommon Kell blood type is vividly illustrated by letters of the eponymous propositus McLeod. The X-linked gene *XK* that underlies McLeod neuroacanthocytosis was discovered in 1994. In 2001 and 2003, the genes *CHAC* and *JPH3* involved in an autosomal-recessive and an autosomal-dominant type of neuroacanthocytosis were identified. Pantothenate kinase associated neurodegeneration can be seen as another neuroacanthocytosis subtype. How is a disorder that mainly affects nerve cells related to the bizarre shape of red blood cells? This question is all the more pressing now that such a variety of origins has been discovered for the one common outcome of neuroacanthocytosis, progressive basal ganglia degeneration.

SEEON 2002

The first ever scientific meeting dedicated to neuroacanthocytosis took place in May 2002 in an idyllic setting in southern Bavaria, the former monastery of Seeon. A small group of researchers with few connections in the past engaged in two days of lively discussion on a topic that usually attracts fleeting interest at best. As will become obvious from this volume, however, the study of neuroacanthocytosis can provide important clues for an understanding of basal ganglia degeneration and of neurodegeneration in general. Advances in

A. Danek (ed.), Neuroacanthocytosis Syndromes, 1–14.

Figure 1. Participants of the meeting in Kloster Seeon, Germany, May 2 to May 5, 2002 ("Neuroacanthocytosis Syndromes: New Perspectives for the Study of Basal Ganglia Degeneration") from left: A. Sano, M. Ho, V. Irvine, G. Irvine, L. Rampoldi, R. Hardie, M. T. Dotti, A.P. Monaco, M. Melone, B. Landwehrmeyer, H.H. Jung, A. Velayos-Baeza, C. Dobson-Stone, T.N. Witt, A. Andreu, A. Weindl, J. Kobal, R. Walker, J.G. de Yébenes, F. Anneser, A. Danek, M. Dose, G. Bosman, M. van den Buuse, A. Storch, F. Tison, M.O. Hengartner, B. Schoser, G. Daniels, B. Gathof, T. Klopstock. Not shown on this photograph are: J. Andrich, T. Brandt, A. Deutschländer, L. de Franceschi, T. Gasser, C.-M. Kosinski, E. Kraft, H. Meierkord, T. Meyer, R. Reilmann, J. Volkmann, U. Wahlländer-Danek.

knowledge are desperately needed to improve the lot of the patients affected by this group of "orphan" syndromes and the relentless progression of neurological impairment with which they are associated.

As organizer of the meeting I feel honored that the participants provided succinct summaries of their contributions, which are collected in this volume. Drawing on a wide variety of backgrounds, they give an overview of neuroacanthocytosis and serve as a reference for future clinical and research work. Additional chapters were generously contributed by experts who could not attend but who cover such important aspects as the connection with lipoprotein disorders and the potentially lethal heart involvement in neuroacanthocytosis syndromes.

Great thanks go to the Fritz Thyssen Stiftung (Cologne, Germany) that provided the main financial support for the meeting. Additional contributions were received from Novartis AG (Nürnberg, Germany), Sanofi-Synthelabo GmbH (Berlin, Germany), Pfizer AG (Karlsruhe, Germany), and the Imperial College Genetic Therapies Centre (London, UK). The "seed" contributions, which gave us the confidence to start came from donors in the United States and Great Britain: Carl H and Elizabeth S Pforzheimer, John and Ellen Buzbee, Francesca Roberts, and Susan and Kirt Mead. John Grooms-Working with Disabled People (London, UK) provided essential financial administration in addition to a financial contribution. Publication of this volume has become only possible with additional resources generously provided by the Jung-Stiftung (Hamburg, Germany), by Novartis AG (Nürnberg, Germany) and by friends and families of patients in Belgium, Canada, Germany, Switzerland, Britain and the US (see page 277). The interest of these institutions and individuals

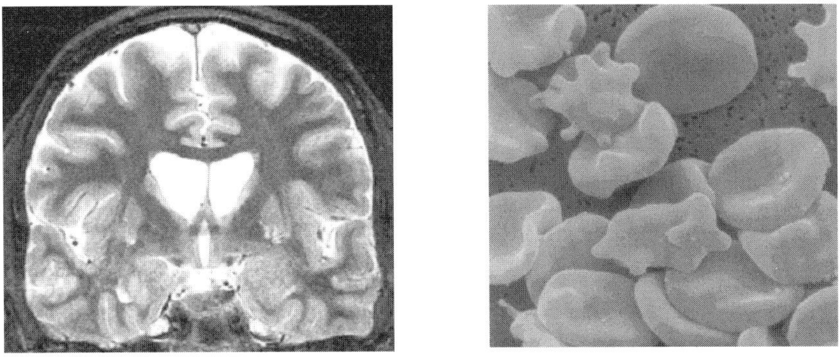

Figure 2. The hallmarks of the neuroacanthocytosis syndromes as evident in a 30 year old female with homozygous mutations in the *CHAC* gene: degeneration of the basal ganglia and acanthocytosis of red blood cells.
Left: The magnetic resonance image taken in the frontal plane six years after onset of motor symptoms shows pronounced atrophy of the heads of the caudate nuclei with subsequent widening of the lateral ventricles as well as incipient atrophy of the putamen, illustrating the mainly striatal pathology of the condition.
Right: Scanning electron microscopy discloses many misshapen erythrocytes in addition to the very few disk-shaped normal cells (discocytes). Acanthocytes are characterized by their thorn-like protrusions (akantha: Greek for thorn).
(Images courtesy of NIH, Bethesda Md, USA: N. Patronas and J. Butman, Clinical Center and V. Ferrans und W. Riemenschneider, National Heart, Lung, and Blood Institute.)

is invaluable, yet without the inspiration, the continuous financial and logistic support, but most of all without the untiring encouragement of Alexandra, Ginger and Glenn Irvine neither the Seeon meeting nor this book would have been realized.

MISSHAPEN BLOOD CELLS AND NEURODEGENERATION

Neuroacanthocytosis is an umbrella term for neurological conditions that occur together with misshapen acanthocytic red cells (Figure 2). In the past, the mode of inheritance of that coincidence had remained ambiguous, with recessive, dominant, and X-linked transmission patterns possible. The definite genetic proof of heterogeneity within neuroacanthocytosis provided the stimulus to convene the Seeon meeting.

The collaboration over patient MT whom we had followed for many years since age 40 [77] helped the group of Tony Monaco identify the X-chromosomal McLeod gene *XK* in 1994 [26]. The numerous DNA samples collected for genotype-phenotype correlations [12] provided the basis for the subsequent linkage of non-McLeod cases of neuroacanthocytosis to chromosome 9 [54] and the eventual cloning of the gene in 2001 [50]. Chorea-acanthocytosis (ChAc) is the name adopted for the autosomal-recessive *CHAC* gene disease (MIM 200150).

The main focus of the Seeon meeting was the characterization and distinction of McLeod syndrome and chorea-acanthocytosis. Both disorders are uncommon, with the prevalence of McLeod syndrome estimated at $0,5 - 1 : 100\,000$ [64]. About 100 patients with McLeod syndrome have been reported and a few additional cases will be known to blood banks. The number of chorea-acanthocytosis cases is estimated in the order of $300 - 500$.

THE STORY OF A MEDICAL TERM

The Syndrome of Levine and Critchley

In 1952, a few years after "acanthocytosis" had been coined to describe the erythrocyte malformation associated in a syndrome with fat malabsorption, ataxia and retinitis pigmentosa (see Chapter 3), Irving M. Levine, working in neurological research for the Boston Veteran's Administration, noticed a new "hereditary neurological disorder with acanthocytosis" which was distinct from Bassen-Kornzweig syndrome. Levine's conference presentations in 1960 and 1964 [38,39] were followed by more extensive reports, including a note on neuropsychiatric involvement [16,40,51], but the pattern of inheritance in the affected family from New England remained ill-defined.

The British neurologist, Edmund Critchley, in 1967 and 1970, respectively, recognized two additional families showing a neurological condition with acanthocytes yet normal lipoproteins, one originating from Kentucky [8,9], the other from the UK [10,11]. All of these families appear to be lost to follow-up.

Few additional cases had been reported in the English literature after Levine's and Critchley's seminal observations [2,3] when the subject was picked up in Japan as "amyotrophic chorea with acanthocytosis". In a short period around 1980 a remarkable number of cases was reported, including a comprehensive review for the health ministry [67]. Unfortunately, only some of these observations are readily accessible [35,56,59,69,78].

Previous Observations?

Reviewing the older literature, Mitchell Brin has proposed that a patient of Cécile and Oskar Vogt's might have suffered from the same condition [6: page 280]. This girl, identified as "Bu42", was diagnosed with "praechorea" in the 1930s. Occurrence of acanthocytes was still unknown at that time. The patient died at the age of 19 and her brain was studied repeatedly because of its distinctive characteristics [72: Figures 210 and 248-250, pages 179 and 434-436; 28,36]. Tissue sections are still kept at the Vogt Institute in Düsseldorf. Amplification of the DNA that we extracted, however, was not successful in Tony Monaco's lab. With genetic analysis in "Bu42" impossible, Brin's speculation about a "Vogt-Hopf syndrome" will stay unproven.

Of some interest, but again inconclusive are five case descriptions that were summarized as Fotopulos syndrome by the French [58]. In 1966 this German-Greek neurologist pointed out an association of chorea and spinal muscular atrophy in a lady whose father had been reported with the same condition a generation previously [18,22]. The most recent report noticed the similarity with neuroacanthocyosis syndromes but a search for acanthocytes failed and a DNA sample is not available [48]. Thus, the status of the syndrome might only be clarified through re-investigation of Fotopulos' original family.

The common occurrence of familial choreatic syndromes with spinal or peripheral neuromuscular changes has been noted repeatedly. Some of these clinical observations are rather typical of Huntington's disease [63] and there is molecular proof for the diagnosis in one family [53] which shows that neuromuscular disease and a chorea syndrome may be associated by chance. In other instances, however, the descriptions are reminiscent of "amyotrophic chorea" [19]. Thus, a neuroacanthocytosis syndrome should always be considered in the differential diagnosis of atypical presentations of chorea.

Neuroacanthocytosis

An observation of "familial tic disorder, parkinsonism, motor neuron disease, and acanthocytosis" [62] led to a controversy over whether "chorea-acanthocytosis", the term commonly used by the Japanese, was truly synonymous with Levine-Critchley syndrome [30,55]. Given the wide range of nervous system manifestations, "neuroacanthocytosis" was subsequently accepted as more appropriate – even if the new term encompasses conditions as diverse as abetalipoproteinemia and mitochondrial cytopathy that show neuromuscular manifestations associated with acanthocytic blood cells.

Existence of an additional subtype of neuroacanthocytosis became obvious, albeit slowly, after the connection with the McLeod phenotype was made. When McLeod, the original propositus, went to his local hospital for minor surgery, it was the hematologist Wimer who first noted acanthocytes in this Kell blood group variant [76]. Lawrence Marsh from the New York Blood Center collaborated on the case and later discovered neurological features in additional cases with the uncommon Kell phenotype, as was reported in a conference abstract [57]. In one instance, Marsh reversed the perspective and specifically asked for a blood sample from a patient suffering from "amyotrophic chorea and acanthocytosis" [17] to test for Kell antigens. His finding of the McLeod phenotype in that case, just briefly mentioned in a book chapter [43], is usually overlooked. It took a while for neurologists to take notice of these observations from the blood transfusion literature. The findings by Marsh, however, made it quite obvious that there is an X-linked form of neuroacanthocytosis, McLeod syndrome, in addition to the commonly described autosomal-recessive and the less common autosomal-dominant forms.

Mr. McLeod is here

McLeod syndrome was named after the index case with the peculiar variant in the Kell blood group system [1]. McLeod's blood was studied many times for elucidation of its unique properties [20,21,44,76], including the confirmation of a mutation in *XK*, the gene affected in subjects with McLeod syndrome [12]. Thanks to him and to the colleagues of the late Lawrence Marsh (notably Colvin Redman), documents are available that illustrate the development of this research too nicely as to be omitted here.

Figure 3. McLeod as portrayed in the university year book. When he had entered as a freshman he was immediately identified as "the possessor of the new phenotype" [1] that came to bear his name. Delineation of the X-linked McLeod subtype of neuroacanthocytosis due to *XK* gene mutations was the final outcome of this discovery in the Boston blood bank. Interestingly, the acanthocytic cell shape was recognized by the hematologist, Wimer, only much later [76].

The original report had introduced McLeod as the "possessor of the new phenotype ... encountered in testing a new class of ... students, who routinely are subgrouped as thoroughly as possible, in search of useful panel donors" [1]. This is the subject's perspective: "I kept getting these notices in my ... mailbox to see Dr. Allen at the Boston Blood Services. I knew that all of the freshmen had had blood drawn for testing. I was sort of leery of what I might find when I did go to his office. ... I finally decided that I would have to face the music at some time, and I went to the Boston Blood Services. I identified myself to the receptionist, she flipped the public address button, and said "Mr. McLeod is here!!". Heads popped up from every desk and around every corner. I was not sure whether I was a celebrity or a freak, with that encounter started a long acquaintance with blood." (personal communication 2000).

While on a lecture trip to New Zealand, Lawrence Marsh had been introduced to a family with acanthocytosis [66] and subsequently wrote to McLeod: "I thought you would like to know that you are no longer unique for we have now tested a large family sent from New Zealand and two of the male members have the red cell McLeod phenotype. ... We have been able to show that some female members in the New Zealand family are carriers of the gene ... We would very much like to test blood samples from your family to confirm that they show the same pattern of inheritance." (letter of November 25, 1975). Marsh then wrote to McLeod's uncle to ask for a blood sample to prove X-linkage: "My laboratory is engaged in research on the human blood groups. For some time we have been studying the blood of your nephew ... The rare type has, in fact, been named after him – the McLeod phenotype. ... Very little is known for sure about inheritance of the McLeod type, but we have some evidence that it may be through ... your sister. ... It would take only a few minutes of your time and your cooperation in this research would be very much appreciated. ... Finally, may I reassure you that if we find that you have this rare type, it should cause you no more concern than the knowledge that you perhaps have blue eyes." (letter of December 30, 1975).

In hindsight Marsh's final statement appears as a naïve euphemism (see Chapters 6 and 16). For example, Patient MT who we had followed for more than two decades died in 2002, at age 63. Although no autopsy was performed, his death quite likely was related to an associated cardiomyopathy.

The recognition of McLeod syndrome took time but its character as a well-defined separate entity is best supported by the observation of a patient with an *XK* point mutation who shows a neuroacanthocytosis syndrome and the peculiar Kell phenotype [13]. Earlier, the association of neuromuscular findings and the red cell phenotype had been interpreted as a contiguous gene syndrome [52]. The distinction of McLeod neuroacanthocytosis as separate was also fairly unclear in the 1991 series of Hardie and collaborators [24]. The X-linkage in their L-family was obscured because it included an exceptional manifesting female gene carrier [27].

Chorea-Acanthocytosis

The eventual cloning of the gene for an autosomal-recessive non-McLeod-type of neuroacanthocytosis in 2001 [50] was immediately confirmed in Japan where patients with the so-called Ehime deletion were observed [70]. The protein involved in chorea-acanthocytosis was called "chorein" (MIM 605978) by these investigators. The status of the majority of the Japanese cases, however, remains to be clarified. The perceived disproportionally high occurrence of neuroacanthocytosis in Japan is yet unexplained. While the observation of three families with homozygous occurrence of the Ehime deletion in the *CHAC* gene could indicate a founder effect, several instances of McLeod syndrome also have been identified among Japanese neuroacanthocytosis patients [23,29,34,71].

SPLITTING UP NEUROACANTHOCYTOSIS

A systematic clinical analysis of chorea-acanthocytosis in patients with one of the many *CHAC* gene mutations found outside of Japan still needs to be completed. While the clinical picture of McLeod syndrome has already been well defined, the process of distinguishing subtypes among the neuroacanthocytosis syndromes is continuing with two additional, recent discoveries.

The unusual family with autosomal-dominant transmission of neuroacanthocytosis that had been followed by Ruth Walker since 1997 [73], was shown in 2002 to suffer from a triplet repeat disease with intranuclear inclusions [74]. Mutations in the *JPH3* gene, initially associated with a Huntington's disease phenocopy, were identified shortly afterwards and subsequently, an additional case with a *JPH3* mutation has revealed acanthocytes in her blood [75].

The neuroacanthocytosis subtype known as the "HARP syndrome" of hypoprebetalipoproteinemia, acanthocytosis, retinitis pigmentosa, and pallidal degeneration [25] has eventually yielded to genetic analysis, too. It has been shown to be an allelic form of pantothenate kinase associated neurodegeneration [7]. The earlier view of Hallervorden-Spatz syndrome as another particular type of neuroanthocytosis [41,65] has thus been confirmed (see Chapter 8). There is only one additional published case of HARP syndrome [47]. Since a test for *PANK2* mutations has proved positive also in this young woman [33], "HARP syndrome" has become an expendable medical term. It would be of great interest to know if patient 17 from the neuroacanthocytosis series of Hardie et al. [24] who presented with acanthocyosis, retinitis pigmentosa and pallidal degeneration [47] – without a lipid abnormality – shows a *PANK2* mutation, too.

Whereas the connection of acanthocytosis with disorders of betalipoproteins is established, "hypoprebetalipoproteinemia" and "aprebetalipoproteinemia" are not commonly used in the lipoprotein field (Hegele, personal communication). While the former is limited to the HARP context, "aprebetalipoproteinemia" has been diagnosed only once [4]. That patient was recently shown to harbor a homozygous nonsense mutation in the *CHAC* gene [5].

Proof of mutations in well-defined genes, *PANK2* and *CHAC*, greatly decreases the importance of the presumed prebetalipoprotein abnormalities for neuroacanthocytosis nosology.

Another instance where genetic analysis has simplified nosology is a group of siblings that had been reported with a picture resembling neuroacanthocytosis but in whom repeated searches for abnormal red cells had failed [31]: heterozygous *CHAC* gene mutations were eventually found in one brother [15].

This observation, of course, casts a shadow on the role that acanthocytes might play. Is a failure to discover them just a problem of sensitivity? Do they take time to develop as had been suggested in another proven instance of ChAc [15,60]? Or are we to accept the existence of a "neuroacanthocytosis without acanthocytes" [42,46]? The availability of new hematologic and genetic tests (see Chapters 9, 19, 20) will eventually solve this awkward problem.

A few older observations must be scrutinized, too, so that unnecessary terminology can be reduced. Foremost among them is Levine's original family in whom the pattern of transmission is not well understood [40]. Also, the reports of "chorea-amyotrophy with chronic hemolytic anemia" [61], of "acanthocytosis and spinocerebellar degeneration" [68] and of "FADAEP", familial autosomal-dominant acanthocytosis with exertion-induced paroxysmal dyskinesias [37], call for re-analysis.

Finally, one further distinct subtype of neuroacanthocytosis could exist in the context of mitochondriopathies, yet the observation of one patient with MELAS syndrome and acanthocytes [45] so far has remained unique. Testing for acanthocytes should systematically be performed in mitochondrial disease in order to elucidate a possible connection.

Further cases of neuroacanthocytosis syndromes will certainly be diagnosed if a dedicated effort is made in patients with Huntington mutation-negative chorea [49] or those diagnosed with "idiopathic HyperCKemia" (see Chapter 10) and in psychiatric conditions that have additional neurological findings such as movement disorders or seizures or show organ involvement such as cardiomyopathy [32].

Summarizing this section, mutations in four different genes (apart from the betalipoprotein disorders, see Chapter 3) are presently known to result in a syndrome of neuroacanthocytosis: the *JPH3* and *PKAN2* genes and most notably *XK*, the McLeod gene, and *CHAC*, the chorein gene.

REASONS FOR LUMPING TOGETHER THE NEUROACANTHOCYTOSIS SYNDROMES

The possible distinctions among neuroacanthocytosis syndromes are important but the vital question remains: Is there a deeper reason that causes a change in shape of blood cells and affects many organ systems, in particular nerve cells? What is the link between these cell types? The possibility that the acanthocytes themselves are responsible for any associated change appears highly unlikely since there is acanthocytosis without nerve cell damage (see Chapters 2 and 3). It is more plausible that there is a superordinate factor causing both observations. The study of cell shape determinants might be particularly helpful for an understanding (see Chapter 18). Eventually, the readily available erythrocyte can serve as the perfect model for changes that affect the inaccessible nerve cells. Studies of erythrocyte properties unfortunately are still sparse in neuroacanthocytosis.

Most fascinating, however, is the similarity in the clinical phenotype of Huntington's disease and its phenocopies with the spectrum of neuroacanthocytosis syndromes. Among the questions that need to be solved soon is the distribution pattern that the proteins coded for by *JPH3*, *PKAN2*, *XK*, and *CHAC* follow in the brain. Given the selective vulnerability of the striatum in McLeod syndrome [14,32] and in ChAc (see Figure 2a) it is an obvious, yet not necessarily a correct, hypothesis that the proteins might specifically be distributed in the basal ganglia.

Similarities in the structures of the affected proteins are unlikely from what is already known, yet it is quite plausible that the different genes affect distinct levels in a cascade of molecular steps that, if dysfunctional, feed into one common pathway to neurodegeneration. An excitotoxic effect related to endothelin neurotransmission in the basal ganglia (see Chapter 23) is perhaps too simple a common denominator.

A variety of complex mechanisms on the level of the protein machinery such as processing, folding, sorting and degradation of proteins can be imagined. Apoptosis processes could be involved if the homology between *XK* and *ced-8* carries through to the functional level (see Chapters 19 and 21).

The main argument for the existence of a common network of metabolic steps involved in the pathogenesis of neuroacanthocytosis syndromes derives not only from clinical similarities but also from their usual onset in adulthood. There might be a substantial redundancy among the various elements of the metabolic cascade that delays disease onset through a certain capacity for mutual substitution.

While it appears possible that clinical work-up might not be performed to the highest levels of differentiation if a diagnosis just of "neuroacanthocytosis" is made, the term should nevertheless be retained. It is an umbrella term that fertilizes research by reminding us of the possibility of a final common pathway that is approached via a variety of molecular mechanisms.

REFERENCES

1. Allen FH, Krabbe SMR, Corcoran PA (1961) A new phenotype (McLeod) in the Kell blood-group system. *Vox Sang* **6:** 555-560.
2. Aminoff MJ (1972) Acanthocytosis and neurological disease. *Brain* **95:** 749-760.
3. Bird TD, Cederbaum S, Valpey RW, Stahl WL (1978) Familial degeneration of the basal ganglia with acanthocytosis: A clinical, neuropathological, and neurochemical study. *Ann Neurol* **3:** 253-258.
4. Bohlega S, Riley W, Powe J, Baynton R, Roberts G (1998) Neuroacanthocytosis and aprebetalipoproteinemia. *Neurology* **50:** 1912-1914.
5. Bohlega S, Al Jishi A, Dobson-Stone C et al (2003) Chorea-acanthocytosis: clinical and genetic findings in three families from the Arabian peninsula. *Mov Disord* **18:** 403-407.
6. Brin MF (1993) Acanthocytosis. In: *Goetz CG, Tanner CM, Aminoff MJ (eds) Handbook of Clinical Neurology, Vol. 19 (63): Systemic Diseases, Part I,* pp 271-299. Elsevier, Amsterdam.
7. Ching KHL, Westaway SK, Gitschier J, Higgins JJ, Hayflick SJ (2002) HARP syndrome is allelic with pantothenate kinase-associated neurodegeneration. *Neurology* **58:** 1673-1674.
8. Critchley EMR, Clark DB, Wikler A (1967) An adult form of acanthocytosis. *Trans Am Neurol Assoc* **92:** 132-137.
9. Critchley EMR, Clark DB, Wikler A (1968) Acanthocytosis and neurological disorder without betalipoproteinemia. *Arch Neurol* **18:** 134-140.

10. Critchley EMR, Betts JJ, Nicholson JT, Weatherall DJ (1970) Acanthocytosis, normolipoproteinemia and multiple tics. *Postgrad Med J* **46:** 698-701.
11. Critchley EMR (1971) Acanthocytosis associated with tics and involuntary movements. *Z Neurol* **200:** 336-340.
12. Danek A, Rubio JP, Rampoldi L et al (2001) McLeod neuroacanthocytosis: Genotype and phenotype. *Ann Neurol* **50:** 755-764.
13. Danek A, Tison F, Rubio J et al (2001) The chorea of McLeod syndrome. *Mov Dis* **16:** 882-889.
14. Danek A, Uttner I, Vogl T, Tatsch K, Witt TN (1994) Cerebral involvement in McLeod syndrome. *Neurology* **44:** 117-120.
15. Dobson-Stone C, Danek A, Rampoldi L et al (2002) Mutational spectrum of the *CHAC* gene in patients with chorea-acanthocytosis. *Eur J Hum Genet* **10:** 773-781.
16. Estes JW, Morley TJ, Levine IM, Emerson CP (1967) A new hereditary acanthocytosis syndrome. *Am J Med* **42:** 868-881.
17. Faillace RT, Kingston WJ, Nanda NC, Griggs RC (1982) Cardiomyopathy associated with the syndrome of amyotrophic chorea and acanthocytosis. *Ann Intern Med* **96:** 616-617.
18. Fotopulos D (1966) Huntington-Chorea und chronisch-progressive spinale Muskelatrophie. *Psychiat Neurol med Psychol (Leipzig)* **18:** 63-71.
19. Frank G, Vuia O (1973) Chorea Huntington – Amyotrophische Lateralsklerose – Spastische Spinalparalyse. Zur Kombination von Systemerkrankungen. *Z Neurol* **205:** 207-220.
20. Galey WR, Evan AP, Van Nice PS et al (1978) Morphology and physiology of the McLeod erythrocyte. *Vox Sang* **34:** 152-161.
21. Glaubensklee CS, Evan AP, Galey WR (1982) Structural and biochemical analysis of the McLeod erythrocyte membrane. *Vox Sang* **42:** 262-271.
22. Grotjahn M (1934) Chronische, progressive Chorea und spinale Muskelatrophie. *Zentralbl gesamte Neurol Psychiatr* **73:** 251-253.
23. Hanaoka N, Yoshida K, Nakamura A et al (1999) A novel frameshift mutation in the McLeod syndrome gene in a Japanese family. *J Neurol Sci* **165:** 6-9.
24. Hardie RJ, Pullon HWH, Harding AE et al (1991) Neuroacanthocytosis: A clinical, haematological and pathological study of 19 cases. *Brain* **114:** 13-49.
25. Higgins JJ, Patterson MC, Papadopoulos NM et al (1992) Hypoprebetalipoproteinemia, acanthocytosis, retinitis pigmentosa, and pallidal degeneration (HARP syndrome). *Neurology* **42:** 194-198.
26. Ho M, Chelly J, Carter N et al (1994) Isolation of the gene for McLeod syndrome that encodes a novel membrane transport protein. *Cell* **77:** 869-880.
27. Ho MF, Chalmers RM, Davis MB, Harding AE, Monaco AP (1996) A novel point mutation in the McLeod syndrome gene in neuroacanthocytosis. *Ann Neurol* **39:** 672-675.
28. Hopf A (1952) Über eine patho-anatomische Sonderform der Chorea. *J Nerv Ment Dis* **116:** 608-618.
29. Ishikawa S, Tachibana N, Tabata KI et al (2000) Muscle CT scan findings in McLeod syndrome and chorea-acanthocytosis. *Muscle Nerve* **23:** 1113-1116.
30. Jankovic J, Killian JM, Spitz MC (1985) Neuroacanthocytosis syndrome and choreoacanthocytosis (Levine-Critchley syndrome). *Neurology* **35:** 1679.
31. Johnson SE, Dahl A, Sjaastad O (1998) Progressive pseudobulbar paresis, early choreiform movements, and later rigidity: Appearance in two sets of dizygotic twins in the same family. *Mov Dis* **13:** 556-562.
32. Jung HH, Hergersberg M, Kneifel S et al (2001) McLeod syndrome: a novel mutation, predominant psychiatric manifestations, and distinct striatal imaging findings. *Ann Neurol* **49:** 384-392.
33. Houlden H, Lincoln S, Farrer M, Cleland PG, Hardy J, Orrell RW (2003) Compound heterozygous *PANK2* mutations confirm HARP and Hallervorden-Spatz syndromes are allelic. *Neurology* **61:** 1423-1426.

34. Kawakami T, Takiyama Y, Sakoe K et al (1999) A case of McLeod syndrome with unusually severe myopathy. *J Neurol Sci* **166:** 36-39.
35. Kito S, Itoga E, Hiroshige Y, Matsumoto N, Miwa S (1980) A pedigree of amyotrophic chorea with acanthocytosis. *Arch Neurol* **37:** 514-517.
36. Lange H, Thorner G, Hopf A, Schröder KF (1976) Morphometric studies of the neuropathological changes in choreatic diseases. *J Neurol Sci* **28:** 401-425.
37. Lerche H, Storch A, Pekrun A et al (2000) A novel form of autosomal dominant neuroacanthocytosis with exertion-induced paroxysmal dyskinesias. *J Neurol* **247 Suppl 3:** III/18.
38. Levine IM, Yettra M, Stefanini M (1960) A hereditary neurological disorder with acanthocytosis. *Neurology* **10:** 425.
39. Levine IM (1964) An hereditary neurological disease with acanthocytosis. *Neurology* **14:** 272.
40. Levine IM, Estes JW, Looney JB (1968) Hereditary neurological disease with acanthocytosis, a new syndrome. *Arch Neurol* **19:** 403-409.
41. Malandrini A, Fabrizi GM, Bartalucci P et al (1996) Clinicopathological study of familial late infantile Hallervorden-Spatz disease: A particular form of neuroacanthocytosis. *Child's Nerv Syst* **12:** 155-160.
42. Malandrini A, Fabrizi GM, Palmeri S et al (1993) Choreo-acanthocytosis like phenotype without acanthocytes: Clinicopathological case report. A contribution to the knowledge of the functional pathology of the caudate nucleus. *Acta Neuropathol (Berl)* **86:** 651-658.
43. Marsh WL (1983) Deleted antigens of the Rhesus and Kell blood groups: Association with cell membrane defects. In: *Garraty G (ed) Blood group antigens and disease,* pp 165-185. American Association of Blood Banks, Arlington, Virginia.
44. Marsh WL, Øyen R, Nichols ME, Allen FH (1975) Chronic granulomatous disease and the Kell blood groups. *Br J Haematol* **29:** 247-262.
45. Mukoyama M, Kazui H, Sunohara N et al (1986) Mitochondrial myopathy, encephalopathy, lactic acidosis, and stroke-like episodes with acanthocytosis: A clinicopathological study of a unique case. *J Neurol* **233:** 228-232.
46. O'Brien CF, Schwarz H, Kurland R (1990) Neuroacanthocytosis without acanthocytes. *Mov Dis* **5 (Suppl. 1):** 98.
47. Orrell RW, Amrolia PJ, Heald A et al (1995) Acanthocytosis, retinitis pigmentosa, and pallidal degeneration: A report of three patients, including the second reported case with hypoprebetalipoproteinemia (HARP syndrome). *Neurology* **45:** 187-192.
48. Pageot N, Vial C, Remy C, Chazot G, Broussolle E (2000) Progressive chorea and amyotrophy without acanthocytes: a new case of Fotopoulos syndrome? *J Neurol* **247:** 392-394.
49. Piccolo I, Defanti CA, Soliveri P et al (2003) Cause and course in a series of patients with sporadic chorea. *J Neurol* **250:** 429-435.
50. Rampoldi L, Dobson-Stone C, Rubio JP et al (2001) A conserved sorting-associated protein is mutant in chorea-acanthocytosis. *Nat Genet* **28:** 119-120.
51. Rovito DA, Pirone FJ (1963) Acanthrocytosis associated with schizophrenia. *Am J Psychiatry* **120:** 182-185.
52. Rowland LP (1988) Clinical concepts of Duchenne muscular dystrophy – the impact of molecular genetics. *Brain* **111:** 479-495.
53. Rubio A, Steinberg K, Figlewicz DA et al (1996) Coexistence of Huntington's disease and familial amyotrophic lateral sclerosis: Case presentation. *Acta Neuropathol (Berl)* **92:** 421-427.
54. Rubio JP, Danek A, Stone C et al (1997) Chorea-acanthocytosis: Genetic linkage to chromosome 9q21. *Am J Hum Genet* **61:** 899-908.
55. Sakai T, Iwashita H, Kakugawa M (1985) Neuroacanthocytosis syndrome and choreoacanthocytosis (Levine-Critchley syndrome). *Neurology* **35:** 1679.

56. Sakai T, Mawatari S, Iwashita H, Goto I, Kuroiwa Y (1981) Choreoacanthocytosis. Clues to clinical diagnosis. *Arch Neurol* **38:** 335-338.
57. Schwartz SA, Marsh WL, Symmans A, Johnson CL, Mueller KA (1982) "New" clinical features of McLeod syndrome. *Transfusion* **22:** 404.
58. Serratrice G, Crémieux G, Pélissier JF, Pouget J (1984) Deux cas de syndrome de Fotopoulos (amyotrophie spinale chronique de la ceinture scapulaire et chorée chronique). *Presse Méd* **13:** 1274.
59. Shibasaki H, Sakai T, Nishimura H et al (1982) Involuntary movements in chorea-acanthocytosis: A comparison with Huntington's chorea. *Ann Neurol* **12:** 311-314.
60. Sorrentino G, De Renzo A, Miniello S, Nori O, Bonvita V (1999) Late appearance of acanthocytes during the course of chorea-acanthocytosis. *J Neurol Sci* **163:** 175-178.
61. Spencer SE, Walker FO, Moore SA (1987) Chorea-amyotrophy with chronic hemolytic anemia: a variant of chorea-amyotrophy with acanthocytosis. *Neurology* **37:** 645-649.
62. Spitz MC, Jankovic J, Killian JM (1985) Familial tic disorder, parkinsonism, motor neuron disease, and acanthocytosis: A new syndrome. *Neurology* **35:** 366-370.
63. Sumner D (1962) Amyotrophy in a family with Huntington's chorea. *World Neurol* **3:** 769-777.
64. Swash M, Schwartz MS, Carter ND et al (1983) Benign X-linked myopathy with acanthocytes (McLeod syndrome), its relationship to X-linked muscular dystrophy. *Brain* **106:** 717-733.
65. Swisher CN, Menkes JH, Cancilla PA, Dodge PR (1972) Coexistence of Hallervorden-Spatz disease with acanthocytosis. *Trans Am Neurol Assoc* **97:** 212-216.
66. Symmans WA, Shepherd CS, Marsh WL et al (1979) Hereditary acanthocytosis associated with the McLeod phenotype of the Kell blood group system. *Br J Haematol* **42:** 575-583.
67. Toyokura Y, Kamakura K, Shimada Y (1982) Familial chorea-acanthocytosis (the Levine-Critchley syndrome) – A review of the reported cases in Japan. In: *The ministry of health and welfare of Japan (ed) Annual report of the research committee of CNS degenerative diseases,* pp 335-351
68. Tsai C-H, Chen R-S, Chang H-C, Lu C-S, Liao K-K (1997) Acanthocytosis and spinocerebellar degeneration: A new association? *Mov Dis* **12:** 456-459.
69. Ueno E, Oguchi K, Yanagisawa N (1982) Morphological abnormalities of erythrocyte membrane in the hereditary neurological disease with chorea, areflexia and acanthocytosis. A study with freeze fracture electron microscopy. *J Neurol Sci* **56:** 89-97.
70. Ueno S, Maruki Y, Nakamura M et al (2001) The gene encoding a newly discovered protein, chorein, is mutated in chorea-acanthocytosis. *Nat Genet* **28:** 121-122.
71. Ueyama H, Kumamoto T, Nagao S et al (2000) A novel mutation of the McLeod syndrome gene in a Japanese family. *J Neurol Sci* **176:** 151-154.
72. Vogt C, Vogt O (1937) *Sitz und Wesen der Krankheiten im Lichte der topistischen Hirnforschung und des Variierens der Tiere. Erster Teil: Befunde der topistischen Hirnforschung als Beitrag zur Lehre vom Krankheitssitz.* J.A.Barth, Leipzig.
73. Walker RH, Melnick A, Walsh M, Davidoff-Feldman B, Brin MF (1997) Dominantly inherited chorea-parkinsonism-dementia-acanthocytosis with red cell protein membrane abnormalities. *Ann Neurol* **42:** 407-408.
74. Walker RH, Morgello S, Davidoff-Feldman B et al (2002) Autosomal dominant chorea-acanthocytosis with polyglutamine-containing neuronal inclusions. *Neurology* **58:** 1031-1037.
75. Walker RH, Rasmussen A, Rudnicki D, Holmes SE, Alonso E, Matsuura T, Ashizawa T, Davidoff-Feldman B, Margolis RL (2003) Huntington's disease-like 2 can present as chorea-acanthocytosis. *Neurology* **61:** 1002-1004.
76. Wimer BM, Marsh WL, Taswell HF, Galey WR (1977) Haematological changes associated with the McLeod phenotype of the Kell blood group system. *Br J Haematol* **36:** 219-224.

77. Witt TN, Danek A, Reiter M et al (1992) McLeod syndrome: A distinct form of neuroacanthocytosis. *J Neurol* **239:** 302-306.
78. Yamamoto T, Hirose G, Shimazaki K et al (1982) Movement disorders of familial neuroacanthocytosis syndrome. *Arch Neurol* **39:** 298-301.

CHAPTER 2

THE DIFFERENTIAL DIAGNOSIS OF NEUROACANTHOCYTOSIS: AN OVERVIEW

François Tison
Service de Neurologie, Hôpital Haut Lévèque, CHU de Bordeaux, Pessac, France

Abstract. The confusing term of neuroacanthocytosis refers to various entities simply defined by the presence of acanthocytosis plus a neurological disorder. In this chapter we will focus on the clinical differential diagnosis. There exist two distinguishable situations to remember: diseases with a primary progressive neurological disorder and systemic diseases with secondary and inconstant neurological symptoms or signs. Primary neurological diseases could then be subclassified in diseases with an abnormal lipid metabolism such as in abetalipoproteinemia and hypobetalipoproteinemia and in diseases without lipid abnormalities such as in chorea-acanthocytosis, in the McLeod syndrome, and in the Hallervorden-Spatz syndrome. The term neuroacanthocytosis should be reserved for the diseases without a lipid metabolism abnormality: Chorea-acanthocytosis and the McLeod syndrome. Besides the molecular biology, numerous clinical and laboratory indices, which are reviewed here, may help the clinician to make the differential diagnosis, particularly in distinction from Huntington's disease or other movement disorders.

INTRODUCTION

As mentioned by Stevenson and Hardie in their recent review [1] the confusing term of neuroacanthocytosis usually refers to various entities simply defined by the presence of acanthocytosis plus a neurological disorder. In this chapter we will focus on the clinical differential diagnosis rather than on pathological and molecular aspects, which are detailed elsewhere. We will present a brief clinical overview to help the clinician find his way to the differential diagnosis and to accurately focus the molecular investigation.

A. Danek (ed.), Neuroacanthocytosis Syndromes, 15–20.

Table 1. Acanthocytosis (adapted from *Brin 1993* [2]).

1) *With prominent neurological signs*
with deficit in lipoproteins
Abetalipoproteinemia (Bassen-Kornzweig disease)
Hypobetalipoproteinemia
HARP syndrome
without deficit in lipoproteins
Chorea-acanthocytosis (Levine-Critchley syndrome)
McLeod syndrome
Pantothenate kinase associated neurodegeneration (Hallervorden-Spatz syndrome)
Mitochondrial cytopathies
2) *With secondary neurological signs due to a primary systemic disease*
Liver cirrhosis
Anorexia nervosa, profound malnutrition
Splenectomy
Myxedema
Wolman's disease
Eales' disease
Psoriasis
Sarcoma

Acanthocytes are abnormal blood cells, normally not circulating in the blood stream. They have to be differentiated from echinocytes or "burr cells" which can be observed as artifacts in routine blood samples or in pathological conditions such as liver failure or splenectomy. Acanthocytes or "spur" cells have less numerous spicules, usually ten to thirty, typically terminated by bulbs.

GENERAL CLASSIFICATION

Many conditions, as proposed by Brin in a review [2], may show an association of acanthocytes and neurological signs and symptoms. For the clinician, however, there exist two distinguishable situations to remember in everyday practice (Table 1): diseases with a primary progressive neurological disorder and systemic diseases with secondary and inconstant neurological symptoms or signs.

Primary neurological diseases may then be subclassified into those with an abnormal lipid metabolism such as abetalipoproteinemia and hypobetalipoproteinemia and into diseases without lipid abnormalities such as chorea-acanthocytosis or Levine-Critchley syndrome, the McLeod syndrome, pantothenate kinase associated neurodegeneration (PKAN, previously known as Hallervorden-Spatz syndrome or as NBIA-1, Neurodegeneration with Brain Iron Accumulation Type 1), and some other entities, for example mitochondrial cytopathies such as the MELAS syndrome.

Many systemic diseases, however, may cause acanthocytes and/or echinocytes to circulate in the blood and might also be associated with some form of neurological involvement, such as for example in hepatic encephalopathies, profound malnutrition syndromes, cancers, thyroid gland disorders, etc. This chapter will focus only on the differential diagnosis of the more common primary neurological diseases associated with acanthocytosis.

ACANTHOCYTES AND PRIMARY PROGRESSIVE NEUROLOGICAL DISEASES

Diseases with abnormal lipid metabolism

Abetalipoproteinemia or Bassen-Kornzweig disease usually is an early onset autosomal-recessive condition (microsomal triglyceride transfer protein gene mutation) presenting with severe spinocerebellar ataxia, more or less masquerading as Friedreich ataxia. Classical other clinical features are: delayed development, axonal neuropathy, retinitis pigmentosa. It usually is easily differentiated from the classical chorea-acanthocytosis. Particular diagnostic clues are provided by clinical chemistry: profound vitamin A, D, E, K deficiency, abetalipoproteinemia (chylomicrons, LDL, HDL), with low cholesterol and triglycerides [3]. Acanthocytes are usually much elevated in the blood. Hypobetalipoproteinemia (autosomal-dominant apolipoprotein B gene mutation) is usually less severe, with later onset spinocerebellar features and detectable but low levels of lipoproteins. There is also the rare condition of hypoprebetalipoproteinemia, acanthocytosis, retinitis pigmentosa and pallidal degeneration syndrome (HARP) [4]. Only two cases of the latter have been described with a variable combination of movement disorders, dementia, retinitis pigmentosa and high levels of acanthocytes (80%), and typical MRI features of pallidal degeneration with iron accumulation, as found in PKAN ("eye of the tiger sign", see Chapter 8).

Diseases without abnormal lipid metabolism

The "True Neuroacanthocytoses": Chorea-Acanthocytosis and the McLeod Syndrome. Chorea-acanthocytosis (Levine-Critchley syndrome) and the McLeod syndrome now are clinically, biologically and molecularly well-defined entities [5,6]. As they constitute the topic of other more detailed chapters of this volume, only the characteristic features are indicated in Table 2. Stevenson and Hardie [1] have proposed to reserve the term neuroacanthocytosis for these diseases without lipid abnormality to resolve nosological confusion.

Their main differential diagnosis is Huntington's disease as indicated in Table 2. Besides molecular biology, numerous clinical and laboratory indices may help the clinician make the differential diagnosis: the movement disor-

Table 2. The neuroacanthocytoses: main clinical features (adapted from *Danek et al 2001* [6]).

	McLeod syndrome	Chorea-acanthocytosis	Huntington's disease
	(MIM 314850)	(MIM 200150)	(MIM 143100)
Age of onset	27-72 years	8-62 years	4-74 years
Movement disorders	chorea	chorea, orofacial dyskinesias, feeding dystonia, lip biting, parkinsonism	chorea, rigidity/ parkinsonism (in juvenile type)
Seizures	Possible	Frequent	In juvenile type
Neuropathy	+	+	−
Myopathy/ Elevated CK	+	+	−
Cardiomyopathy	+	+	−
Blood	Acanthocytes, McLeod phenotype	Acanthocytes, Normal Kell phenotype	Normal Kell phenotype
Gene	*XK* (chromosome X)	*CHAC* (chromosome 9)	*IT15* (chromosome 4)
Transmission	X-linked	Autosomal-recessive	Autosomal-dominant

der with classical lip biting and with feeding dystonia, neuromuscular involvement, particularly an axonal neuropathy, presence of acanthocytes, elevated CK and/or weak expression of Kell blood antigens in the case of McLeod syndrome [5,6].

Pantothenate Kinase Associated Neurodegeneration (PKAN). PKAN is an autosomal-recessive severe disease due to mutations in a newly discovered pantothenate kinase gene (*PANK2*) [7,8]. Pantothenate kinases regulate coenzyme A synthesis. The disorder is somewhat heterogeneous but the classical picture is: onset during the first two decades of life, extrapyramidal signs (dystonia, rigidity, choreoathetosis), pyramidal signs/spasticity, progressive intellectual impairment, retinitis pigmentosa/optic atrophy, inconstant cytosomes in lymphocytes and sea-blue histiocytes in the bone marrow, and inconstant acanthocytes. The "eye of the tiger" sign on MRI, which represents the classical finding of iron accumulation in the medial globus pallidus permits the diagnosis during life (see page 68). Within this broad phenotype, late onset or adult forms may be misleading because they may present with late-onset parkinsonism, dementia, dystonia or as choreatic forms. Particularly, these may imitate chorea-acanthocytosis. MRI, however, usually reveals the diagnosis if the classical "eye of the tiger" sign is remembered [8].

Other Movement Disorders

Huntington's disease is probably the more difficult differential diagnosis to discuss with the neuroacanthocytoses, particularly when a typical autosomal-dominant family history is not present or cannot be obtained. Along with *dentato-rubro-pallido-Luysian atrophy* (*DRPLA*), chorea-acanthocytosis is probably the most frequent alternative diagnosis to look for in a patient with typical features of Huntington's disease. For this differential diagnosis, the search for circulating acanthocytes and elevated CK are simple screening tests in blood before a molecular diagnosis can be obtained.

Wilson's disease can mimic any young-onset movement disorder and may present with orofacial and choreic movements. *Non Wilsonian hepato-lenticular degeneration* may also present as a choreic and/or parkinsonian movement disorder, possibly with acanthocytes or echinocytes depending upon the nature of the underlying liver disease. *Gilles de la Tourette syndrome*, because of the prominence of tics and compulsive disorders has also to be discussed. The rare *Lesh-Nyhan syndrome* with atypical late onset with chorea, tics and self mutilations can also imitate chorea-acanthocytosis. In these disorders a careful clinical examination supported by a search for typical signs on imaging and laboratory or molecular markers should lead the clinician to the diagnosis.

CONCLUSION

In clinical practice, the mode of transmission and the features at examination, such as the presence of peculiar orofacial movements with tongue biting, neuropathy, cardiomyopathy, MRI findings, presence of acanthocytes, CK elevation, and the pattern of red cell Kell antigens may help the neurologist to find his way through the differential diagnosis. However, once a diagnosis is clinically suspected the clinician must definitively confirm his working hypothesis by asking for the appropriate molecular test which will be the final diagnostic step.

REFERENCES

1. Stevenson VL, Hardie RJ (2001) Acanthocytosis and neurological disorders. *J Neurol* **48:** 87-97.
2. Brin MF. Acanthocytosis. Handbook of Clinical Neurology, Systemic Diseases, Part I (1993) *19(63)* 272-299
3. Bohlega S, Riley W, Powe J, Baynton R, Roberts G (1998) Neuroacanthocytosis and aprebetalipoproteinemia. *Neurology* **50:** 1912-1914.
4. Orrell RW, Amrolia PJ, Heald A, Cleland PG, Owen JS, Morgan-Hughes JA, Harding AE, Marsden CD (1995) Acanthocytosis, retinitis pigmentosa, and pallidal degeneration: a report of three patients, including the second reported case with hypoprebetalipoproteinemia (HARP syndrome). *Neurology* **45:** 487-492.
5. Danek A, Rubio JP, Rampoldi L, Ho M, Dobson-Stone C, Tison F, Symmans WA, Oechsner M, Kalckreuth W, Watt JM, Corbett AJ, Hamdalla HHM, Marshall AG, Sutton I,

Dotti MT, Malandrini A, Walker RH, Daniels G, Monaco AP (2001) McLeod neuro-acanthocytosis: Genotype and phenotype. *Ann Neurol* **50:** 755-764.

6. Danek A, Tison F, Rubio J, Oechsner M, Kalckreuth W, Monaco AP (2001) The chorea of McLeod syndrome. *Mov Dis* **16:** 882-889.
7. Halliday W (1995) The nosology of Hallervorden-Spatz disease *J Neurol Sci* **134:** 84-91.
8. Racette BA, Perry A, D'Avossa G, Perlmutter JS (2001) Late-onset neurodegeneration with brain iron accumulation type 1: Expanding the clinical spectrum. *Mov Dis* **16:** 1148-1152.

CHAPTER 3

ACANTHOCYTES AND DISORDERS OF LIPOPROTEIN METABOLISM

Khalid Al-Shali and Robert A. Hegele
Blackburn Cardiovascular Genetics Laboratory,
Robarts Research Institute, London, Ontario, Canada

Abstract. Acanthocytosis and disordered lipoprotein metabolism are inextricably linked in medical history. The term "acanthocytosis" was itself coined a half-century ago to describe the dysmorphic erythrocytes seen in abetalipoproteinemia (ABL), a complex metabolic disorder that is characterised by fat malabsorption, atypical retinitis pigmentosa and spinocerebellar ataxia. ABL results from mutation in the gene encoding microsomal triglyceride transfer protein (MTP), which results in failure to synthesize apolipoprotein B-containing lipoproteins, producing very low plasma concentrations of cholesterol and triglycerides, and fat soluble vitamin deficiency. A phenotypically similar condition called homozygous hypobetalipoproteinemia results from mutation in the gene encoding apo B itself, and is clinically very similar to ABL. These conditions are both rare, with fewer than 1 affected subject in 100 000 in most populations. In addition, some neurological disorders with acanthocytosis are associated with depressed plasma lipoproteins.

INTRODUCTION: COEXISTENCE OF LIPOPROTEIN ANOMALIES WITH ACANTHOCYTOSIS

The two archetypal disorders of lipoprotein metabolism with acanthocytosis are abetalipoproteinemia (ABL; MIM 200100) and homozygous familial hypobetalipoproteinemia (FHBL; MIM 107730). In addition, Anderson disease, HARP syndrome and an atypical variant of Wolman disease have been noted to have both reduced plasma lipoproteins and acanthocytosis. These disorders are summarized in Table 1. The elucidation of the molecular basis of both ABL and FHBL were seen as important milestones in lipoprotein research.

A. Danek (ed.), Neuroacanthocytosis Syndromes, 21–30.

Table 1. Lipoprotein anomalies with acanthocytosis.

Disease	MIM	comments
Abetalipoproteinemia	200100	- classical disorder with acanthocytes due to mutant MTP gene - associated findings include fat malabsorption, atypical retinitis pigmentosa, spinocerebellar ataxia - normal plasma apo B in heterozygotes
Familial hypobetalipoproteinemia	107730	- very similar phenotype to ABL - due to mutant *APOB* gene - reduced plasma apo B in heterozygotes
Anderson disease	246700	- chylomicron retention disease - molecular defect unknown - acanthocytosis is mild and sometimes absent
HARP	607236	- hypo-prebetalipoproteinemia, acanthocytosis, retinitis pigmentosa, and pallidal degeneration - *PANK2* mutation (as in Hallervorden-Spatz disease)
Atypical Wolman disease	278100	- atypical variant with hypolipoproteinemia and acanthocytosis

ABETALIPOPROTEINEMIA

Discovery and Clinical Features

The world's index ABL patient was an 18-year-old female with a history of steatorrhea since childhood, Friedreich-type ataxia and atypical retinitis pigmentosa whose peripheral blood film showed erythrocytes with bizarre shapes [4]. Her 9-year-old brother had similar looking erythrocytes and early pigmentary changes in the retina. The syndrome comprising fat malabsorption, abnormal erythrocytes, Friedreich-like ataxia and atypical retinitis pigmentosa was given the name "Bassen-Kornzweig syndrome" [4]. Two years later, the term "acanthrocytosis" was suggested to describe the peculiar erythrocytes in this condition [36] and this was later changed to "acanthocytosis" [14]. The link with low plasma lipoproteins was made in 1960 [32], when beta-migrating lipoproteins were noted to be absent from serum, leading to the proposed name of "abetalipoproteinemia" for the disease. Early microscopic examination of the peripheral nerves from ABL subjects showed extensive central and peripheral demyelination [37]. Improvements during the 1960's with respect to the qualitative and quantitative assessment of plasma lipoproteins resulted in narrowing of the fundamental defect in ABL to the inability to synthesize lipoprotein particles containing apolipoprotein (apo) B, namely chylomicrons, very low density lipoprotein (VLDL) and low density lipoprotein (LDL).

At birth, infants with ABL are asymptomatic. With a diet that is rich in lipids, gastrointestinal symptoms and signs develop during the months after birth. The initial presentation resembles celiac disease with diarrhea, vomiting, and abdominal swelling. Later on, the gastrointestinal signs subside, in part because the patients themselves learn to avoid fatty foods [2,17,23]. The chronic malabsorption of lipids also leads to fat-soluble vitamin deficiency because the plasma transport and delivery of these vitamins to tissues depends almost exclusively (for vitamin E and beta-carotene) or in part (for vitamins A, D, and K) on intact synthesis and secretion of apo B-containing lipoproteins.

Without any intervention, the vitamin deficiencies result in neuro-ophthalmologic complications by the third decade of life, which dominate the clinical picture and determine the morbidity of ABL. However, there is heterogeneity in disease presentation. Almost all reported ABL patients > 20 years of age who had not received vitamin E supplementation developed neuro-retinal complications. Some were blind and bedridden. In $\sim 25\%$ of reported cases, the diagnosis was made after age 20 [5]. This clinical heterogeneity is unexplained, although there may be some correlation with the severity of the molecular defect [38]. The affected organ systems and tissues are described in detail below.

Gastrointestinal Consequences

With a normo-lipidemic diet, steatorrhea is invariably present. This symptom reflects lipid malabsorption and is attenuated or fully relieved after introduction of a low-fat diet. The lipid malabsorption affects height and weight gain, and may lead to secondary malabsorption of other nutrients. Endoscopic examination of the intestine reveals a "gelee blanche" or white hoar frosting appearance [2,13]. This coating of the duodenum and jejunum reflects the infiltration of the mucosa by lipids. Intestinal cells become engorged with fat, which provided an early clue that the metabolic defect in ABL prevented the normal secretion of dietary fat from enterocytes into the plasma through the intestinal lymphatics.

Hepatic Consequences

As in the intestine, the livers of ABL subjects can show marked lipid accumulation. This hepatic steatosis is occasionally associated with elevation of transaminases either with or without hepatomegaly [3]. Rarely, there is evolution to fibrosis and occasionally progression to cirrhosis, requiring transplantation [7,8].

Neurological Consequences

The initial neurological sign is often the diminution and then the loss of deep tendon reflexes followed by a progressive loss of position and vibratory senses, a spinocerebellar syndrome, and muscular weakness. Also, slowed intellectual development is present in up to one-third of patients. Neuropathology reveals axonal degeneration of the spinocerebellar tracts and demyelination of the fasciculus cuneatus and gracilis [37]. Vitamin E deficiency was first recognized in patients with ABL in 1965 [24] and is now considered to be the cause of the spinocerebellar degeneration [31]. The myopathy results from both neural degeneration and an intrinsic myositis. Although the clinical course is variable without treatment, it leads progressively to impaired mobility, and some patients become wheelchair-bound or even bedridden. The severe effects on the central nervous system are the ultimate cause of death in most patients with ABL, which (before the advent of high-dose vitamin E therapy) often occurred by the fifth decade [22]. However, early administration of vitamin E has been shown to cause objective arrest of the usually progressive neuropathy and myopathy [18].

Ophthalmological Consequences

Initially, ABL patients complain of decreased night and color vision, followed by a decrease in visual acuity. The visual field shows a concentric contraction. If left untreated, virtual blindness occurs by the fourth decade. Fundoscopic examination shows an atypical pigmentation of the retina. Pathologically, the retina shows reduced numbers of photoreceptor cells and accumulation of lipofuscin [11]. In some cases, angioid streaks have been reported [15]. Ophthalmoplegia, ptosis, and anisocoria have also been described [22].

Hematological Signs

As mentioned above, acanthocytes were first named in ABL subjects. It has been speculated that the membrane deformation results from decreased membrane fluidity caused by changes in lipid composition, specifically an increase in the ratio of sphingomyelin to lecithin and changes to the fatty acid composition of the phospholipids [39]. In addition to acanthocytosis, patients with ABL may have a moderate to severe anemia that results from hemolysis and shortening of the erythrocyte half-life [35]. Abnormalities in coagulation (elevated prothrombin time) caused by deficiency in vitamin K-dependent coagulation factors may also be seen in ABL patients. This may be symptomatic, leading to bruising or hemorrhage [9].

Molecular Genetics

Approximately one-third of the cases of ABL result from consanguineous marriages, suggesting an autosomal-recessive mode of transmission. Typically, obligate heterozygotes have normal plasma lipid levels [5]. Disease frequency is < 1 in 100 000. ABL is caused by mutations in the gene encoding the microsomal triglyceride transfer protein (MTP; MIM 157147) [28,33,38,41] on chromosome 4q22-24. MTP was a candidate gene for sequencing because Wetterau et al originally reported that liver and intestinal MTP mediated the intracellular transport of membrane-associated lipids [39]. They further suggested that MTP was required for normal assembly and secretion of apo B-containing lipoproteins [40]. MTP is a heterodimer composed of the multifunctional enzyme protein disulfide isomerase (PDI), and a unique 97 kDa subunit. PDI appears to be necessary to maintain the structural integrity of MTP, however no mutations in PDI have been reported in subjects with ABL. Wetterau et al demonstrated that MTP activity and the large subunit of MTP were present in intestinal biopsy samples from 8 control persons but were absent in 4 abetalipoproteinemic subjects [41]. The findings proved that MTP is defective in ABL and is required for lipoprotein assembly [41]. In the absence of MTP, apo B cannot be properly lipidated and undergoes rapid presecretory degradation [12,26].

Treatment of ABL

Early diagnosis and treatment is essential to prevent growth retardation and neuro-ophthalmological complications secondary to chronic lipid malabsorption and deficits in the lipid-soluble vitamins.

Low Fat Diet. The steatorrhea and vomiting caused by the lipid malabsorption leads to secondary deficiencies in carbohydrates and proteins. A low-fat diet eliminates these symptoms allowing normal absorption of carbohydrates and proteins. To provide an adequate amount of total calories, the proportion of protein and carbohydrate in the diet must be increased to allow resumption of growth in height and weight. The lipid-poor diet should provide the daily requirements in essential fatty acids in the form of vegetable oils. Oral medium-chain triglycerides (MCTG) have been used to provide dietary fatty acids for absorption through the portal circulation, thus bypassing the defective MTP-mediated assembly of apo B-containing lipoproteins in ABL. However, this treatment is controversial, with suggestion of induced hepatic fibrosis in an ABL patient [20].

Fat Soluble Vitamin Replacement. The main function of vitamin E is thought to be prevention of lipid peroxidation. Therefore, its deficiency leads to an increase in the peroxidation of polyunsaturated fatty acids in photoreceptor cells, myelin, and cell membranes in general. ABL subjects require lifetime therapy with vitamin E in large oral doses of 100-300 mg/kg per day to prevent this complication [27]. Such high doses of vitamin E can be absorbed through the portal system. Plasma levels of vitamin E rarely exceed 10%-30% of normal even after long-term therapy [5]. Nevertheless, the levels in adipose tissue [11,18], hepatic tissue [6], and erythrocytes [11] almost always increase and even normalize with large doses of vitamin E administered by oral and/or parenteral routes.

Vitamin A is thought to stabilize photoreceptor membranes and the pigmented epithelial cells of the retina. In ABL patients, vitamin A deficiency is easily compensated for by oral supplementation because after intestinal absorption and transport to the liver, vitamin A (unlike vitamin E) has its own transport system, which is independent of lipoproteins. Daily doses two- to four-fold of the normal recommended doses are generally required to normalize the levels of vitamin A [6], or its surrogate analyte, beta-carotene.

Vitamin K administration whether orally or parenterally rapidly corrects coagulation factors. The deficit in vitamin K is exacerbated when large doses of vitamin E are absorbed and therefore it is important to administer vitamin K prophylactically when beginning vitamin E therapy.

Vitamin D deficiency is not classically described in ABL because the metabolism of vitamin D does not depend much on apo B-containing lipoproteins, since there is partial absorption via the portal path and specific vitamin D transport proteins. However the development of rickets and osteomalacia has been reported [25] and therefore prophylaxis should be instituted in infants during growth.

Thus, early treatment with Vitamin E and vitamin A appears to prevent the onset of neuroretinal complications of ABL [7,27]. However, vitamin therapy does not typically reverse clinical features if patients are treated too late and neurological and ophthalmological signs are already established [25].

FAMILIAL HYPOBETALIPOPROTEINEMIA (FHBL)

Clinical Features and Pathogenesis

Homozygous FHBL shares most of the above clinical attributes of ABL, with the main distinguishing factor of half-normal plasma concentrations of plasma apo B-containing lipoproteins in heterozygote parents, compared to normal levels in the parent of ABL subjects. Homozygous FHBL is rare, occurring in perhaps 1 in 10^6 persons. As in ABL homozygotes may be detected at a young age because of fat malabsorption and reduced plasma cholesterol levels. Fat malabsorption results from an inability to form chylomicrons in the intestine and a subsequent failure to absorb fats and fat-soluble vitamins. The failure to form chylomicrons is a direct effect of the absence of apoB. Cholesterol absorption is probably also impaired, as demonstrated by a transgenic mouse that lacks intestinal apoB expression and chylomicron formation [42]. Fat malabsorption may lead to a progressive neurologic degenerative disease resulting from vitamin E deficiency. It may also cause retinitis pigmentosa and acanthocytosis. Despite the low plasma cholesterol levels, steroidogenesis appears to be normal except when demands are quite high [21]. Homozygotes who produce enough of a truncated isoform of apoB to facilitate some fat absorption may have a milder phenotype.

Molecular Genetics

FHBL segregates as an autosomal-codominant trait. Homozygotes have clinical and biochemical findings similar to those of ABL with LDL cholesterol in the lowest fifth percentile. However, while obligate heterozygotes for ABL were clinically and biochemically normal, obligate heterozygotes for FHBL were noted to have hypocholesterolemia secondary to low LDL cholesterol concentrations which are usually in the lowest tenth percentile. Apart from hypocholesterolemia, obligate heterozygotes for FHBL are healthy and usually have no difficulty absorbing fat. Genetic linkage analyses in the 1980's indicated that the defect in some cases of FHBL was within the *APOB* gene on chromosome 2p24, which was distinct from the ABL locus. The *APOB* gene defects were mostly truncation-producing mutations of apo B of the nonsense or frameshift variety. However the molecular defects underlying the majority of FHBL cases are not known [34]. Pulai et al reported a four-generation family with FHBL in which linkage to the *APOB* gene was excluded [30]. Affected members in this family were designated to have FHBL2 (MIM 605019). Yuan et al conducted a genome-wide search in this kindred and found strong evidence for linkage of FHBL2 to 3p22-p21.1 [43].

Treatment

The treatment of homozygous FHBL is similar to that of ABL. No specific treatment is indicated for heterozygotes, but dietary supplementation with fat-soluble vitamins (especially vitamin E) is reasonable. Heterozygotes should be informed that if their spouse also has a very low plasma cholesterol level, the possibility exists that offspring could have homozygous or compound heterozygous hypobetalipoproteinemia; in this situation, subjects should be referred to a lipid clinic for genetic counseling.

CHYLOMICRON RETENTION DISEASE

Patients with chylomicron retention disease, also known as Anderson disease, resemble those with ABL with respect to dietary fat malabsorption and its consequences [34]. These patients are able to synthesize the full length form of apo B in the liver (called apoB-100), but not the intestinal form of apo B (called apoB-48). The intestinal transcription of apoB-100 mRNA and its editing to apoB-48 mRNA appear to proceed at normal rates, and both apoB and dietary triglycerides are found in enterocytes, suggesting that enterocytes can produce elements of chylomicrons. Nevertheless, chylomicrons are not secreted [34].

Aguglia et al described 2 brothers, aged 19 and 12 years, with Marinesco-Sjögren syndrome (MIM 248800), both of whom also had very low serum vitamin E concentrations with an absence of postprandial chylomicrons [1]. Findings on electron microscopy of the intestinal mucosa were consistent with a chylomicron retention disease (MIM 246700). Aguglia et al suggested that both chylomicron retention disease and Marinesco-Sjögren syndrome are related to defects in a gene crucial for the assembly or secretion of the chylomicron particles, leading to very low serum levels of vitamin E.

HARP SYNDROME

Orrell et al described three unrelated individuals with orobuccolingual dystonia, mild cognitive impairment, and acanthocytosis [29]. All three of these patients had hypointense signal in the globus pallidus within a high signal region, giving the appearance of the 'eye of the tiger' sign. The authors presented these cases as a variant of Hallervorden-Spatz disease (MIM 234200). In one of the three cases, lipoprotein electrophoresis demonstrated absence of a prebeta band. The authors referred to this case as being the second reported example of the HARP syndrome (MIM 607236) (hypoprebetalipoproteinemia, retinitis pigmentosa, and pallidal degeneration). This term had been introduced by Higgins et al [19]. However, the significance of the lipid abnormality was not clear, inasmuch as the patient's sister and mother had a similar lipid disorder but no retinal or neurologic disease. Recently, this condition was shown to be part of the spectrum of diseases caused by mutation in pantothenate kinase 2 (*PANK2*) [10], thus confirming the earlier hypothesis of a close relationship with Hallervorden-Spatz disease (see also Chapter 8).

ATYPICAL VARIANT OF WOLMAN DISEASE

Eto and Kitagawa described a disorder with features of malabsorption of lipid, vomiting, growth failure, and adrenal calcification [16]. The presence of hypolipoproteinemia and acanthocytosis suggested this to be distinct from Wolman disease (MIM 278000).

SUMMARY

Largely based upon cellular and molecular studies in ABL and FHBL, it is now understood that assembly of apo B-containing lipoproteins requires functional MTP, which integrates lipids into the evolving lipoprotein particle built on a backbone of structurally intact apo B. Homozygosity for a defect either in apo B or MTP cripples this process, and results in failure to secrete apo B-containing lipoproteins into the plasma, with consequent deficiencies of fat-soluble vitamins.

REFERENCES

1. Aguglia U, Annesi G, Pasquinelli G, Spadafora P, Gambardella A, et al (2000) Vitamin E deficiency due to chylomicron retention disease in Marinesco-Sjögren syndrome. *Ann Neurol* **47:** 260-264.
2. Akamatsu K, Sakaue H, Mizukami Y, Yamaguchi S, Tanaka A, Ohta Y (1983) A case report of abetalipoproteinemia (Bassen-Kornzweig syndrome): The first case in Japan. *Jpn J Med* **22:** 231-236.

3. Avigan MI, Ishak KG, Gregg RE, Hoofnagle JH (1984) Morphologic features of the liver in abetalipoproteinemia. *Hepatology* **4:** 1223.
4. Bassen FA, Kornzweig AL (1950) Malformation of the erythrocytes in a case of atypical retinitis pigmentosa. *Blood* **5:** 381-387.
5. Berriot-Varoqueaux N, Aggerbeck LP, Samson-Bouma ME, Wetterau JR (2000) The role of the microsomal triglyceride transfer protein in abetalipoproteinemia. *Annu Rev Nutr* **20:** 663-697.
6. Bieri JG, Hoeg JM, Ernst JS, Zech LA, Brewer HB (1984) Vitamin A, vitamin E replacement in abetalipoproteinemia. *Ann Int Med* **100:** 238-239.
7. Black DD, Rick VH, Rohwer-Nutter PL, Ellinas H, Stephens JK, et al (1991) Intestinal and hepatic apolipoprotein B gene expression in abetalipoproteinemia. *Gastroenterology* **101:** 520-528.
8. Braegger CP, Belli DC, Mentha G, Steinmann B (1998) Persistence of the intestinal defect in abetalipoproteinaemia after liver transplantation. *Europ J Pediat* **157:** 576-578.
9. Caballero FM, Buchanan GR (1980) Abetalipoproteinemia presenting as severe K deficiency. *Pediatrics* **65:** 161-163.
10. Ching KH, Westaway SK, Gitschier J, Higgins JJ, Hayflick SJ (2002) HARP syndrome is allelic with pantothenate kinase-associated neurodegeneration. *Neurology* **58:** 1673-1674.
11. Cogan DG, Rodrigues M, Chu FC, Schaefer EJ (1984) Ocular abnormalities in abetalipoproteinemia. *Ophthalmology* **91:** 991-998.
12. Davis R, Thrift RN, Wu CC, Howell KE (1990) Apolipoprotein B is both integrated into and translocated across the endoplasmic reticulum membrane. *J Biol Chem* **265:** 10005-10011.
13. Delpre G, Kadish U, Glantz I, Aividor I (1978) Endoscopic assessment in abetalipoproteinemia. *Endoscopy* **10:** 59-62.
14. Druez G (1959) Un nouveau cas d'acanthocytose: Dysmorphie erythrocytaire congenitale avec retinite, troubles nerveux et stigmates degeneratifs. *Rev Hemat* **14:** 3-11.
15. Duker JS, Belmont J, Bosley TM (1987) Angioid streaks associated with abetalipoproteinemia. Case report. *Arch Ophthalmol* **105:** 1173-1174.
16. Eto Y, Kitagawa T (1970) Wolman's disease with hypolipoproteinemia and acanthocytosis: Clinical and biochemical observations. *J Pediat* **77:** 862-867.
17. Friedman IS, Cohn H, Zymaris M, Goldner MG (1960) Hypocholesterolemia in idiopathic steatorrhea. *Arch Int Med* **105:** 112-121.
18. Hegele RA, Angel AA (1985) Arrest of neuropathy and myopathy in abetalipoproteinemia with high-dose vitamin E therapy. *Can Med Assoc J* **132:** 41-44.
19. Higgins JJ, Patterson MC, Papadopoulos NM, Brady RO, Pentchev PG, Barton NW (1992) Hypoprebetalipoproteinemia, acanthocytosis, retinitis pigmentosa, and pallidal degeneration (HARP syndrome). *Neurology* **42:** 194-198.
20. Illingworth DR, Connor WE, Miller RG (1980) Abetalipoproteinemia. Report of two cases and review of therapy. *Arch Neurol* **37:** 659-662.
21. Illingworth DR, Kenny TA, Orwoll ES (1982) Adrenal function in heterozygous and homozygous hypobetalipoproteinemia. *J Clin Endocrinol Metab* **54:** 27-33.
22. Kane JP, Havel RJ (1995) Disorders of the biogenesis and secretion of lipoproteins containing the B apolipoproteins. In: *Scriver CR, Beaudet AL, Sly WS, Valle D (eds) The Metabolic and Molecular Basis of Inherited Disease.* McGraw-Hill, New York, pp 1853-1885.
23. Kayden HJ (1980) Is essential fatty acid deficiency part of the syndrome of abetalipoproteinemia? *Nutr Rev* **38:** 244-246.
24. Kayden HJ, Silber R (1965) The role of vitamin E deficiency in the abnormal autohemolysis of acanthocytosis. *Trans Assoc Am Phys* **78:** 334-342.
25. Lazaro RP, Dentinger MP, Rodichok LD, Barron KD, Satya-Murti S (1986) Muscle pathology in Bassen-Kornzweig syndrome and vitamin E deficiency. *Am J Clin Pathol* **86:** 378-387.

26. Leiper J, Bayliss JD, Pease RJ, Brett DJ, Scott J, Shoulders CC (1994) Microsomal triglyceride transfer protein, the abetalipoproteinemia gene product, mediates the secretion of apolipoprotein B-containing lipoproteins from heterologous cells. *J Biol Chem* **269:** 21951-21954.
27. Muller DPR, Lloyd JK, Wolff OH (1985) The role of vitamin E in the treatment of the neurological features of abetalipoproteinemia and other disorders of fat absorption. *J Inherit Metab Dis* **8:** 88-92.
28. Narcisi TME, Shoulders CC, Chester SA, Read J, Brett DJ, et al (1995) Mutations of the microsomal triglyceride transfer protein gene in abetalipoproteinemia. *Am J Hum Genet* **57:** 1298-1310.
29. Orrell RW, Amrolia PJ, Heald A, Cleland PG, Owen JS, Morgan-Hughes JA, Harding AE, Marsden CD (1995) Acanthocytosis, retinitis pigmentosa, and pallidal degeneration: A report of three patients, including the second reported case with hypoprebetalipoproteinemia (HARP syndrome). *Neurology* **45:** 487-492.
30. Pulai JI, Neuman RJ, Groenewegen AW, Wu J, Schonfeld G (1998) Genetic heterogeneity in familial hypobetalipoproteinemia: Linkage and non-linkage to the apoB gene in Caucasian families. *Am J Med Genet* **76:** 79-86.
31. Rader DJ, Brewer HB (1993) Abetalipoproteinemia: New insights into lipoprotein assembly and vitamin E metabolism from a rare genetic disease. *J Am Med Assoc* **270:** 865-869.
32. Salt HB, Wolff OH, Lloyd JK, Fosbrooke AS, Cameron AH, Hubble DV (1960) On having no beta-lipoprotein: A syndrome comprising abetalipoproteinaemia, acanthocytosis and steatorrhoea. *Lancet* 2: 325-329
33. Sharp D, Blinderman L, Combs KA, Kienzle B, Ricci B, Wager-Smith K, Gil CM, Turck CW, Bouma ME, Rader DJ, Aggerbeck LP, Gregg RE, Gordon DA, Wetterau JR (1993) Cloning and gene defects in microsomal triglyceride transfer protein associated with abetalipoproteinaemia. *Nature* **365:** 65-69.
34. Schonfeld G (1995) The hypobetalipoproteinemias. *Annu Rev Nutr* **15:** 23-34.
35. Simon ER, Ways P (1964) Incubation hemolysis and red cell metabolism in acanthocytosis. *J Clin Invest* **43:** 1311-1321.
36. Singer K, Fisher B, Perlstein MA (1952) Acanthrocytosis: A genetic erythrocytic malformation. *Blood* **7:** 577-591.
37. Sobrevilla LA, Goodman ML, Kane CA (1964) Demyelinating central nervous system disease, macular atrophy and acanthocytosis (Bassen-Kornzweig syndrome). *Am J Med* **37:** 821-832.
38. Wang J, Hegele RA (2000) Microsomal triglyceride transfer protein (MTP) gene mutations in Canadian subjects with abetalipoproteinemia. *Hum Mutat* **15:** 294-295.
39. Ways P, Reed CF, Hanahan DJ (1963) Red-cell and plasma lipids in acanthocytosis. *J Clin Invest* **42:** 1248-1260.
40. Wetterau JR, Aggerbeck LP, Laplaud PM, McLean LR (1991) Structural properties of the microsomal triglyceride-transfer protein complex. *Biochemistry* **30:** 4406-4412.
41. Wetterau JR, Aggerbeck LP, Bouma ME, Eisenberg C, Munck A, Hermier M, Schmitz J, Gay G, Rader DJ, Gregg RE (1992) Absence of microsomal triglyceride transfer protein in individuals with abetalipoproteinemia. *Science* **258:** 999-1001.
42. Young SG, Cham CM, Pitas RE, et al (1995) A genetic model for absent chylomicron formation: Mice producing apolipoprotein B in the liver, but not in the intestine. *J Clin Invest* **96:** 2932-2946.
43. Yuan B, Neuman R, Duan SH, Weber JL, Kwok PY, Saccone NL, Wu JS, Liu KY, Schonfeld G (2000) Linkage of a gene for familial hypobetalipoproteinemia to chromosome 3p21.1-22. *Am J Hum Genet* **66:** 1699-1704.

CHAPTER 4

LEVINE-CRITCHLEY SYNDROME OF NEUROACANTHOCYTOSIS: A CLINICAL REVIEW

RICHARD J. HARDIE
Department of Neurology, Atkinson Morley's Hospital, Copse Hill, London, UK

Abstract. The term Levine-Critchley syndrome has been applied to an inherited multi-system degenerative neurological disorder associated with acanthocytosis in the absence of any lipid abnormality. Involuntary movements particularly affect the orofacial region and may cause dysarthria, dysphagia and vocalisations as well as tongue and lip biting which, when present, are virtually diagnostic. Psychiatric features, seizures, peripheral neuropathy and elevated plasma creatine kinase levels are all common. However, recent advances in molecular biology have shown the syndrome to be genetically heterogeneous, with mutations in at least two separate genes, both the *CHAC* and McLeod genes being associated with very similar phenotypes.

INTRODUCTION AND HISTORICAL BACKGROUND

The association of chorea with acanthocytosis was initially described in 1967 in two separate North American kindreds [3,8], and fewer than 70 cases have since been reported, at least in the English literature. The condition is actually an inherited multi-system degenerative neurological disorder with acanthocytosis in the absence of any lipid abnormality, and seems to be more common in Japan (see Chapter 5).

Descriptive terms variously applied to this association have included *familial amyotrophic chorea with acanthocytosis* or *chorea-acanthocytosis*, emphasizing the commonest features. Because chorea does not occur universally, other designations have included *familial neuroacanthocytosis* or the *Levine-*

A. Danek (ed.), Neuroacanthocytosis Syndromes, 31–37.

Critchley syndrome. Many cases have affected relatives, but various authors have proposed conflicting patterns of inheritance over the last 35 years.

Acanthocytosis was also reported in 1977 to be a feature, both in males and in female carriers [24], in a rare X-linked abnormality of expression of the Kell blood group antigens known as the McLeod phenotype (see Chapter 6). Males have elevated serum creatine kinase (CK) levels and a mild myopathy, and obscure reference had been made in 1983 to the coexistence of chorea and areflexia in certain male McLeod subjects [17].

The New England Family Described by Levine

In 1967 a *new hereditary acanthocytosis syndrome* was described in 9 out of 15 members amongst three generations in a family from New England, and autosomal-dominant inheritance was postulated [8]. The clinical features of the 50-year-old male proband were detailed by Levine et al [15]. The development at 27 years of progressive weakness, wasting and involuntary movements of the limbs was followed by seizures and dementia. Chorea also involved the face and tongue. Tendon reflexes were absent and sensation was impaired distally. Muscle biopsy was compatible with denervation. His brother was similarly affected, and both had normal lipoproteins, elevated CK levels and $> 10\%$ acanthocytosis.

The proband's 5-year-old daughter was the only other family member with that level of abnormal red cells, and was classified as neurologically abnormal on the basis of "hyporeflexia, slight weakness left peroneal muscles". There are discrepancies between these 2 and other even earlier reports that appeared elsewhere. For example, acanthocytosis was indicated in the proband's mother to be first absent [8] and then present [15]. Moreover, she was reported to be "the healthiest member of the family", but also to have mild left-sided hyperreflexia and mild spastic drag on the left, mild digital tremor, whilst her sister was said to have chorea [15].

The Critchley Cases

A syndrome of acanthocytosis, involuntary vocalizations and mutilating orofacial dyskinesia was also reported by Critchley et al [3,4] in a 29-year-old man from Kentucky. He frequently bit his lips and tongue, was dysarthric and made grunting and sucking noises. A wide spectrum of involuntary movements also affected the limbs including tics, dystonic and choreiform movements, and he had originally been given a tentative diagnosis of Tourette syndrome. Hypotonia and areflexia were also present and vitamin E and lipid studies were normal. Three siblings had already died young having had similar involuntary movements, and a fourth was areflexic with mild chorea and acanthocytosis. 2 of these siblings had had seizures. Critchley later described very similar findings in a young English woman [5].

Two male siblings of the first proband had asymptomatic acanthocytosis [4]. A 25-year-old daughter of one of these brothers had a neurological disorder resembling Friedreich's disease with ataxia, deafness and a cardiomyopathy but no acanthocytosis, and so considered to be coincidental.

CLINICAL FEATURES

A review in 1991 reported the wide spectrum of clinical features encountered in a series of 19 English cases with neuroacanthocytosis and also reviewed the 30 or so previously reported patients [11]. We identified 7 non-familial cases, together with a sibling pair and 10 affected members from two other families. Two examples of the McLeod phenotype were identified amongst 6 affected cases in one of the families, including one female. All other living cases were negative for the McLeod phenotype, and pedigrees were consistent with autosomal-recessive or X-linked inheritance. There is only one convincingly dominant pedigree which was in itself unusual because of the occurrence of hypobetalipoproteinemia in the proband and son [14]. Another atypical family with different clinical, genetic and neuropathological features is described elsewhere (see Chapter 7).

The commonest age of onset tends to be early in the fourth decade although cases as young as 8 and as old as 62 at onset have been described. The main clinical features were neuropsychiatric and movement disorders, with a subclinical axonal polyneuropathy. Extensor plantar responses and sensory deficits were seldom found. Treatment is purely symptomatic and similar to that of Huntington's disease (HD) (see Chapter 27).

Neuropsychiatric Features

Cognitive impairments, psychiatric features, behaviour disturbance and organic personality change consistent with frontal lobe dysfunction are common (see Chapter 12). Psychiatric symptoms are usual and may include depression, anxiety and obsessive-compulsive disorders. In a neuropsychological review 9 of the 10 patients studied were impaired on tests of executive skills consistent with subcortical dementia. Seizures are also common, occurring in more than one-third of reported cases [11] and were sometimes the presenting feature.

Movement Disorders

A full range of movement disorders may be seen including tics, dystonia and akinetic-rigid syndromes. Chorea affecting the limbs resembles that seen in HD. Dystonia was sometimes the presenting and predominant movement disorder [11]. The combination of chorea, dystonia, and exaggerated extensor posturing can result in a bizarre gait that may be mistaken for a functional gait disorder. Progression from a hyperkinetic state to parkinsonism has been observed as well as both together. Rarely, parkinsonism may be the presenting feature [11,20].

The most striking clinical feature is that of the orofaciolingual movement disorder, characteristically a combination of dystonia and chorea, and marked pseudobulbar disturbance [1,11]. Dysarthria and dysphagia are common and sometimes progress to anarthria and severe feeding difficulty requiring gastrostomy. Dystonic protrusion of the tongue induced by chewing and attempted swallowing is sometimes referred to as feeding dystonia. It can result in mutilating lip and tongue biting. Some patients develop the habit of holding objects, typically the corner of a handkerchief or something similar, both to protect the buccal and tongue mucosa and presumably inhibit the involuntary movements. It may be helped somewhat by botulinum toxin injection into the tongue base (personal observation). Involuntary vocalisations caused by phonic tics such as grunting and clicking are also common.

Hematology and Biochemistry

True acanthocytes should probably never be present in normal blood. However, examination of fresh blood for acanthocytes by an experienced hematologist is necessary for detection and also to avoid false positives. Repeated blood film examination may be necessary to confirm the diagnosis because acanthocytes can be elusive, until late in the disease course [22]. Indeed, two cases of neuroacanthocytosis without acanthocytes have even been published [16,18]. This and other hematological aspects are reviewed in more detail in other chapters.

Serum lipoprotein electrophoresis is by definition normal, to exclude disorders such as abetalipoproteinemia with secondary vitamin E deficiency, an important differential diagnosis. Most cases have mildly elevated levels of plasma creatine kinase (CK) [11], strongly suggestive of a membrane defect but the exact cause is still uncertain.

Cranial Neuroimaging

Both computerised tomography and magnetic resonance imaging (MRI) have been reported to show caudate and more generalised cerebral atrophy similar to HD. Increased signal on T2-weighted MRI in the caudate and putamen is also seen [6,11]. Positron emission tomography (PET) studies have demonstrated reduced blood flow and oxygen metabolism throughout the brain but particularly in the caudate nucleus, putamen and frontal cortex. PET studies utilising [^{18}F]dopa have demonstrated reduced posterior putamen uptake comparable to levels seen in patients with Parkinson's disease [2,20,23].

Peripheral Nerve, Muscle and Neuropathology

About half the reported cases have an associated neuropathy, usually characterised clinically by reduced or absent tendon reflexes. Electrophysiologically, sensory nerve action potentials are reduced but nerve conduction velocities are always normal. Peripheral nerve examinations have indicated predominant involvement of the large diameter myelinated fibres and histological changes of a chronic axonal neuropathy [11]. Some reports have suggested a demyelinating process, but these changes may have been secondary to axonal degeneration.

Electromyography (EMG) may show evidence of denervation. There have also been some reports of prominent muscle wasting with weakness but, despite elevated serum CK levels, muscle biopsies are consistently normal apart from neurogenic changes.

Postmortem studies of the brain have been described in at least 11 patients (reviewed in Chapter 11).

PHENTOTYPE-GENOTYPE CORRELATION AND DIFFERENTIAL DIAGNOSIS

The series reported over a decade ago by this author and his colleagues were presumed to share a common diagnosis of neuroacanthocytosis [11]. However, the passage of time and the application of molecular genetics have shown these 19 cases to have a number of different genotypes, and the same is presumably true of other published cases. Family B and several other sporadic cases have been found to have mutations of the *CHAC* gene [7,21]. Case 17 later developed an abnormality of lipid metabolism and to conform to the phenotype known as HARP (hypoprebetalipoproteinemia and retinitis pigmentosa) [19]. Family L had already been identified to contain at least 2 cases of McLeod phenotype including the first ever manifesting female. Subsequent work identified the X-linked gene mutations in these cases [12]. It has thus been suggested that the term Levine Critchley syndrome should be reserved purely for McLeod negative cases.

However, there are a number of problems with this. The neurological features are very similar, apart from less prominent orofacial dyskinesia and pseudobulbar disturbance and an absence of dystonia and parkinsonism in McLeod families. The McLeod phenotype was not tested for in the original Levine or Critchley kindreds, and was only documented to be negative in 2 pre-1991 case reports [10,20]. Even neurologically affected family members with proven heterozygous frameshift McLeod mutations may express Kell antigens sufficiently strongly not to be classified as having the classical blood group serotype.

Indeed, Levine's pedigree is strikingly similar to Family L and could well have had the McLeod syndrome [15]. One niece of Critchley's first proband had a phenotype resembling Friedreich's ataxia, for which the gene is also on chromosome 9q, and co-segregation with the *CHAC* gene could explain this observation.

Whilst of historical interest, therefore, the use of the label Levine-Critchley syndrome cannot be recommended, and the terms chorea- or neuro-acanthocytosis with or without the McLeod serotype are preferable for affected cases with normal lipoproteins and vitamin E levels. In the future, genotyping will be essential to permit a precise diagnosis and accurate genetic counselling. We eagerly await further advances in molecular biology to offer more specific and effective treatment for the unfortunate victims of this yet incurable disorder.

REFERENCES

1. Brin MF (1993) Acanthocytosis. In: Vinken PJ, Bruyn GW, Klawans HL (eds) *Handbook of Clinical Neurology.* Elsevier, Amsterdam, pp 271-299
2. Brooks DJ, Ibanez V, Playford ED et al (1991) Presynaptic and postsynaptic striatal dopaminergic function in neuroacanthocytosis: A positron emission tomographic study. *Ann Neurol* **30:** 166-171.
3. Critchley EMR, Clark DB, Wikler A (1967) An adult form of acanthocytosis. *Trans Amer Neurol Assoc* **92:** 132-137.
4. Critchley EMR, Clark DB, Wikler A (1968) Acanthocytosis and neurological disorder without abetalipoproteinemia. *Arch Neurol* **18:** 134-140.
5. Critchley EMR, Betts JJ, Nicholson JT, Weatherall DJ (1970) Acanthocytosis, normolipoproteinaemia and multiple tics. *Postgrad Med J* **46:** 698-701.
6. Dervin JE, Kendall BK, Hardie RJ (1991) Neuroacanthocytosis: Correlation of clinical and neuroimaging abnormalities. *Neuroradiology* **33(Suppl):** 575-577.
7. Dobson-Stone C, Danek A, Rampoldi L, Hardie R, Chalmers R, Wood N et al (2002) Mutational spectrum of the *CHAC* gene in patients with chorea-acanthocytosis. *Eur J Hum Genet* **10:** 773-781.
8. Estes JW, Morley TJ, Levine IM, Emerson CP (1967) A new hereditary acanthocytosis syndrome. *Amer J Med* **42:** 868-881.
9. Faillace RT, Kingston WJ, Nanda NC, Griggs RC (1982) Cardiomyopathy associated with the syndrome of amyotrophic chorea and acanthocytosis. *Ann Intern Med* **96:** 616-7.
10. Gross KB, Skrivanek JA, Carlson KC, Kaufman DM (1985) Familial amyotrophic chorea with acanthocytosis. New clinical and laboratory investigations. *Arch Neurol* **42:** 753-756.
11. Hardie RJ, Pullon HWH, Harding AE et al (1991) Neuroacanthocytosis. A clinical, haematological and pathological study of 19 cases. *Brain* **114:** 13-49.
12. Ho MF, Chalmers RM, Davis MB, Harding AE, Monaco AP (1996) A novel point mutation in the McLeod syndrome gene in neuroacanthocytosis. *Ann Neurol* **139:** 672-675.
13. Kartsounis LD, Hardie RJ (1996) The pattern of cognitive impairments in neuroacanthocytosis. *Arch Neurol* **53:** 77-80.
14. Kito S, Itoga E, Hiroshige Y, Matsumoto N, Miwas S (1980) A pedigree of amyotrophic chorea with acanthocytosis. *Arch Neurol* **37:** 514-517.
15. Levine IM, Estes JW, Looney JM (1968) Hereditary neurological disease with acanthocytosis: A new syndrome. *Arch Neurol* **19:** 403-409.
16. Malandrini A, Fabrizi GM, Palmeri S et al (1993) Choreo-acanthocytosis like phenotype without acanthocytes: Clinicopathological case report. A contribution to the knowledge of the functional pathology of the caudate nucleus. *Acta Neuropathol* **86:** 651-658.
17. Marsh WL (1983). Deleted antigens of the Rhesus and Kell blood groups: association with cell membrane defects. In: *Garratty G (ed) Blood Group Antigens and Disease. American Assoc Blood Banks.* Arlington, pp 165-185.
18. O'Brien CF, Schwarz J, Kurlan R (1990) Neuroacanthocytosis without acanthocytes. *Mov Dis* **5:** 98.
19. Orrell RW, Amrolia PJ, Heald A et al (1995) Acanthocytosis, retinitis pigmentosa, and pallidal degeneration: a report of three patients, including the second reported case with hypoprebetalipoproteinaemia (HARP syndrome). *Neurology* **45:** 487-492.
20. Peppard RF, Lu CS, Chu N, Martin WRW, Calne DB (1990) Parkinsonism with neuroacanthocytosis. *Can J Neurol Sci* **17:** 298-301.

21. Rubio JP, Danek A, Stone C et al (1997) Chorea-acanthocytosis: genetic linkage to chromosome 9q21. *Am J Hum Genet* **61:** 899-908.
22. Sorrentino G, De Renzo A, Miniello S, Nori O, Bonavita V (1999) Late appearance of acanthocytes during the course of chorea-acanthocytosis. *J Neurol Sci* **163:** 175-178.
23. Tanaka M, Hirai S, Kondo S et al (1998) Cerebral hypoperfusion and hypometabolism with altered striatal signal intensity in chorea-acanthocytosis: a combined PET and MRI study. *Mov Dis* **13:** 100-107.
24. Wimer BM, Marsh WL, Taswell HF, Galey WR (1977) Haematological changes associated with the McLeod phenotype of the Kell blood group system. *Br J Haematol* **36:** 219-224.

CHAPTER 5

CHOREA-ACANTHOCYTOSIS WITH THE EHIME-DELETION MUTATION

SHU-ICHI UENO[1], KAZUE KAMAE[1], YORIAKI YAMASHITA[2], YOSHIKO MARUKI[1], YUKO TOMEMORI[1], MASAYUKI NAKAMURA[1], MANABU IKEDA[1], HIROTAKA TANABE[1], AND AKIRA SANO[1,3]

[1]Department of Neuropsychiatry, Ehime University School of Medicine;
[2]Department of Neurology, Matsuyama Red Cross Hospital, Ehime;
[3]present address: Department of Psychiatry, Kagoshima University Graduate School of Medical and Dental Sciences, Kagoshima; Japan

Abstract. Chorea-acanthocytosis (ChAc) is a hereditary neurodegenerative disease showing Huntington disease-like neuropsychiatric symptoms and peripheral blood red cell acanthocytosis. Recently, we have identified the gene, *CHAC*, encoding a newly discovered protein, chorein, in which a deletion mutation was found in three Japanese ChAc families. Although four patients possessed the same mutation homozygously, their clinical characteristics varied, which means that multifactorial effects on the pathogenesis are present. Even some of the heterozygous carriers in the families showed a slight degree of acanthocytosis and psychiatric features including emotional lability or cognitive disturbance. We found heterozygous carriers of the deletion-mutation allele in the group of patients with mood disorder. The *CHAC* gene product, chorein, may function as an important protein in the brain not only for motor but also for mental function.

Chorea-acanthocytosis (ChAc) is a hereditary neurodegenerative disorder showing Huntington disease-like neuropsychiatric symptoms and peripheral blood red cell acanthocytosis [1]. Recently, we have identified the gene, *CHAC*, encoding a newly discovered protein, chorein, in which a deletion mutation was found in three Japanese ChAc families [2].

All four ChAc patients from three families had the same Ehime-deletion mutation described in the following, however, the clinical pictures were quite different. The age of onset of the disease varies from 16 to 36. The first

A. Danek (ed.), Neuroacanthocytosis Syndromes, 39–43.

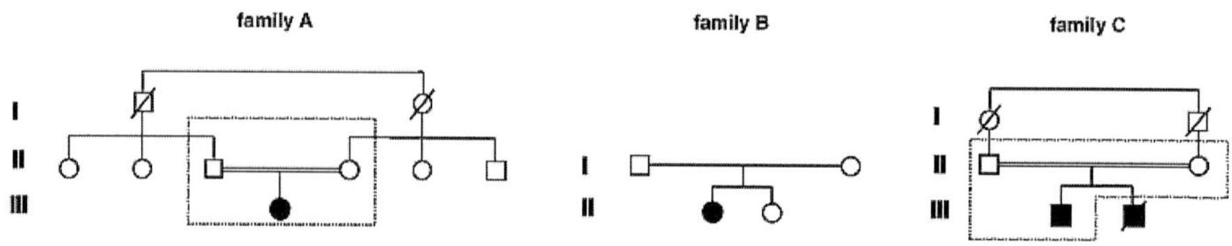

Figure 1. Pedigrees of the four ChAc patients. All three pedigrees originated in Ehime prefecture, a specific small area in Japan.

symptom seen in Case 1 was autistic tendency, and the diagnosis at the time was schizophrenia. She had been admitted to a psychiatric hospital. Epileptic seizures occurred in two patients out of four. All of the patients showed oral involuntary movements like spitting, clicking their tongues, and protrusion of their tongues. Two of them showed severe oral self-mutilation.

The *CHAC* cDNA consists of 9,859 bp including an open reading frame of 9,288 bp (accession number AB054005) [2]. The deduced protein, chorein, consists of 3,095 amino acid residues, and the 260 bp-deletion mutation (Ehime deletion-mutation) in the cDNA from the disease allele leads to a frame shift resulting in the production of a truncated protein. The 260 bp-deletion in the cDNA is due to the 5,937 bp-deletion including two exons in genomic DNA of the disease allele.

Some of the ChAc patients' parents, obligate carriers who were heterozygous for the deletion mutation, showed a slight degree of acanthocytosis in their

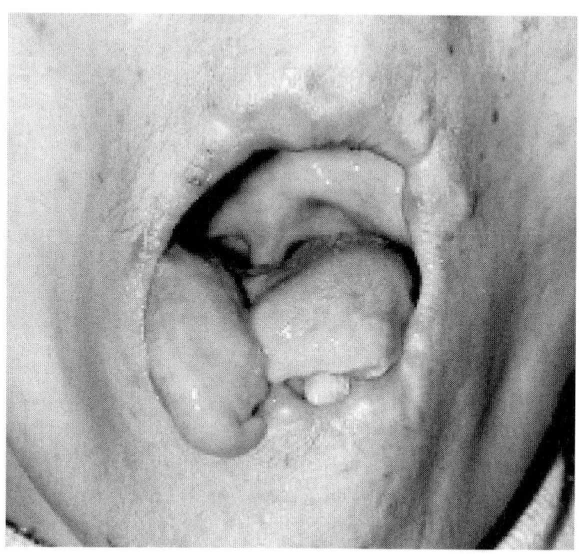

Figure 2. Oral self-mutilation [2]. Severe oral self-mutilation of left side of tongue and lips was observed in Case 1.

Table 1. Clinical summary of four ChAc patients.

	Family A	Family B	Family C	
	Case 1	Case 2	Case 3	Case 4
Age	39	33	40*	45
Sex	female	female	male	male
Consanguinity	+	unknown	+	+
Age at onset	16	27	21	36
First symptom	psychiatric autistic	epileptic seizure	oral dyskinesia/ dystonia	gait instability & epileptic seizure
Chorea	+	+	+	+
Oral dyskinesia-dystonia	+	+	+	+
Oral self-mutilation	+	−	+	−
Epilepsy	−	+	−	+
Psychiatric symptoms	autistic, psychotic	subcortical dementia (frontal lobe deficits dominant)	severe dementia	severe dementia
Elevated CK	+	+	+	+

*age at death

peripheral blood and/or psychiatric features like emotional lability or cognitive disturbance. This fact indicates that the mode of inheritance in this disease is semidominant with a relatively low penetrance. Neurological evaluation of the ChAc-non-affected family members has been performed, and none of the members showed ChAc-specific neurological abnormalities except for one parent's cryptogenic abducens palsy. On psychiatric evaluation, one parent showed a tendency to emotional lability, and another showed slight cognitive impairment. Hosler et al recently performed the linkage of familial amyotrophic lateral sclerosis with frontotemporal dementia (FALS/FTD), and they obtained evidence suggesting that a defective gene located in the chromosome 9q21-q22 region may be linked to FALS/FTD [3], in which the *CHAC* gene is located [4,5]. These facts indicate that the *CHAC* gene mutation might be responsible for some types of mental or neurodegenerative disorders.

Then we analyzed the deletion mutation in a series of patients with psychiatric diseases, including Alzheimer's disease (AD), frontotemporal dementia, and epilepsy. Normal healthy Japanese and Caucasians were used as controls. No deletion allele was identified in Japanese and Caucasian controls and in patients with AD, schizophrenia, idiopathic epilepsy and hereditary spinocerebellar ataxia. However, two patients with mood disorder were found to be heterozygous for the deletion allele ($p = 0.25$, $\chi^2 = 1.34$). Both suffered from monopolar, recurrent depressive disorder.

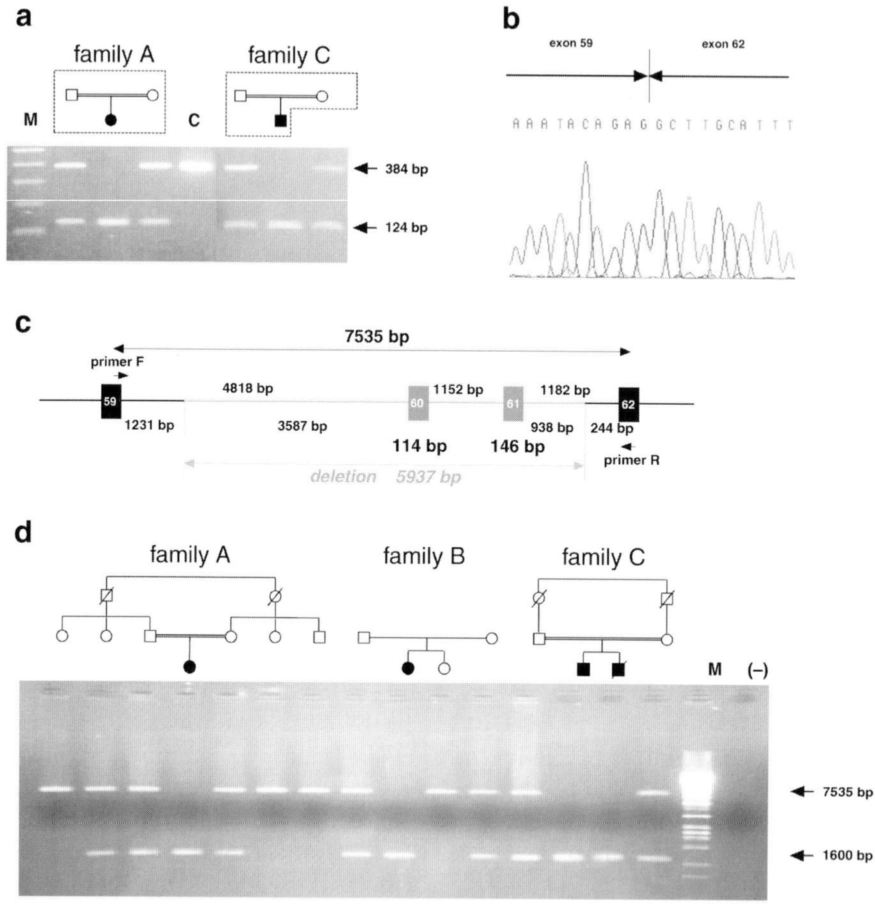

Figure 3. Identification of the Ehime-deletion mutation [2].
a, Reverse transcription-polymerase chain reaction (RT-PCR) of leukocyte cDNA from patients and their parents in families A and C. Hatched line, individuals examined in the pedigree trees in a. M, 100-bp ladder marker; C, PCR product of a normal control subject.
b, Frame-shift mutation in the leukocyte cDNA clone from a patient with ChAc.
c, Deletion mutation in genomic DNA in the three ChAc pedigrees. Primers F and R were used for amplification.
d, Genomic PCR of the deletion region in the three ChAc families. Normal allele 7535 bp, deleted allele 1600 bp. The same 5937-bp segment including exons 60 and 61 is deleted in three ChAc families. The deletion allele has two bases inserted (thymine and cytosine) at the deleted site. M, λ phage cut with the Eco130I marker; (-), PCR product without template.

It is of much interest that the deletion mutant allele was found only in patients with mood disorder. Many organic brain syndromes show mood changes as part of the symptoms, and it is highly possible that mood disorder consists of various diseases due to different etiology. Both mood disorder patients

Table 2. Frequency of the 260 bp-deletion (Ehime-deletion mutation) allele.

	number of chromosomes	number of mutated alleles
Caucasian controls	148	0
Japanese controls	200	0
Schizophrenia	140	0
Mood disorder	296	2
Alzheimer's disease	334	0
Frontotemporal dementia	22	0
Idiopathic epilepsy	50	0

who had the *CHAC* deletion were diagnosed as recurrent depressive disorder. They showed no physical symptoms and no acanthocytes were detected in their peripheral blood samples. They responded well to antidepressant medication.

Further studies, including experiments at the molecular level are needed to understand the real function of chorein in the nervous system.

REFERENCES

1. Brin M (1993) *Handbook of Clinical Neurology* **19:** 271-299 Elsevier Science Publishers, Amsterdam, The Netherlands.
2. Ueno S, Maruki Y, Nakamura M, Tomemori Y, Kamae K, Tanabe H, Yamashita Y, Matsuda S, Kaneko S, Sano A (2001) The gene encoding a newly discovered protein, chorein, is mutated in chorea-acanthocytosis. *Nature Genet* **28:** 121-122.
3. Hosler BA, Siddique T, Sapp PC, Sailor W, Huang MC, Hossain A, Daube JR, Nance M, Fan C, Kaplan J, Hung W-Y, McKenna-Yasek D, Haines JL, Pericak-Vance MA, Horvitz HR, Brown RH (2000) Linkage of familial amyotrophic lateral sclerosis with frontotemporal dementia to chromosome 9q21-q22. *JAMA* **284:** 1664-1669.
4. Rubio JP, Danek A, Stone C, Chalmers R, Wood N, Verellen C, Ferrer X, Malandrini A, Fabrizi GM, Manfredi M, Vance J, Pericak-Vance M, Brown R, Rudolf G, Picard F, Alonso E, Brin M, Nemeth AH, Farrall M, Monaco AP (1997) Chorea-acanthocytosis: genetic linkage to chromosome 9q21. *Am J Hum Genet* **61:** 899-908.
5. Rampoldi L, Dobson-Stone C, Rubio JP, Danek A, Chalmers RM, Wood NW, Verellen C, Ferrer X, Malandrini A, Fabrizi GM, Brown R, Vance J, Pericak-Vance, M, Rudolf G, Carre S, Alonso E, Manfredi M, Nemeth AH, Monaco AP (2001) A conserved sorting-associated protein is mutant in chorea-acanthocytosis. *Nature Genet* **28:** 119-120.

CHAPTER 6

MCLEOD SYNDROME: A CLINICAL REVIEW

HANS H. JUNG

Department of Neurology, University Hospital Zürich, Switzerland

Abstract. McLeod syndrome is an X-linked multi-system disorder that is classified with the neuroacanthocytosis syndromes. It is caused by mutations of the *XK* gene encoding the XK protein, a membrane transport protein of yet unknown function. Hematologically, McLeod syndrome is characterized by an absent Kx erythrocyte antigen, weak expression of Kell antigens, acanthocytosis, and compensated hemolysis. Asymptomatic male McLeod carriers have elevated serum creatine kinase levels, and are prone to develop neurological symptoms at a mean onset ranging from 30 to 40 years. Neuromuscular manifestations include myopathy, sensory motor axonal neuropathy, and cardiomyopathy. Central nervous system manifestations resemble Huntington's disease, with a choreatic movement disorder, subcortical cognitive deficits, psychiatric abnormalities, and generalized seizures.

INTRODUCTION

The McLeod blood group phenotype was detected by routine screening for new antibodies and was named after the first propositus [1]. It is characterized by absent expression of Kx erythrocyte antigen, weak expression of Kell glycoprotein antigens, and X-linked inheritance [29]. Marsh and colleagues demonstrated that male carriers of the McLeod blood group phenotype have elevated serum levels of creatine kinase (CK) indicating a muscle cell abnormality [21]. Subsequently, they observed that carriers of the McLeod blood group phenotype had a "neurological disorder characterized by involuntary dystonic or choreiform movements, areflexia, wasting of limb muscles, elevated CK, and congestive cardiomyopathy" [26]. This first short description as well as the subsequent clinical observations define the McLeod syndrome as a multi-system disorder with hematological, neuromuscular, and central nervous system involvement [6,8,12].

A. Danek (ed.), Neuroacanthocytosis Syndromes, 45–53.

In 1994, Ho and co-workers discovered the causative *XK* gene [10]. The *XK* gene encodes the XK protein, a membrane transport protein of yet unknown function, that carries the Kx erythrocyte antigen [10]. The vast majority of disease-causing mutations comprise deletions, nonsense mutations, or splice site mutations predicting an absent or truncated XK protein devoid of the Kell protein binding site [6,10,12]. The XK protein is linked to the Kell glycoprotein, and the two proteins most probably form a functional complex [25]. Although the precise role of the XK-Kell-complex in brain and muscle is not known, preliminary data suggest an important role in muscle and nerve cell physiology [17,27].

DIAGNOSIS OF MCLEOD SYNDROME

Diagnosis of McLeod syndrome is based on the immunohematologic findings of absent Kx antigen and weakened Kell glycoprotein antigens [1]. In virtually all McLeod patients, associated laboratory findings such as erythrocyte acanthocytosis, compensated hemolysis, and elevated serum CK levels are present [6,8,12,29]. In addition, about a third of McLeod patients have a hepatosplenomegaly [6]. In most carriers of the McLeod blood group phenotype, hematological abnormalities are not symptomatic. Boys with chronic granulomatous disease carrying the McLeod blood group phenotype who receive multiple transfusions may have serious transfusion hazards due to allogenic antibody production. However, transfusion hazards have to be considered in all carriers of the McLeod blood group phenotype.

Consequently, many carriers of the McLeod blood group phenotype are recognized in blood banks. In most McLeod blood group phenotype carriers reported in the hematological literature, data concerning possible neurological involvement are not available, and no cognitive, neuromuscular, cardiac, and striatal imaging examinations were performed [1,18]. However, long term follow-up in some primarily asymptomatic McLeod blood group phenotype carriers, including the initial index patient, demonstrated development of neurological manifestations during the disease course [6]. Onset of symptoms in McLeod patients ranges between 18 years and 61 years, with a mean onset age of about 35 years (Table 1) [6,8,12]. There is converging evidence that the clinical penetrance of McLeod syndrome is high, in particular after the age of 50 years. However, clinical diagnosis is frequently delayed due to the gradual onset and wide diversity of signs and symptoms, often leading to misdiagnosis such as familial hemolytic anemia, Huntington's disease, Tourette's syndrome, myopathy, or spinal muscular atrophy.

CENTRAL NERVOUS SYSTEM INVOLVEMENT

CNS manifestations of McLeod syndrome resemble Huntington's disease. Symptoms comprise the prototypic triad of a progressive neuro-degenerative

Table 1. Signs and symptoms of 35 McLeod patients at disease onset (reported in refs. 6, 8, and 12).

Sign or symptom	Percentage of Patients (Number)	Mean Age (Range; in years)
Kell antigen phenotype	23% (8)	n.a.
Movement disorder	31% (11)	37.6 (18 – 57)
Psychiatric abnormalities	20% (7*)	33.3 (25 – 51)
Seizures	20% (7)	42.1 (22 – 61)
Muscle atrophy/weakness	3% (1)	26
Cardiopathy	3% (1)	40
Hepato-splenomegaly	6% (2)	21; 34

*second symptom in 2 patients

basal ganglia disease including 1) choreiform movement disorder, 2) subcortical cognitive impairment, and 3) psychiatric symptoms.

Movement Disorder

Involuntary dystonic or choreiform movements were noted in the first clinical description of McLeod syndrome [26]. In subsequent reports, choreiform movements are the presenting symptom in about 30% of McLeod patients (Table 1) [5,6,8,12]. A slight generalized motor restlessness with frequent changes of posture or tic-like movements may be the first manifestations of a movement disorder [5,12]. Remarkably, motor restlessness may remain the sole manifestation of a movement disorder over many years [12]. Later, the majority of McLeod patients develop a choreiform movement disorder [6,8,12]. Additional involuntary movements in McLeod patients include facial dyskinesia, dysarthria, and involuntary vocalizations (Table 2) [5,6,8,12]. The inter- and intrafamilial variability of the type and the severity of the movement disorder is considerable [12]. In contrast to autosomal-recessive chorea-acanthocytosis (Levine-Critchley syndrome), only a minority of McLeod patients have habitual lip or tongue biting, dysphagia, dystonia, or extrapyramidal symptoms (Table 2) [3,6,8,19].

Cognitive Impairment

Cognitive impairment is not a major presenting symptom in McLeod syndrome. During the course of the disease, however, about 50% of McLeod patients develop cognitive decline [4,6,8,12,15]. Available data suggest a subcortical pattern of cognitive deficits similar to Huntington's disease (see Chapter 12). Severity of cognitive alterations shows a remarkable intrafamilial variability ranging from slight memory impairment to evident dementia [12].

Table 2. Clinical findings of 35 McLeod patient (reported in refs. 6, 8, and 12).

Finding	Frequency (%)	Number of Patients Affected/Examined
Weak Kell antigens	100	33/33
Acanthocytosis	100	33/33
Serum CK elevation	100	33/33
Splenomegaly	33	5/15
Cardiac disease	56	14/25
Areflexia	90	29/32
Reduced vibration sense	28	6/21
Muscle weakness or atrophy	44	14/32
Seizures	40	9/23
Psychiatric abnormalities	68	17/25
Cognitive impairment	46	11/24
Limb chorea	79	22/28
Facial hyperkinesia	60	15/25
Involuntary vocalizations	50	9/18
Habitual tongue/lip biting	5	1/22
Dysarthria	52	13/25
Dystonia	21	5/24
Parkinsonism	13	3/24

Psychiatric Abnormalities

In 1963, Rovito and Pirone described for the first time the association of acanthocytosis with schizophrenia [24]. Since no Kx/Kell serology or determination of serum CK levels are reported from the initial propositus, it is not known if he suffered from McLeod syndrome or autosomal-recessive chorea-acanthocytosis. Consecutive reports demonstrated that about 20% of McLeod patients manifest with psychiatric abnormalities including personality disorder, anxiety, depression, obsessive-compulsive disorder, bipolar disorder, or schizo-affective disorder (Table 1) [6,8,12,20]. Psychiatric manifestations may be the predominant initial symptom in certain families [12]. During the disease course, however, a majority of McLeod patients develop psychiatric abnormalities (Table 2) [6,8,12]. The spectrum of neuropsychiatric symptoms in McLeod syndrome is comparable to other neurodegenerative basal ganglia disorders, including Huntington's disease [23] (see Chapter 12).

Seizures

Generalized seizures may be the presenting symptom in about 20% of McLeod patients [6], however, up to 40% have seizures later in the course of the disease [6,8]. Electroencephalographic findings seem not to be specific [8].

Pathophysiological Basis

In contrast to Huntington's disease, neuropathological changes and striatal imaging alterations in McLeod syndrome are restricted to the basal ganglia [9]. Neuropathological examination of one McLeod patient revealed marked neuronal loss and astrocytic gliosis in caudate nucleus and putamen. In contrast, no alterations were found in cortex, thalamus, subthalamic nucleus, brainstem and cerebellum [8]. The absence of cortical alterations indicates a subcortical seizure origin. Corresponding to the pathological findings, computed tomography and magnetic resonance imaging studies demonstrate atrophy of caudate nucleus and putamen, particularly in McLeod patients with advanced disease [6,8,12]. ^{18}F-fluorodeoxyglucose (FDG)-positron emission tomography studies demonstrate impaired striatal glucose metabolism in all McLeod patients examined so far, even in carriers of the McLeod blood group phenotype without clinical signs of cerebral involvement [12,22]. With the help of quantified FDG-positron emission tomography in comparison to data from a large control group, no impairment of glucose metabolism was found in the cerebral cortex [12]. Consequently, McLeod syndrome might be a "model disease" for the understanding of neuropsychiatric symptoms in basal ganglia disorders [2].

NEUROMUSCULAR INVOLVEMENT

Axonopathy

Almost all McLeod patients show absence of deep tendon reflexes, at least at the ankles [6,8,12]. About a third of McLeod patients have a reduced vibration sense at the feet, and only a minority has sensory symptoms [6,8,12]. Therefore, axonopathy seems to be of minor clinical relevance. Electroneurography as well as nerve biopsy demonstrate an underlying sensory-motor axonal neuropathy [8]. Electromyography, in contrast, may reveal myopathic as well as neurogenic changes [6,8].

Myopathy

Areflexia, wasting of limb muscles, elevated CK, and congestive cardiomyopathy were already noted in the first clinical description of McLeod syndrome [26]. Subsequent reports demonstrated elevated serum CK levels rarely exceeding 4000 U/l as a sign of subclinical or manifest myopathy in virtually all carriers of the McLeod blood group phenotype [6,8,12,28,29]. About 50% of McLeod patients have a clinically relevant muscle weakness or atrophy (Table 2). The deterioration rate of the myopathy is slow, and only a minority of McLeod patients develop severe weakness [6,8,12,16,28,29]. One McLeod patient is reported to develop a severe rhabdomyolysis with renal insufficiency after a prolonged period of motor restlessness due to an agitated psychotic state [14]. This observation indicates that McLeod myopathy is a predisposing factor for severe rhabdomyolysis, in particular if excessive movement disorder or neuroleptic medication is present.

Cardiomyopathy

Congestive cardiomyopathy was noted in the first clinical descriptions of McLeod patients [7,26]. Additional cardiac manifestation of McLeod syndrome include dilated cardiomyopathy, atrial fibrillation, and tachyarrhythmia [20,30] About 60% of McLeod patients are reported to develop cardiac manifestations during the course of the disease [6]. Although long-term follow-up is lacking in many McLeod patients, cardiac problems might be an important cause of death [6]. Due to these possibly serious complications and the availability of treatment, McLeod patients should be monitored carefully for cardiac disease.

Pathophysiological Basis

Histological findings of skeletal muscle in McLeod patients include fiber type grouping, type 1-fiber predominance, type 2-fiber atrophy, increased variability in fiber size, and increased central nucleation [13,28]. In normal skeletal muscle, XK immunohistochemistry reveals a type 2 fiber-specific intracellular staining most probably confined to the sarcoplasmic reticulum. XK staining is absent in McLeod myopathy. This finding correlates to the type 2-fiber atrophy observed in McLeod myopathy, and suggests that the XK protein represents a crucial factor for the maintenance of normal muscle structure and function [13].

RARE MANIFESTATIONS

A minority of McLeod patients has additional signs and symptoms concerning CNS involvement, sleep behavior, and autonomous nervous system involvement. Some McLeod patients have cerebellar symptoms, and neuroradiological evaluation demonstrated extended white matter alterations in 2 other cases [31]. Three McLeod patients had a sleep apnea syndrome [6]. As a probable manifestation of autonomous nervous system involvement, one patient had profuse sweating, and 2 others developed fecal incontinence in late stages of the disease [6]. Although no clinico-pathological comparisons are available due to the limited number of post mortem examinations performed up to date, these findings support the multi-system nature of McLeod syndrome.

MANIFESTING FEMALE MCLEOD CARRIERS

Two *female heterozygotes* were reported to develop the typical McLeod phenotype [8,16]. The most probable reason for this finding is an unfavorable inactivation of the X chromosome carrying the normal *XK* gene, as it is demonstrated in one case [8,11]. Some other studies suggest that *XK* mutations may be pseudo-dominant. In several female heterozygotes, a blood group mosaicism of the Kx/Kell system is found [12,29]. In one McLeod family, female McLeod mutation carriers had slight cognitive deficits [12]. In addition, reduction of striatal glucose uptake was demonstrated with quantified FDG-positron emission tomography indicating slight CNS involvement in female heterozygotes who do not manifest a movement disorder [12]. Although interpretations have to be made with caution because of the limited number of cases, cognitive testing, flow cytometry with Kell antibodies and quantified positron emission tomography could be useful for the determination of the probable carrier status in asymptomatic females from McLeod families who have not been genetically tested.

DISEASE COURSE AND THERAPY

Almost all clinical observations indicate a slowly progressive disease course [6,8,12]. Activities of daily living may be impaired due to the movement disorder, psychiatric symptoms, cognitive impairment, or autonomic dysfunction. Because it is difficult to determine the exact time of disease onset, little reliable data about the disease duration are available. Intervals between reported disease onset and death range from 7 years to 51 years [8,12]. In a heterogeneous series of 8 McLeod patients, mean age of death was 53 years, ranging from 31 to 69 years [8,12,20] Cardiac problems, in particular tachyarrhythmia, are suggested as a major cause of death in young McLeod patients [6,8]. Cardiovascular causes, epileptic seizures, and aspiration pneumonia might be the major causes of death in older McLeod patients [6,8,12].

Up to date, no causal therapy is available to prevent or slow down the progression of McLeod syndrome. Therefore, recognition and treatment of cardiac problems and seizures might be the most important issue in the clinical follow-up of McLeod patients [6,8]. In order to recognize rhabdomyolysis, serum CK levels should be carefully monitored in patients with McLeod syndrome, in particular if excessive movement disorder or neuroleptic medication are present [14]. Psychiatric problems should be treated according to the clinical presentation. Dopamine antagonists such as tiapride, clozapine, or quetiapine can ameliorate the choreatic movement disorder. Last but not least, an extended and continuous multidisciplinary psychosocial support should be provided for the patients and their families.

REFERENCES

1. Allen FH, Krabbe SMR, Corcoran PA (1961) A new phenotype (McLeod) in the Kell blood-group system. *Vox Sang* **6:** 555-560.
2. Bunney WE, Bunney BG (2000) Evidence for a compromised dorsolateral prefrontal cortical parallel circuit in schizophrenia. *Brain Res Brain Res Rev* **31:** 138-146.

3. Critchley EM, Clark DB, Wikler A (1968) Acanthocytosis and neurological disorder without betalipoproteinemia. *Arch Neurol* **18:** 134-140.
4. Danek A, Uttner I, Vogl T, Tatsch K, Witt TN (1994) Cerebral involvement in McLeod syndrome. *Neurology* **44:** 117-120.
5. Danek A, Tison F, Rubio J, Oechsner M, Kalckreuth W, Monaco AP (2001) The chorea of McLeod syndrome. *Mov Dis* **16:** 882-889.
6. Danek A, Rubio JP, Rampoldi L, Ho M, Dobson-Stone C, Tison F, Symmans WA, Oechsner M, Kalckreuth W, Watt JM, Corbett AJ, Hamdalla HHM, Marshall AG, Sutton I, Dotti MT, Malandrini A, Walker RH, Daniels G, Monaco AP (2001) McLeod neuroacanthocytosis: genotype and phenotype. *Ann Neurol* **50:** 755-764.
7. Faillace RT, Kingston WJ, Nanda NC, Griggs RC (1982) Cardiomyopathy associated with the syndrome of amyotrophic chorea and acanthocytosis. *Ann Intern Med* **96:** 616-617.
8. Hardie RJ, Pullon HW, Harding AE, Owen JS, Pires M, Daniels GL, Imai Y, Misra VP, King RH, Jacobs JM (1991) Neuroacanthocytosis. A clinical, haematological and pathological study of 19 cases. *Brain* **114:** 13-49.
9. Hedreen JC, Peyser CE, Folstein SE, Ross CA (1991) Neuronal loss in layers V and VI of cerebral cortex in Huntington's disease. *Neurosci Lett* **133:** 257-261.
10. Ho M, Chelly J, Carter N, Danek A, Crocker P, Monaco AP (1994) Isolation of the gene for McLeod syndrome that encodes a novel membrane transport protein. *Cell* **77:** 869-880.
11. Ho MF, Chalmers RM, Davis MB, Harding AE, Monaco AP (1996) A novel point mutation in the McLeod syndrome gene in neuroacanthocytosis. *Ann Neurol* **39:** 672-675.
12. Jung HH, Hergersberg M, Kneifel S, Alkadhi H, Schiess R, Weigell-Weber M, Daniels G, Kollias S, Hess K (2001) McLeod syndrome: a novel mutation, predominant psychiatric manifestations, and distinct striatal imaging findings. *Ann Neurol* **49:** 384-392.
13. Jung HH, Russo D, Redman C, Brandner S (2001) Kell and XK immunohistochemistry in McLeod myopathy. *Muscle Nerve* **24:** 1346-1351.
14. Jung HH, Brandner S (2002) Malignant McLeod myopathy. *Muscle Nerve* **26:** 424-427.
15. Kartsounis LD, Hardie RJ (1996) The pattern of cognitive impairments in neuroacanthocytosis. A frontosubcortical dementia. *Arch Neurol* **53:** 77-80.
16. Kawakami T, Takiyama Y, Sakoe K, Ogawa T, Yoshioka T, Nishizawa M, Reid ME, Kobayashi O, Nonaka I, Nakano I (1999) A case of McLeod syndrome with unusually severe myopathy. *J Neurol Sci* **166:** 36-39.
17. Lee S, Lin M, Mele A, Cao Y, Farmar J, Russo D, Redman C (1999) Proteolytic processing of big endothelin-3 by the kell blood group protein. *Blood* **94:** 1440-1450.
18. Lee S, Russo D, Redman C (2000) Functional and structural aspects of the Kell blood group system. *Transfus Med Rev* **14:** 93-103.
19. Levine IM, Estes JW, Looney JM (1968) Hereditary neurological disease with acanthocytosis. A new syndrome. *Arch Neurol* **19:** 403-409.
20. Malandrini A, Fabrizi GM, Truschi F, Di Pietro G, Moschini F, Bartalucci P, Berti G, Salvadori C, Bucalossi A, Guazzi G (1994) Atypical McLeod syndrome manifested as X-linked chorea-acanthocytosis, neuromyopathy and dilated cardiomyopathy: report of a family. *J Neurol Sci* **124:** 89-94.
21. Marsh WL, Marsh NJ, Moore A, Symmans WA, Johnson CL, Redman CM (1981) Elevated serum creatine phosphokinase in subjects with McLeod syndrome. *Vox Sang* **40:** 403-411.
22. Oechsner M, Buchert R, Beyer W, Danek A (2001) Reduction of striatal glucose metabolism in McLeod choreoacanthocytosis. *J Neurol Neurosurg Psychiatry* **70:** 517-520.
23. Ring HA, Serra-Mestres J (2002) Neuropsychiatry of the basal ganglia. *J Neurol Neurosurg Psychiatry* **72:** 12-21.
24. Rovito DA, Pirone FJ (1963) Acanthrocytosis associated with schizophrenia. *Am J Psychiatry* **120:** 182-185.

25. Russo D, Redman C, Lee S (1998) Association of XK and Kell blood group proteins. *J Biol Chem* **273:** 13950-13956.
26. Schwartz SA, Marsh WL, Symmans A, Johnson CL, Mueller KA (1982) New clinical features of McLeod syndrome. *Transfusion* **22:** 404.
27. Stanfield GM, Horvitz HR (2000) The ced-8 gene controls the timing of programmed cell deaths in *C. elegans*. *Mol Cell* **5:** 423-433.
28. Swash M, Schwartz MS, Carter ND, Heath R, Leak M, Rogers KL (1983) Benign X-linked myopathy with acanthocytes (McLeod syndrome). Its relationship to X-linked muscular dystrophy. *Brain* **106:** 717-733.
29. Wimer BM, Marsh WL, Taswell HF, Galey WR (1977) Haematological changes associated with the McLeod phenotype of the Kell blood group system. *Br J Haematol* **36:** 219-224.
30. Witt TN, Danek A, Reiter M, Heim MU, Dirschinger J, Olsen EG (1992) McLeod syndrome: a distinct form of neuroacanthocytosis. Report of two cases and literature review with emphasis on neuromuscular manifestations. *J Neurol* **239:** 302-306.
31. Nicholl DJ, Sutton I, Dotti MT, Supple SG, Danek A, Lawden M (2002) Florid white matter abnormalities on MRI in neuroacanthocytosis. *Mov Dis* **17:** S174-S175.

CHAPTER 7

AUTOSOMAL-DOMINANT CHOREA-ACANTHOCYTOSIS: REPORT OF A FAMILY AND NEUROPATHOLOGY

RUTH H. WALKER[1,2], SUSAN MORGELLO[3], BARBARA DAVIDOFF-FELDMAN[4], ARI MELNICK[5], MICHAEL J. WALSH[2], P. SHASHIDHARAN[2], AND MITCHELL F. BRIN[2,6]

[1]Department of Neurology, Veterans Affairs Medical Center, Bronx, NY; Departments of [2]Neurology, [3]Pathology and [5]Medicine (Division of Hematology), Mount Sinai School of Medicine, New York, NY; [4]Genetics Ambulatory Service, Nassau County Medical Center, East Meadow, NY; [6]Allergan Pharmaceuticals, Irvine, CA

Abstract. We report a family with autosomal-dominant inheritance of chorea-acanthocytosis. Clinical and hematological evaluations were performed on all available family members and neuropathologic examination was performed on one patient. There were variable clinical features of chorea or parkinsonism, and marked cognitive changes. On hematologic analysis there were abnormalities of band 3. Neuropathologic examination revealed severe neuronal loss in the caudate-putamen and intranuclear inclusion bodies throughout the cerebral cortex. These inclusion bodies were immunoreactive for ubiquitin, expanded polyglutamine repeats, and torsinA. This family extends the genetic spectrum of chorea-acanthocytosis to include autosomal-dominant inheritance, possibly due to expanded trinucleotide repeats.

INTRODUCTION

The term chorea-acanthocytosis (C-A, Levine-Critchley syndrome, neuroacanthocytosis) [7,16,17] refers to a group of genetically and phenotypically heterogeneous disorders. Inheritance typically appears to be autosomal-recessive [1-4,9,24,26,28,31], unless the syndrome is associated with the X-linked McLeod phenotype. A small number of families have been reported in which there may have been autosomal-dominant inheritance with reduced penetrance [7,15-17]

A. Danek (ed.), *Neuroacanthocytosis Syndromes*, 55–65.

but this form of inheritance has not been unequivocally demonstrated. The abnormalities of red blood cell (RBC) membrane proteins which characterise this group of disorders have been determined in McLeod syndrome [11] and in a small number of cases of apparently autosomal-recessive C-A [5,13,14], yet the relationship between the acanthocytosis and the neurodegenerative process remains obscure. Neuropathologic descriptions of C-A cases are phenotypically and probably genotypically heterogeneous [1,3,4,8-10,21,24,28,31], and demonstrate neuronal loss and gliosis in the striatum and pallidum, and variably in the substantia nigra. We describe a kindred with an autosomal-dominantly inherited form of C-A with RBC band 3 abnormalities, and document the presence of ubiquitinated, polyglutamine-containing, intranuclear neuronal inclusions in an affected family member [30].

CASE REPORTS

Patient 1

The proband (IV:1) was a 54-year-old African-American man, initially diagnosed with Huntington's disease (HD), who came to our attention after his daughter sought genetic counseling. The family is of African-American, Cherokee and Caucasian origin.

The patient's mother (III:2) and maternal grandmother (II:2) were described as nervous, "glassy-eyed", forgetful, shaky, and having jerky movements (Figure 1). The patient's birth, growth, and development were normal until age 34, when he developed a slowly progressive deterioration in memory and personality, becoming antisocial and withdrawn, with non-purposeful and non-repetitive involuntary movements of the face and hands.

On examination at age 54 years, the patient was not able to stand. He had no spontaneous speech and would occasionally obey very simple commands. He had difficulty holding up his trunk and had mild, continuous choreiform movements of head, trunk, and both upper extremities, with intermittent rapid jerks. He held both hands fisted with extension of the thumbs. He had marked salivation and anterior protrusion of the jaw. Eye movements were normal, with no blink on initiation of saccadic gaze. Power was normal, with no muscle atrophy or fasciculations, and tone was decreased. Reflexes at biceps and triceps were trace bilaterally, absent in the brachioradialis, but with spread to the fingers and positive Hoffmann signs bilaterally. Reflexes were increased at both knees and ankles with two to three beats of clonus on the left and bilateral flexor plantar responses.

MRI scan of the brain showed diffuse atrophy of the cortex, and marked atrophy of the caudate nuclei.

The following tests were normal or negative: creatine kinase, serum vitamin E, cholesterol, triglycerides, apolipoprotein B, ceruloplasmin, rapid plasma reagent (RPR), erythrocyte sedimentation rate, thyroid function tests, ammonia, immunoelectrophoresis, cerebrospinal fluid protein, glucose, and cell

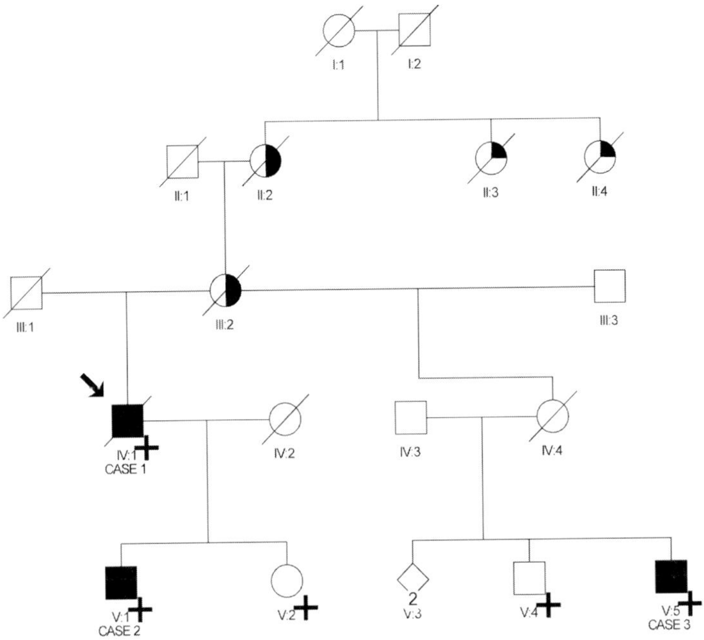

Figure 1. Pedigree of family. Blackened symbols indicate individuals with acanthocytosis, neurological abnormalities and dementia; half-blackened symbols indicate individuals affected by history; quarter-blackened symbols indicate individuals possibly affected by history. Arrow indicates patient 1. A plus (+) sign indicates that an individual was examined. Reproduced from [30] with permission Lippincott, Williams & Wilkins ©2002.

count, urinary copper, mercury and lead excretion. Vitamin B_{12} was less than 100 pg/ml (range 190-900 pg/ml), attributed to poor oral intake. Genetic testing for HD and dentato-rubro-pallido-Luysian atrophy (DRPLA) was negative. Red cell Kell antigen testing for McLeod phenotype was negative. Fragile X testing showed 57 repeats, in the premutation range. Genetic testing for linkage to the site on chromosome 9p21 involved in autosomal-recessive C-A [22] (ChAc; MIM 200150) was negative. Transthoracic echocardiography showed normal cardiac size and function. CSF contained normal levels of 5-hydroxyindoleacetic acid and 3-O-methyldopa. Homovanillic acid was slightly decreased at 131 nM (range 145-324 nM). Electromyography and nerve conduction studies were normal. Muscle biopsy of the quadriceps revealed diffusely small fibers without structural abnormalities, type 2 predominant (approximately 76%) with a mean myofiber diameter of 57 μm (range 44-88 μm). There was no fiber type grouping, and no structural abnormalities were detected on NADH-TR, modified trichrome, ATPase pH 4.2 and 9.4, neuron specific enolase, and alkaline and acid phosphatase stains. Sural nerve biopsy was unremarkable.

The patient expired at age 57. After a post mortem interval of 510 minutes, one cerebral hemisphere was immersed in 10% formalin for neuropathologic examination and the other frozen at −80° C.

Patient 2

The son of patient 1 (V:1) was diagnosed with fragile X syndrome (660 triplet repeats). He was 26 years old at the time of initial examination. He was moderately mentally retarded, with frequent behavioral problems, including aggression. No gait difficulty, abnormal movements or seizures were noted. On initial evaluation there was no evidence of hyper- or hypokinetic movements, and neurologic examination was consistent with his diagnosis of fragile X syndrome. MRI of the brain was normal.

Three years later his gait deteriorated, and he started falling. On examination there was mild chorea of lower face, trunk, and both upper limbs, with frequent dystonic posturing of both upper extremities, especially on walking. Reflexes were increased symmetrically with spread, absent snout and glabellar response, and flexor plantar responses. Tone and coordination were normal. Mastication was performed at the front of the mouth but there was no extrusion of food. He walked with a shuffling, slightly lurching, wide-based gait with feet turned out. On postural reflex testing he collapsed at the knees, or took extra steps backwards. Non-contrast head CT showed generalized cerebral and caudate atrophy.

Patient 3

The nephew of patient 1 was examined (V:5). His mother (IV:4) was the victim of homicide at the age of 25 years; on post mortem examination her brain weighed 1250 g and was reported to show no gross abnormalities on serial section.

The patient's birth, growth and early development were normal. At the age of 28 he started to become socially withdrawn with deteriorating memory, was no longer able to work and stopped taking care of himself. In the months before examination he developed generalized motor slowing, chewing movements of his mouth and tremor of his hands, which improved with clonazepam.

At the time of examination he was 32 years old. He had marked difficulty in obeying complex commands and scored 20/57 on the Modified Mini-Mental State Exam. There was moderate facial masking; glabellar, snout and jaw jerk responses were positive. All cranial nerves including eye movements were normal. Posture holding in supine, prone and wing-beating positions showed a mild postural tremor, more marked on the left than the right. Finger-to-nose testing showed occasional small amplitude jerking movements. Rapid alternating and successive movements were performed with mild clumsiness. Performance of movements with his feet induced synkinetic movements of the ipsilateral hand. Power and sensation were normal, with mildly increased tone throughout. Reflexes were abnormally brisk in all extremities with a positive Hoffmann sign and 3-4 beats of clonus at the ankles, and flexor plantar responses. On walking, he had mild generalized bradykinesia but normal arm swing and normal postural stability.

An MRI scan performed two years prior to this evaluation showed generalized atrophy. Metabolic work-up including ceruloplasmin, plasma and urinary amino acids was negative. Testing for fragile X showed 56 repeats, in the premutation range.

RESULTS

Hematologic Analyses

In all three patients, peripheral blood smear and scanning electron microscopy showed 30-35% acanthocytosis. SDS-PAGE of RBC membrane extract showed breakdown of band 3 proteins [30].

Neuropathology: Patient 1

Autopsy was limited to examination of the brain and spinal cord. The brain weighed 970 g and externally displayed severe, diffuse cortical atrophy. On coronal section, the externally noted cortical atrophy appeared to involve all lobes (with particular severity in the cingulum), apart from the hippocampal formation and medial temporal lobe, which were strikingly spared. There was severe atrophy of the caudate and putamen, which were discolored, granular, and contracted. The pallidum appeared relatively preserved. The substantia nigra and locus coeruleus were pale. A 2 × 2 × 2 centimeter firm, white intracortical mass was identified in the right anterior mid and inferior frontal gyri which on microscopic examination revealed a markedly fibrotic focus of meningioangiomatosis.

On microscopic examination, the caudate and putamen demonstrated severe neuronal loss and gliosis, with scattered residual large-calibre neurons. A subpopulation of these residual neurons displayed immunoreactivity for somatostatin, confirming their identity as interneurons. Small foci of hemosiderin deposition and rare spheroids were identified. Lesser degrees of neuronal loss were present in all areas of neocortex sampled, as well as in globus pallidus, substantia nigra, locus coeruleus and amygdala. In neocortex, many large pyramidal neurons in layer III appeared atrophic, with amphophilic/purple cytoplasm and contracted nuclei.

Immunohistochemical stains demonstrated the presence of round to oval, ubiquitinated intranuclear neuronal inclusions scattered throughout all areas of cortex, substantia nigra, and thalamus (Figure 2A). The inclusions appeared

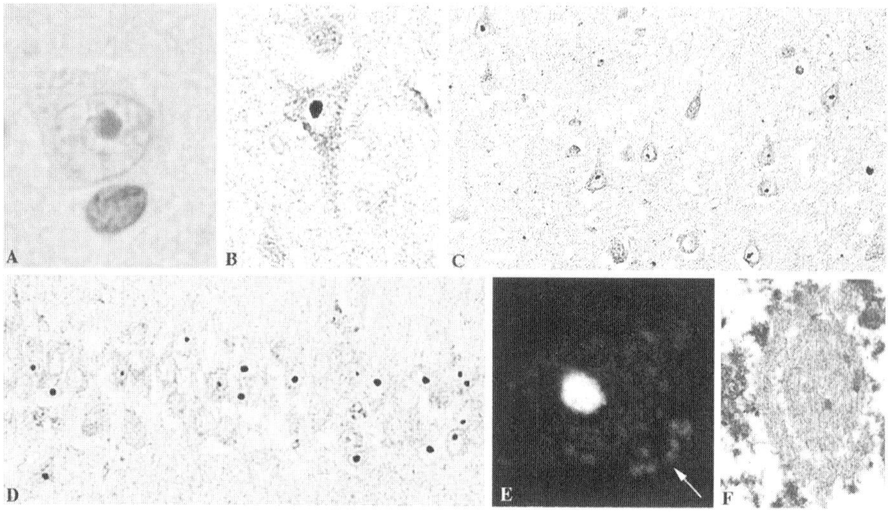

Figure 2. Intranuclear neuronal inclusions in cortical neurons immunolabeled using DAB for ubiquitin (counterstained with hematoxylin) (original magnification ×100) (A), and torsinA (original magnification ×60) (B). Low power views of torsinA immunoreactivity in neocortex (C) and hippocampal granule cell layer (D) (original magnification ×20). Confocal laser microscopy of fluorochrome-labeled immunoreactivity to polyglutamine repeats (original magnification ×100); arrow indicates lipofuscin granules at periphery of cell body (E). Electron micrograph of inclusion body showing filamentous structure (original magnification ×20,000) (F). Reproduced from [30] with permission Lippincott, Williams & Wilkins ©2002.

most numerous in the insula. Inclusions were not identified in basal ganglia and locus coeruleus. The inclusions also demonstrated immunoreactivity for torsinA [25] (Figure 2B, C, D), and expanded polyglutamine repeats (Figure 2E) but did not label with antibodies directed against tau, phosphorylated neurofilament, α-synuclein, or p53. Nuclear immunoreactivity to torsinA was decreased in the presence of an inclusion body, suggesting redistribution of the protein. Portions of formalin-fixed cortex were processed for examination by electron microscopy. Several intranuclear inclusions were identified, and consisted of relatively compact aggregates of filamentous material, with no separation from the surrounding nucleoplasm (Figure 2F).

DISCUSSION

The patients in this family show evidence of an autosomal-dominantly inherited neurodegenerative disorder, which we describe as chorea-acanthocytosis. The nomenclature of this disorder has been debated and is likely to be eventually determined by genotype. A number of features of the family are distinct from the patients previously described. In particular we do not see the marked orobucco-lingual dyskinesia, self-mutilation, or seizures which have been associated with this condition. Elevated creatine kinase is reported in some [4,6,26], but not all, cases [6], but we did not find this, nor evidence of peripheral neuropathy.

Autosomal-recessive inheritance appears to be the inheritance pattern of the majority of patients [1-4,9,24,26,28,31], with the gene localized to chromosome 9q21 [23,27]. Several reports are suggestive of autosomal-dominant inheritance with reduced penetrance, although the specificity of the neurological abnormalities and their relationship to the presence of acanthocytosis is unclear. In the family reported by Levine and coworkers [16,17], chorea was documented in the proband, a sibling, and a maternal aunt. A cousin had chorea gravidarum. Many other family members were found to have hyporeflexia, slight weakness, mild tremor, or other neurological features which are hard to interpret. Acanthocytosis was a variable feature, present in patients both with and without neurological abnormalities. Lipoproteins were normal; McLeod syndrome was not excluded, but male-male transmission of some features makes this diagnosis unlikely. The family reported by Critchley and coworkers is less convincing, with several affected siblings, and the niece of the proband who did not have any involuntary movements, but a syndrome resembling Friedreich's ataxia with cardiac involvement and a few possible acanthocytes [7]. Acanthocytosis was seen in another neurologically normal second generation child. Kito and coworkers reported chorea, vocal tics and seizures in a paternal uncle and a sibling of the proband, along with acanthocytosis and hypo-β-lipoproteinemia in the proband's son, raising the possibility of a lipoprotein disorder [15].

The presence of immunoreactivity to expanded polyglutamine repeats in the ubiquitinated intranuclear inclusions suggests that this disease may be due to expanded trinucleotide repeats, as are a number of other disorders characterised by intranuclear inclusions, such as HD and the spinocerebellar ataxias. We did not note any clear evidence of anticipation, as is often seen in the autosomal-dominant trinucleotide repeat diseases. Disease onset in patient 1 was at age 34, in his son (patient 2) at age 29, and his nephew (patient 3) with maternally-inherited disease, at age 28.

The association of fragile X carrier status with C-A is intriguing, however, it is the most common type of familial mental retardation affecting boys, with a prevalence of approximately $1 : 2,000$, thus we suspect that the presence of expanded trinucleotide repeats in the fragile X gene is a chance association, and unrelated to the neurological abnormalities in this family (see [30] for more discussion).

The striking RBC morphological abnormalities characterizing the heterogeneous group of disorders known as C-A have generated optimism that molecular characterization of the RBC defect might clarify the basis for neurodegeneration. As in cases which appear to have had autosomal-recessive C-A [5,14], we also find abnormalities of band 3, a major transmembrane glycoprotein of the RBC. The protein implicated in autosomal-recessive chorea-acanthocytosis has recently been identified [19,27], and may play an important role in protein sorting and trafficking, with dysfunction impairing plasma membrane structure [19]. Abnormalities of another red cell membrane protein, XK, in McLeod syndrome [11], also characterized by chorea and acanthocytosis, may suggest a common pathophysiology of neurodegeneration.

Neuropathologic studies in previously reported cases of C-A are phenotypically and genetically heterogeneous, and include probable autosomal-recessive [1,4,9,10,24,28,31] or sporadic [10,12,28] cases, and a female with McLeod phenotype [10]. The predominant findings are of degeneration of the basal ganglia, in particular of the caudate nucleus, and sometimes the substantia nigra [20], with the cortex being much less affected and the total brain weights being more nearly normal than in our patient [1,3,4,8-10,12,20,24,28,31]. Inclusion bodies have not previously been reported in C-A.

TorsinA is a 332 amino acid protein of unknown function, mutation of which is associated with one form of autosomal-dominantly-inherited childhood-onset dystonia [18]. It is found ubiquitously in brain and peripheral tissues [25,29], and has homology with the heat shock family of proteins [18]. Other proteins which colocalise to inclusions are members of the heat shock protein family which play an important role in ubiquitin-related proteolysis, implicating a role for the stress response in the formation of these bodies. TorsinA has also been demonstrated in Lewy bodies in idiopathic Parkinson's disease and in dementia with Lewy bodies [32] and in intranuclear inclusion bodies in HD and spinocerebellar ataxia type 3 [36]. The presence of torsinA in intranuclear neuronal inclusions may be suggestive of a role in protein folding or degradation, or may just reflect protein aggregation in a malfunctioning neuron.

Further work will determine the significance of the expanded polyglutamine repeats in the inclusion bodies, and the relationship of this to the genetic etiology of this autosomal-dominant disease.

Note Added in Press

It has been determined that the genetic mutation in this family is identical to that found in Huntington's disease-like 2 (HDL2). This mutation is a trinucleotide repeat expansion within the junctophilin-3 gene, which encodes for a protein involved in junctional membrane structures [33,34]. Patient 1 had 51/13 trinucleotide repeats, patient 2 had 58/14, and patient 3 57/15 (normal < 40). One additional patient with HDL2 and acanthocytosis has been identified [35]. The presence of acanthocytosis in HDL2, the function of junctophilin-3, and its relationship to striatal degeneration and acanthocytosis, are the subject of ongoing research.

Acknowledgements

The authors thank the family for facilitating these studies. We also thank Keith Hyland, PhD (Institute for Metabolic Disease, Baylor Research Institute, Dallas, TX) for performing the CSF analysis; Justin Rubio, PhD and Anthony Monaco, PhD (Wellcome Trust Centre for Human Genetics, Headington, UK) for performing the genetic testing for autosomal-recessive chorea-acanthocytosis; Jacinta Murray, BS, and Daniela Sandu, MS, for technical assistance; J.T. Wilkinson, MD (Temple, TX) for facilitating fragile X testing of patient 3; and Dana Doheny, MS, for assistance with the pedigree. The neuropathology autopsy and analysis were performed under the auspices of the Manhattan HIV Brain Bank (R24MH59724). This work was supported in part by the Bachmann-Strauss Dystonia & Parkinson Foundation, Inc. Confocal laser scanning microscopy was performed at the MSSM-Microscopy Center, supported with funding from an NSF Major Research Instrumentation grant (DBI-9724504).

REFERENCES

1. Alonso ME, Teixeira F, Jimenez G, Escobar A (1989) Chorea-acanthocytosis: report of a family and neuropathological study of two cases. *Can J Neurol Sci* **16:** 426-431.
2. Aminoff MJ (1972) Acanthocytosis and neurological disease. *Brain* **95:** 749-760.
3. Bharucha EP, Bharucha NE (1989) Choreo-acanthocytosis. *J Neurol Sci* **89:** 135-139.
4. Bird TD, Cederbaum S, Valpey RW, Stahl WL (1978) Familial degeneration of the basal ganglia with acanthocytosis: a clinical, neuropathological, and neurochemical study. *Ann Neurol* **3:** 253-258.
5. Bosman GJCGM, Bartholomeus IGP, Degrip WJ, Horstink MWIM (1994) Erythrocyte anion transporter and antibrain immunoreactivity in chorea-acanthocytosis – A contribution to etiology, genetics, and diagnosis. *Brain Res Bull* **33:** 523-528.
6. Brin MF, Bressman SB, Fahn S, Resor SR, Weitz J, Sagman DL (1985) Chorea-acanthocytosis: Clinical and laboratory features in 5 cases. *Neurology* **35 (Suppl 1):** 110.
7. Critchley EM, Clark DB, Wikler A (1968) Acanthocytosis and neurological disorder without betalipoproteinemia. *Arch Neurol* **18:** 134-140.
8. de Yebenes JG, Vazquez A, Martínez A, Mena MA, del Rio RM, De Felipe C, Del Rio J (1988) Biochemical findings in symptomatic dystonias. *Adv Neurol* **50:** 167-175.
9. Gross KB, Skrivanek JA, Carlson KC, Kaufman DM (1985) Familial amyotrophic chorea with acanthocytosis. New clinical and laboratory investigations. *Arch Neurol* **42:** 753-756.
10. Hardie RJ, Pullon HW, Harding AE, Owen JS, Pires M, Daniels GL, Imai Y, Misra VP, King RH, Jacobs JM (1991) Neuroacanthocytosis. A clinical, haematological and pathological study of 19 cases. *Brain* **114:** 13-49.
11. Ho M, Chelly J, Carter N, Danek A, Crocker P, Monaco AP (1994) Isolation of the gene for McLeod syndrome that encodes a novel membrane transport protein. *Cell* **77:** 869-880.
12. Iwata M, Fuse S, Sakuta M, Toyokura Y (1984) Neuropathological study of chorea-acanthocytosis. *Jpn J Med* **23:** 118-122.
13. Kay MMB, Goodman J, Goodman S, Lawrence C (1990) Membrane protein band 3 alteration associated with neurologic disease and tissue-reactive antibodies. *Exp Clin Immmunogenet* **7:** 181-199.

14. Kay MMB, Goodman J, Lawrence C, Bosman G (1990) Membrane channel protein abnormalities and autoantibodies in neurological disease. *Brain Res Bull* **24:** 105-111.
15. Kito S, Itoga E, Hiroshige Y, Matsumoto N, Miwa S (1980) A pedigree of amyotrophic chorea with acanthocytosis. *Arch Neurol* **37:** 514-517.
16. Levine IM (1964) An hereditary neurologic disease with acanthocytosis. *Neurology* **14:** 272.
17. Levine IM, Estes JW, Looney JM (1968) Hereditary neurological disease with acanthocytosis. A new syndrome. *Arch Neurol* **19:** 403-409.
18. Ozelius LJ, Hewett JW, Page CE, Bressman SB, Kramer PL, Shalish C, de Leon D, Brin MF, Raymond D, Jacoby D, Penney J, Risch NJ, Fahn S, Gusella JF, Breakefield XO (1998) The gene (DYT1) for early-onset torsion dystonia encodes a novel protein related to the Clp protease/heat shock family. *Adv Neurol* **78:** 93-105.
19. Rampoldi L, Dobson-Stone C, Rubio JP, Danek A, Chalmers RM, Wood NW, Verellen C, Ferrer X, Malandrini A, Fabrizi GM, Brown R, Vance J, Pericak-Vance M, Rudolf G, Carre S, Alonso E, Manfredi M, Nemeth AH, Monaco AP (2001) A conserved sorting-associated protein is mutant in chorea-acanthocytosis. *Nat Genet* **28:** 119-120.
20. Rinne JO, Daniel SE, Scaravilli F, Harding AE, Marsden CD (1994) Nigral degeneration in neuroacanthocytosis. *Neurology* **44:** 1629-1632.
21. Rinne JO, Daniel SE, Scaravilli F, Pires M, Harding AE, Marsden CD (1994) The neuropathological features of neuroacanthocytosis. *Mov Dis* **9:** 297-304.
22. Rubio JP, Danek A, Stone C, Chalmers R, Wood N, Verellen C, Ferrer X, Malandrini A, Fabrizi GM, Manfredi M, Vance J, Pericak-Vance M, Brown R, Rudolf G, Picard F, Alonso E, Brin M, Nemeth AH, Farrall M, Monaco AP (1997) Chorea-acanthocytosis: genetic linkage to chromosome 9q21. *Am J Hum Genet* **61:** 899-908.
23. Rubio JP, Levy ER, Dobson-Stone C, Monaco AP (1999) Genomic organization of the human galpha14 and Galphaq genes and mutation analysis in chorea-acanthocytosis (CHAC). *Genomics* **57:** 84-93.
24. Sakai T, Mawatari S, Iwashita H, Goto I, Kuroiwa Y (1981) Choreoacanthocytosis. Clues to clinical diagnosis. *Arch Neurol* **38:** 335-338.
25. Shashidharan P, Kramer BC, Walker RH, Olanow CW, Brin MF (2000) Immunohistochemical localization and distribution of torsinA in normal human and rat brain. *Brain Res* **853:** 197-206.
26. Spitz MC, Jankovic J, Killian JM (1985) Familial tic disorder, parkinsonism, motor neuron disease, and acanthocytosis: a new syndrome. *Neurology* **35:** 366-370.
27. Ueno S, Maruki Y, Nakamura M, Tomemori Y, Kamae K, Tanabe H, Yamashita Y, Matsuda S, Kaneko S, Sano A (2001) The gene encoding a newly discovered protein, chorein, is mutated in chorea-acanthocytosis. *Nat Genet* **28:** 121-122.
28. Vance JM, Pericak Vance MA, Bowman MH, Payne CS, Fredane L, Siddique T, Roses AD, Massey EW (1987) Chorea-acanthocytosis: a report of three new families and implications for genetic counselling. *Am J Med Genet* **28:** 403-410.
29. Walker RH, Brin MF, Sandu D, Gujjari P, Olanow CW, Shashidharan P (2001) Distribution and immunohistochemical characterization of torsinA immunoreactivity in rat brain. *Brain Res* **900:** 348-354.
30. Walker RH, Morgello S, Davidoff-Feldman B, Melnick A, Walsh MJ, Shashidharan P, Brin MF (2002) Autosomal-dominant chorea-acanthocytosis with polyglutamine-containing neuronal inclusions. *Neurology* **58:** 1031-1037.
31. Yamamoto T, Hirose G, Shimazaki K, Takado S, Kosoegawa H, Saeki M (1982) Movement disorders of familial neuroacanthocytosis syndrome. *Arch Neurol* **39:** 298-301.
32. Shashidharan P, Good PF, Hsu A, Perl D, Brin MF, Olanow CW (2000) TorsinA accumulation in Lewy bodies in sporadic Parkinson's disease. *Brain Res* **877:** 379-381.
33. Holmes SE, O'Hearn E, Rosenblatt A, Callahan C, Hwang HS, Ingersoll-Ashworth RG, Fleisher A, Stevanin G, Brice A, Potter NT, Ross CA, Margolis RL (2001) A repeat

expansion in the gene encoding junctophilin-3 is associated with Huntington disease-like 2. *Nat Genet* **29:** 377-378.

34. Margolis RL, O'Hearn E, Rosenblatt A, Willour V, Holmes SE, Franz ML, Callahan C, Hwang HS, Troncoso JC, Ross CA (2001) A disorder similar to Huntington's disease is associated with a novel CAG repeat expansion. *Ann Neurol* **50:** 373-380.
35. Walker RH, Rasmussen A, Rudnicki D, Holmes SE, Alonso E, Matsuura T, Ashizawa T, Davidoff-Feldman B, Margolis RL (2003) Huntington's disease-like 2 can present as chorea-acanthocytosis. *Neurology* **61:** 1002-1004.
36. Walker RH, Good PF, Shashidharan P (2003) TorsinA immunoreactivity in inclusion bodies in trinucleotide repeat diseases. *Mov Dis* **18:** 1041-1044.

CHAPTER 8

ACANTHOCYTES IN PANTOTHENATE KINASE ASSOCIATED NEURODEGENERATION

THOMAS KLOPSTOCK[1], MATTHIAS ELSTNER[1], AND ALESSANDRO MALANDRINI[2]

[1]Neurologische Klinik und Poliklinik, Ludwig-Maximilians-Universität, München, Germany, [2]Institute of Neurological Sciences, University of Siena, Italy

Abstract. Hallervorden-Spatz syndrome (HSS) and HARP syndrome (hypoprebetalipoproteinemia, acanthocytosis, retinitis pigmentosa, and pallidal degeneration) were shown to be due to mutations in the gene for pantothenate kinase 2 (*PANK2*). It was suggested that this group of disorders should now be referred to as pantothenate kinase associated neurodegeneration (PKAN). *PANK2* is a key regulatory enzyme in the biosynthesis of coenzyme A, which in turn is pivotal in phospholipid biosynthesis and membranogenesis. Thus, defective membrane repair may be the molecular basis for acanthocyte formation in PKAN.

HALLERVORDEN-SPATZ SYNDROME (HSS)

Clinical Findings

Classical HSS begins in late childhood or early adolescence and progresses slowly over 10 to 20 years. Death usually occurs before age 30. Clinically, the condition is characterized by extrapyramidal (rigidity, dystonia, choreoathetosis) and pyramidal signs. Both involuntary movements and rigidity may involve muscles supplied by cranial nerves, resulting in difficulties in articulation and swallowing. In most cases, mental deterioration is present. Ataxia, myoclonus, epilepsy, and optic atrophy have also been mentioned in a few reports. In a review by Dooling, pigmentary retinopathy was found in 11 of 42 patients [2]. The hallmark of the disease is large deposits of iron in the globus pallidus, caudate and substantia nigra, which reveal the diagnosis post mortem and on

A. Danek (ed.), Neuroacanthocytosis Syndromes, 67–70.

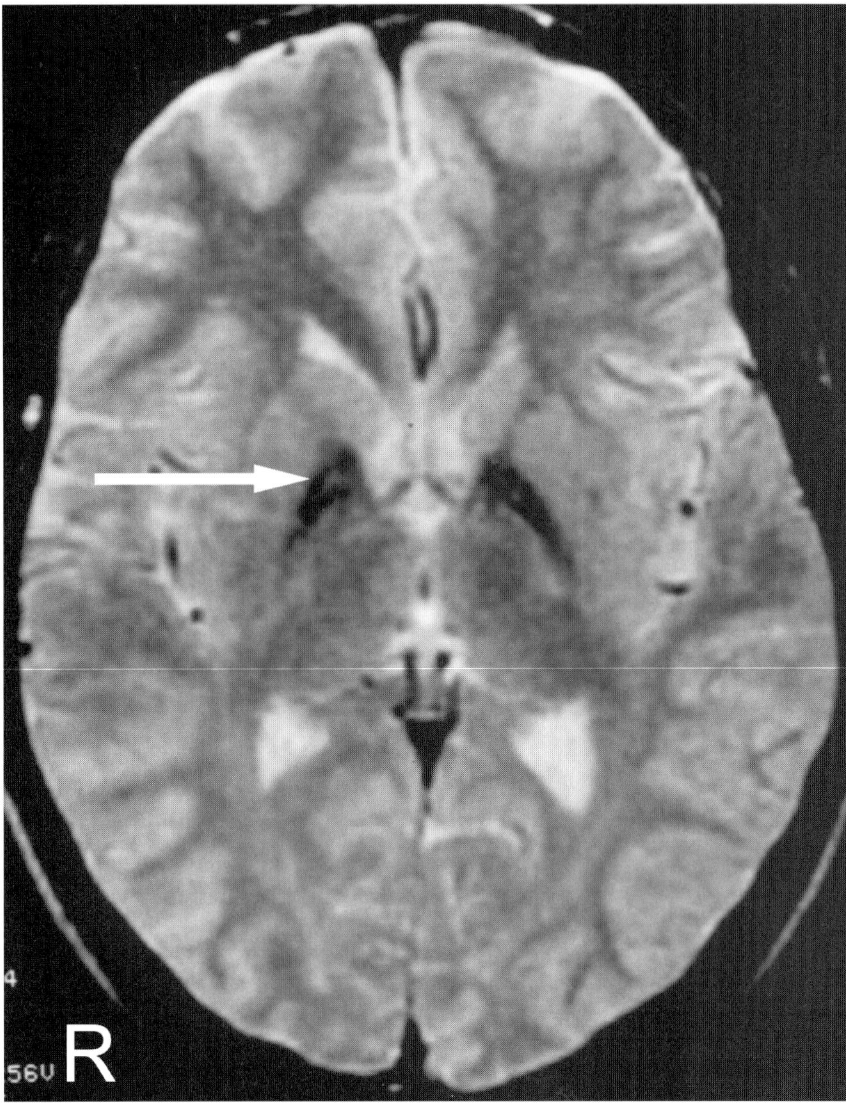

Figure 1. Brain MR showing the "tiger's eye" appearance of the globus pallidus. Note the bilateral hypointensity of the globus pallidus and the small central area of hyperintensity on the right side (arrow). The latter is thought to resemble the eye of a tiger.

magnetic resonance imaging (MRI). The classical MRI appearance is known as the "eye of the tiger" sign (Figure 1). Atypical HSS is diagnosed in individuals who may have a somewhat different phenotype yet have radiographic or pathologic evidence of increased basal ganglia iron.

Molecular Findings

The HSS gene has been linked to chromosome 20p13 [11], and recently it was shown that HSS is due to mutations in the gene for pantothenate kinase 2 (*PANK2*) [13]. Mutations were found not only in classical HSS, but also in atypical cases. *PANK2* is a key regulatory enzyme in the biosynthesis of coenzyme A (CoA) from pantothenate (vitamin B5). Subsequently, the name pantothenate kinase associated neurodegeneration (PKAN; MIM 234200) was proposed [13].

HARP SYNDROME

Only two patients have been reported with HARP syndrome (hypoprebetalipoproteinemia, acanthocytosis, retinitis pigmentosa, and pallidal degeneration; MIM 200150) [3,9]. The syndrome resembles HSS by MRI appearance and clinical findings, but is distinguished by the lipoprotein abnormality. However, it is not clear if the hypoprebetalipoproteinemia is indeed an integral part of the phenotype. Other authors failed to demonstrate lipid abnormalities in patients with sporadic and familial HARP-like syndrome [6,8,9,12]. Recently, a homozygous nonsense mutation was found in exon 5 of the *PANK2* gene in the original HARP patient [1]. This finding confirms the clinical assumption that HARP is part of the PKAN disease spectrum.

ACANTHOCYTES IN PKAN

Apart from HARP syndrome, acanthocytes have been frequently reported in HSS [4,5,7,8,9,12]. The relation between PKAN and acanthocyte formation is yet to be determined. As mentioned above, *PANK2* is a key regulatory enzyme in the biosynthesis of CoA from pantothenate (vitamin B5). CoA in turn plays a central role in several important pathways, including phospholipid biosynthesis and, therefore, membranogenesis. As erythrocytes frequently undergo trauma, defective membrane repair may be involved in acanthocyte formation [10].

NOSOLOGICAL ASPECTS

The term neuroacanthocytosis was coined to define the association of neurological syndromes and acanthocytosis. It comprises three fairly distinct syndromes: Abetalipoproteinemia Bassen-Kornzweig, McLeod syndrome and chorea-acanthocytosis. In these disorders, acanthocytosis is detected in a high percentage of patients, and is regarded as a distinctive feature. Although acanthocytosis is obligatory in HARP and rather frequent in HSS, these two disorders were never ranked among the neuroacanthocytosis syndromes.

Due to the fact that the molecular basis is now known for abetalipoproteinemia, McLeod syndrome, chorea-acanthocytosis, and PKAN, the term neuroacanthocytosis in our opinion is rendered unnecessary. It will be interesting to define the frequency and, hopefully, the pathomechanism of acanthocytosis in a defined population of PKAN patients.

REFERENCES

1. Ching KH, Westaway SK, Gitschier J, Higgins JJ, Hayflick SJ (2002) HARP syndrome is allelic with pantothenate kinase associated neurodegeneration. *Neurology* **58:** 1673-4.
2. Dooling EC, Schoene WC, Richardson EP (1974) Hallervorden-Spatz syndrome. *Arch Neurol* **30:** 70-83.
3. Higgins JJ, Patterson MC, Papadopoulos NM, Brady RO, Pentchev PG, Barton NW (1992) Hypoprebetalipoproteinemia, acanthocytosis, retinitis pigmentosa, and pallidal degeneration (HARP syndrome). *Neurology* **42:** 194-8.
4. Köhler B (1989) Hallervorden-Spatz Syndrom mit Akanthozytose. *Monatsschr Kinderheilkd* **137:** 616-9.
5. Luckenbach MW, Green WR, Miller NR, Moser HW, Clark AW, Tennekoon G (1983) Ocular clinicopathologic correlation of Hallervorden-Spatz syndrome with acanthocytosis and pigmentary retinopathy. *Am J Ophthalmol* **95:** 369-82.
6. Malandrini A, Cesaretti S, Mulinari M et al (1996) Acanthocytosis, retinitis pigmentosa, pallidal degeneration. Report of two cases without serum lipid abnormalities. *J Neurol Sci* **140:** 129-31.
7. Malandrini A, Fabrizi GM, Bartalucci P et al (1996) Clinicopathological study of familial late infantile Hallervorden-Spatz disease: A particular form of neuroacanthocytosis. *Childs Nerv Syst* **12:** 155-60.
8. Muthane UB, Shetty R, Panda K, Yasha TC, Jayakumar PN, Taly AB. (1999) Hallervorden Spatz disease and acanthocytes. *Neurology* **53:** 32A.
9. Orrell RW, Amrolia PJ, Heald A et al (1995) Acanthocytosis, retinitis pigmentosa, and pallidal degeneration: a report of three patients, including the second reported case with hypoprebetalipoproteinemia (HARP syndrome). *Neurology* **45:** 487-92.
10. Stevenson VL, Hardie RJ (2001) Acanthocytosis and neurological disorders. *J Neurol* **248:** 87-94.
11. Taylor TD, Litt M, Kramer P et al (1996) Homozygosity mapping of Hallervorden-Spatz syndrome to chromosome 20p12.3-p13. *Nat Genet* **14:** 479-81.
12. Tripathi RC, Tripathi BJ, Bauserman SC, Park JK (1992) Clinicopathologic correlation and pathogenesis of ocular and central nervous system manifestations in Hallervorden-Spatz syndrome. *Acta Neuropathol (Berl)* **83:** 113-9.
13. Zhou B, Westaway SK, Levinson B, Johnson MA, Gitschier J, Hayflick SJ (2001) A novel pantothenate kinase gene (*PANK2*) is defective in Hallervorden-Spatz syndrome. *Nat Genet* **28:** 345-9.

CHAPTER 9

DIAGNOSTIC TEST FOR NEUROACANTHOCYTOSIS: QUANTITATIVE MEASUREMENT OF RED BLOOD CELL MORPHOLOGY

Alexander Storch[1] and Johannes Schwarz[2]
[1]Department of Neurology, University of Ulm, Ulm, and [2]Department of Neurology, University of Leipzig, Leipzig, Germany

Abstract. Acanthocytosis is defined as an increased amount of abnormally shaped red blood cells with multiple spicules, which are irregular in shape, orientation and distribution. These morphological changes occur normally because of ultrastructural abnormalities of the erythrocyte membrane due to lipid or membrane skeleton alterations. Since genetic analyses and/or specific laboratory tests are available only for a minority of disorders associated with acanthocytosis, quantitative measurements of abnormal erythrocytes play a pivotal role in the diagnosis of neurological disorders associated with acanthocytosis. This review focuses on standardized procedures to quantify red blood cell morphology.

INTRODUCTION

Acanthocytes ($\alpha\kappa\alpha\nu\theta\alpha$/akantha = thorn or spine in Greek) are mature red blood cells (RBC) with multiple protrusions or spicules, often with terminal bulbs, which are irregular in shape, orientation, and distribution. Consequently, acanthocytosis is defined as an increased amount of such misshaped RBCs in peripheral blood. This characteristic morphology may be associated with a wide spectrum of symptoms including hemolysis, splenomegaly, seizures, amyotrophy, myopathy, chorea, dystonia and Parkinsonism [4,9,18].

Etiologies comprise: (i) hereditary disorders such as autosomal-recessive chorea-acanthocytosis (ChAc) with mutations in the *CHAC* gene, X-linked McLeod syndrome, dyslipoproteinemias (such as abetalipoproteinemia or Bassen-Kornzweig disease) or mutations in the erythrocyte anion transporter gene

A. Danek (ed.), Neuroacanthocytosis Syndromes, 71–77.

[6,16,18], and (ii) secondary acanthocytosis due to hypovitaminosis, malnutrition, kidney or liver diseases [4] (see also Chapter 2).

Since genetic analyses and/or specific laboratory tests are available only for a minority of these disorders, testing for acanthocytes remains the first-line approach to establish the diagnosis. This review focuses on procedures to quantify acanthocyte levels in peripheral blood and gives a range for normal values as well as data on specificity and sensitivity.

RED BLOOD CELL MORPHOLOGY

A normal RBC has the shape of a biconcave disc or discocyte. Several changes both *in vivo* and *in vitro* can result in crenation or deformation of discocytes forming the spiculated cell type called echinocytes. *In vivo*, crenation might be caused by uremia, liver failure or splenectomy [4]. On the other hand, there are several known conditions to cause echinocytic RBC formation *in vitro*, such as high pH value, ATP depletion, incubation with various anionic drugs, and contact to glass surfaces [8,13]. These changes may remain reversible depending on the degree of echinocyte formation [3]. In contrast to echinocytes, acanthocytes show less numerous spicules on an irregular cell surface, which are of variable size and frequently demonstrate terminal bulbs [1,3,12,17]. The distinction between acanthocytes and echinocytes may be obscured in dried blood smears, but is facilitated in wet blood preparations and scanning electron microscopy [11]. Acanthocytes are not seen in bone marrow preparations. Transformation of normal RBCs into acanthocytes has not been demonstrated *in vitro*.

In 1973, Bessis introduced a classification of RBC morphology based on scanning electron microscopy findings, which was modified by Redman and coworkers in 1989 [1,17]. These studies classified normal RBCs as category I, irregularly contoured disks with protrusions or wavy membranes as AI, "true" acanthocytes – flat cells with spicules – as AII, cupped-shaped cells as category II, and obvious stomatocytes as category III.

Pathophysiology of Acanthocytosis

The pathophysiological mechanisms underlying acanthocyte formation have remained unclear. Most likely, a defect of cell membrane function possibly due to altered erythrocyte membrane lipids or abnormalities of the cytoskeleton contribute to acanthocyte formation. Abnormalities of the membrane lipid composition and/or vitamin E content can explain acanthocytosis in patients with hypolipoproteinemias, i.e. changes in plasma lipids lead to an increased content of sphingomyelin in the outer leaflet of the membrane bilayer resulting in decreased membrane fluidity with subsequent acanthocyte formation [2,5,9,18]. In most studies of chorea-acanthocytosis, the RBC membrane lipid composition was grossly normal [9,14], only some Japanese studies report altered phospholipid content in RBC membrane preparations [15]. On the other hand, freeze-fracture electron microscopy showed ultrastructural abnormalities of the membranous skeleton, and there is evidence for altered phosphorylation of membrane proteins and abnormal ion fluxes through the membrane [14].

INVESTIGATIONS OF RED BLOOD CELL MORPHOLOGY

True acanthocytes (type AII according to the Redman classification [17]) should never be present in normal blood. In an earlier study of 19 cases suffering from neuroacanthocytosis, a proportion of > 3 % of crenated erythrocytes in dried blood smears of ethylen-diamine-tetracetic acid (EDTA) blood samples was regarded as significantly abnormal [9]. However, there are several reports of cases with neuroacanthocytosis without significant acanthocytosis in this kind of blood investigations [8,10] and repeated blood smear examination was recommended [9]. Feinberg and co-workers (1991) introduced a method utilizing the phenomenon that acanthocytes show a dramatic increase in their sensitivity to echinocytic stress of isotonic dilution of the blood sample [8,13]. The development of > 15 % of echinocytes on a wet preparation of blood diluted 1 : 2 with isotonic saline or a significant percentage of true acanthocytes in scanning electron microscopy was considered abnormal [8]. Indeed, two patients with clinically suspected ChAc without acanthocytosis in dried smears of EDTA blood showed significantly increased levels of echinocytic RBCs [8], suggesting a higher sensitivity of this method compared to the standard method using dried EDTA blood smears. However, this study only investigated 16 controls and did not report normal values or the specificity/sensitivity of the testing procedure.

We performed a prospective, randomized trial in healthy volunteers and patients with movement disorders to establish a defined method for measuring acanthocyte levels and provide a range for normal values as well as specificity and sensitivity of the test procedure [19]. Thus, we used the method introduced by Feinberg and coworkers (1991) with modifications [8]. Differentiation between acanthocytes (type AI and AII according to Redman and co-workers [17]) and echinocytes is only possible with the help of scanning electron microscopy [4,11]. Since we wanted to establish an easy and inexpensive screening method, we used phase-contrast microscopy to investigate both the dry and wet blood preparations and we counted both acanthocytic and echinocytic red blood cells. The main difference between these two morphologies of red blood cells is the reversibility of the abnormality in echinocytes. The reversibility of echinocyte formation was demonstrated by adding levomepromazine at a final concentration of 1 mM to each blood sample.

Quantitative Measurement of Acanthocytosis – Test Procedure

The detailed test procedure and study design is described elsewhere [19]. In brief, blood was drawn into a commercially available syringe pre-filled with potassium-EDTA-solution (final concentration: 1.6 mg per ml blood; Sarstedt, Nümbrecht, Germany) and into a 10 ml plastic syringe, respectively. Blood samples without EDTA (5 ml) were immediately mixed with 5 ml of heparinized saline solution (isotonic (0.9%) sodium chloride containing 10 units of heparin per ml of saline). For reversion of acanthocyte formation levomepromazine (Neurocil, Bayer Pharmaceuticals, Leverkusen, Germany) at a final concentration of 1 mmol/l saline was added. These mixtures were incubated at room temperature for 30 to 120 minutes on a shaker. Then, dry blood smears were prepared on glass object slides and stained using the standard Pappenheim staining procedure (Merck KG, Darmstadt, Germany). Wet unfixed blood preparations were prepared by putting a drop of blood mixture on a glass object slide and carefully placing a glass cover slip on top of the blood drop to achieve a bubble-free monolayer of blood cells. Dry blood smears and wet cell monolayers, respectively, were investigated using a phase-contrast microscope (Axiophot 125; Zeiss, Jena, Germany) at a magnification of 1000× (oil immersion) and at least five microphotographs were taken from each preparation. All red blood cells of the photographs were analyzed by a blinded investigator. Cells with spicules, which were irregular in shape and orientation/distribution (corresponding to type AI/AII acanthocytes and echinocytes according to Redman's classification using electron scanning microscopy [17]) were counted as abnormal. The amounts of acanthocytic/echinocytic red blood cells were expressed as percent of total erythrocytes.

Every third patient seen in our movement disorder clinic from October 1997 to June 1999 was asked to give informed consent and participate in the study. Patients were diagnosed according to published standard diagnostic criteria. In addition, we studied 31 patients with movement disorders in whom the diagnosis remained unclear. A total of 37 healthy controls (female/male: 19/18; age 49 ± 20 years, mean $\pm$ SD), 100 patients with defined movement disorders (female/male: 39/61; age 62 ± 14 years), and 31 patients with undiagnosed movement disorders (female/male: 12/18; age 44 ± 27 years) participated in the study. Patients with defined movement disorders suffered from Parkinson's disease ($n = 68$), Huntington's disease ($n = 5$), essential tremor ($n = 5$), Lewy body disease ($n = 4$), Multiple System Atrophy ($n = 3$), Progressive Supranuclear Palsy ($n = 3$) and various other movement disorders ($n = 12$). Patients in whom diagnosis remained unclear suffered from a hyperkinetic movement disorder ($n = 12$), Parkinsonism ($n = 8$), and other movement disorders ($n = 11$). Thus, diagnoses in the defined movement disorder group comprise all major differential diagnoses of hereditary neuroacanthocytosis, such as Huntington's disease and parkinsonian syndromes [4,6,18].

In controls, the levels of echinocytic/acanthocytic RBC were between 0.2% in dried smears of EDTA blood and 1.8% in wet preparations of isotonically diluted blood samples. Similar results were obtained in patients with defined movement disorders with echinocyte levels of between 0.1% in dried smears of EDTA blood and 1.9% in wet preparations of diluted blood samples.

Table 1. Parameters of the standardized acanthocyte screening test (adapted from *Storch et al, 2003*).

Blood/Smear Type	Normal value[a]	Specificity	Sensitivity[b]
EDTA/dry smear	< 1.2 %	0.99	Low (1/3)
EDTA/wet preparation	< 3.7 %	0.98	Low (1/3)
Diluted/dry smear	< 3.0 %	0.99	Middle (2/3)
Diluted/wet preparation	< 6.3 %	0.98	High (3/3)

[a]99th percentile of healthy controls and defined movement disorder patients
[b]Number of detected patients per genetically confirmed ChAc patients

In general, echinocytes/acanthocytes were significantly more frequent following isotonic dilution as well as in wet preparations of blood [19]. There was no statistical difference between controls and patients with defined movement disorders in all test conditions. As reported by Feinberg and co-workers [8], addition of levomepromazine (final concentration: 1 mM) nearly completely reverses the formation of acanthocytes in all blood preparations [19]. We found no significant correlation of acanthocyte levels in both groups of volunteers in all blood samples and preparations with age, gender or diagnosis [19]. Thus, to define normal values we combined the results of the two groups of volunteers and calculated the 99th percentile. Table 1 indicates these normal values as well as the specificity and sensitivities of the tests.

In the group of patients with undefined movement disorders, 10 of 31 patients displayed echinocyte/acanthocyte levels above the normal range in diluted blood samples and wet preparations (10.5 to 72.3%) [19]. Five of these ten patients displayed typical symptoms of ChAc including chorea, dystonia, amyotrophy, oral self-mutilations, CK elevation and one affected sibling. Two other patients showed unknown neurological disorders, but positive family history including pathologic levels of acanthocytes above the 99th percentile in at least one additional family member. Therefore, a diagnosis of hereditary neuroacanthocytosis was made in a total of seven patients. After we had closed the prospective phase of the study, two additional patients and one family member were tested positive for ChAc using genetic analysis [7]. The amounts of acanthocytes in patients with ChAc did not correlate with clinical presentation or with severity or time course of symptoms [9,19]. As reported by Feinberg and co-workers (1991) [8], two of our patients with genetically confirmed ChAc showed normal levels of acanthocytes in standard dry blood smears of EDTA blood samples, but had greatly abnormal values in wet blood preparations (38 and 43%, respectively).

These data indicate that standard EDTA blood samples to prepare dry smears have only limited sensitivity to detect neuroacanthocytosis. Unfortunately, most laboratories use this technique to search for acanthocytes if this diagnosis is suspected. On the other hand, unfixed wet blood preparations of isotonically diluted blood samples seem to bear the highest sensitivity for acanthocytosis since all genetically confirmed patients of our cohort were correctly identified (Table 1).

CONCLUSIONS

Genetic analyses and/or specific laboratory tests are available only for a minority of the disorders associated with acanthocytosis in the peripheral blood. Thus, quantitative measurement of acanthocytic erythrocytes is pivotal for the diagnosis of these disorders. The use of standard hematological techniques (dried smears of EDTA blood samples) shows low sensitivity for the detection of acanthocytosis, at least in neurological disorders [8,10,19]. We systematically evaluated a method of Feinberg and co-workers (1991) [8,19], which utilizes the phenomenon that acanthocytes show a dramatic increase in their sensitivity to echinocytic stresses [8,13]. This technique uses isotonically diluted blood and unfixed wet preparations to detect acanthocytosis, with a normal range of $< 6.3\,\%$ acanthocytes among total erythrocyte number after an incubation period of the samples of 0.5 to 3 h at room temperature. The interval between blood drawing and smear preparation is crucial, since the amount of acanthocytes and echinocytes increased with time [8,19]. Furthermore, it is suggested to include a healthy volunteer as well as a levomepromazine control in each test series. In our experience, our method is useful to identify acanthocytosis with high sensitivity and specificity in large numbers of patients with suspected neuroacanthocytosis.

REFERENCES

1. Bessis M (1973) Red cell shapes: An illustrated classification and its rationale. In: *Bessis M, Weed R, LeBlond SF (eds) Red cell shape.* Springer Verlag, New York, pp 1-25.
2. Bird TB, Cederbaum S, Valpey RW, Stahl WL (1978) Familial degeneration of the basal ganglia with acanthocytosis: a clinical, neuropathological and neurochemical study. *Ann Neurol* **3:** 253-258.
3. Brecher G, Bessis M (1972) Present status of spiculated red cells and their relationship to the discocyte-echinocyte transformation: a critical review. *Blood* **40:** 333-344.
4. Brin MF (1993) Acanthocytosis. In: *Goetz C, Tanner CM, Aminoff MJ (eds) Handbook of Clinical Neurology.* Elsevier, Amsterdam, pp 271-299.
5. Clarke MR, Aminoff MJ, Chiu DT-Y, Kuypers FA, Friend DS (1989) Red cell deformability and lipid composition in two forms of acanthocytosis: enrichment of acanthocytic populations by density gradient centifugation. *J Lab Clin Med* **113:** 469-481.
6. Danek A, Rubio JP, Rampoldi L, Ho M, Dobson-Stone C, Tison F, Symmans WA, Oechsner M, Kalckreuth W, Watt JM, Corbett AJ, Hamdalla HHM, Marshall AG, Sutton I, Dotti MT, Malandrini A, Walker RH, Daniels G, Monaco AP (2001) McLeod neuroacanthocytosis: Genotype and phenotype. *Ann Neurol* **50:** 755-764.
7. Dobson-Stone C, Danek A, Rampoldi L, Hardie R, Chalmers RM, Wood NW, Bohlega S, Dotti MS, Federico A, Shizuka M, Tanaka M, Watanabe M, Ikeda Y, Brin M, Goldfarb LG, Karp BI, Mohiddin S, Storch A, Fryer AE, Maddison P, Sibon I, Trevisol-Bittencourt

PC, Singer C, Requena I, Aasly JO, Schmierer KO, Zeviani M, Meiner V, Lossos A, Johnson S, Mercado FC, Sorrentino G, Rouleau G, Volkmann J, Lees A, Geraud G, Chouinard S, Németh A, Monaco AP (2002) Mutational spectrum of the *CHAC* gene in patients with chorea-acanthocytosis. *Eur J Hum Gen,* **10:** 773-781.

8. Feinberg TE, Cianci CD, Morrow JS, Pehta JC, Redman CM, Huima T, Koroshetz WJ (1991) Diagnostic tests for chorea-acanthocytosis. *Neurology* **41:** 1001-1006.
9. Hardie RJ, Pullon HW, Harding AE, Owen JS, Pires M, Daniels GL, Imai Y, Misra VP, King RHM, Jacobs JM, Tippett P, Duchen LW, Thomas, PK, Marsden CD (1991) Neuroacanthocytosis: a clinical, hematological and pathological study of 19 cases. *Brain* **114:** 13-49.
10. Johnson SE, Dahl A, Sjaastad O (1998) Progressive pseudobulbar paresis, early choreiform movements, and later rigidity: appearance in two sets of dizygotic twins in the same family. *Mov Dis* **13:** 556-562.
11. Kayden HJ, Bessis M (1970) Morphology of normal erythrocyte and acanthocyte using Nomarski optics and the scanning electron microscope. *Blood* **36:** 427-436.
12. Lessin LS, Klug PP, Jensen WN (1976) Clinical implication of red cell shape. *Adv Intern Med* **21:** 451-500.
13. Morrow J, Andersen R (1986) Shaping the too fluid bilayer. *Lab Invest* **54:** 237-239.
14. Olivieri O, De Franceschi L, Bordin L, Manfredi M, Miraglia del Giudice E, Perrotta S, De Vivo M, Guarini P, Corrocher R (1997) Increased membrane protein phosphorylation and anion transport activity in chorea-acanthocytosis. *Haematologica* **82:** 648-53.
15. Oshima M, Osawa Y, Asano K, Saito T (1985) Erythrocyte membrane abnormalities in patients with amyotrophic chorea with acanthocytosis. Part 1. Spin labeling studies and lipid analyses. *J Neurol Sci* **68:** 147-160.
16. Rampoldi L, Dobson-Stone C, Rubio JP, Danek A, Chalmers RM, Wood NW, Verellen C, Ferrer X, Malandrini A, Fabrizi GM, Brown R, Vance J, Pericak-Vance M, Rudolf G, Carre S, Alonso E, Manfredi M, Nemeth AH, Monaco AP (2001) A conserved sorting-associated protein is mutant in chorea-acanthocytosis. *Nat Genet* **28:** 119-120.
17. Redman CM, Huima T, Robbins E, Lee S, Marsh WL (1989) Effects of phosphatidylserine on the shape of McLeod red cell acanthocytes. *Blood* **74:** 1826-1835.
18. Stevenson VL, Hardie RJ (2001) Acanthocytosis and neurological disorders. *J Neurol* **248:** 87-94.
19. Storch A, Kornhass M, Schwarz J (2003) Acanthocytosis in movement disorders: How much is normal? *Submitted*

CHAPTER 10

DIFFERENTIAL DIAGNOSIS OF SERUM CREATINE KINASE ELEVATION

BENEDIKT G.H. SCHOSER[1] AND THOMAS N. WITT[2]
[1]Department of Neurology, Friedrich-Baur-Institut, and [2]Department of Neurosurgery; Ludwig-Maximilians-Universität, Munich, Germany

Abstract. Routine blood chemistry profiles normally include measurements of creatine kinase (CK) activity. Therefore, an increasing number of patients with elevated creatine kinase levels are identified. General practitioners, anesthesiologists, surgeons and neurologists might be confronted with such an unexplained non-cardiac hyperCKemia. Although this problem is not uncommon, it has been the subject of only a few studies. We provide a clinical review of the enzyme CK in humans and of 343 case reports of "idiopathic" persistently elevated CK serum levels. At present, altogether nine studies provide retrospective data on patients admitted to neuromuscular centers for a work-up of "idiopathic" hyperCKemia and/or susceptibility to malignant hyperthermia. After complete examination, including a muscle biopsy, a subclinical myopathy could be delineated in up to 50% of the patients.

INTRODUCTION

At least 4 functionally active genes and a number of pseudogenes code for human creatine kinase. The gene products are termed CK-MM (chromosome 19q13), CK-MB and CK-BB, which originate from skeletal muscle, from the heart and from brain, respectively [MIM 123310, 123295, 123290, 123280, 123270].

In general, creatine kinase (CK) is the enzyme which catalyses the reversible reaction

$$\text{ATP} + \text{Cr} \Longleftrightarrow \text{ADP} + \text{PCr}$$

(adenosine triphosphate+creatine ⇔ adenosine diphosphate+phosphocreatine)

A. Danek (ed.), Neuroacanthocytosis Syndromes, 79–86.

Table 1. Causes of hyperCKemia unrelated to systemic neuromuscular disorders.

a.	muscle trauma, muscle hematoma, compartment syndrome
b.	intramuscular injections, needle EMG
c.	acute psychosis, epileptic seizures, stroke, cerebral hemorrhage, head trauma
d.	burns, hyperthermia, septic shock
e.	macro-CKemia, hematopathies
f.	pregnancy
g.	drugs:
	β-blockers;
	antipsychotics: clozapine, loxapine, haloperidol, melperone, risperidone, olanzapine;
	cholesterol-lowering agents: statins, clofibrate;
	various: AZT, amiodarone, chloroquin, cyclosporine, doxorubicin, emetine, interferons, isotretinoin, perhexiline, tumor necrosis factor alpha, tryptophan, vincristine, etc.

Mice deficient in cytosolic and mitochondrial CK [19] showed blocked phosphocreatine (PCr) to ATP transphosphorylation in skeletal muscle. In vivo, mutant muscle showed significantly impaired tetanic force output, increased relaxation times, altered mitochondrial volume and location, and conspicuous tubular aggregates of sarcoplasmic reticulum membranes. In depolarised myotubes, absence of CK influenced both the release and sequestration of calcium ion. These data linked the CK-PCr system to the regulation of calcium ion flux during excitation and relaxation phases of muscle contraction [19].

The measurement of serum CK activity encompasses the cytoplasmic dimeric isoenzymes CK-MM, CK-MB, their forms modified after synthesis and the activity of macro-creatine kinase (macro-CK). Macro-CK is a serum enzyme with a molecular mass higher than that of the corresponding enzyme normally found in serum [8,20]. There are two different macro-CKs in humans: macro-CK type 1, a postsynthetic isoenzyme of CK-BB; and macro-CK type 2, a mitochondrial CK [20]. In a healthy person, 96% of the total CK in the human body is contained in the skeletal muscle. Normal values for CK activity are below 70 U/l in women and below 80 U/l in men, with some variation depending upon the test kit [6].

Without a relation to generalized neuromuscular disorders, CK levels may be elevated due to presence of macro-CK and to trauma, burns, hyperthermia, pregnancy, hematopathies, muscle hematoma, intramuscular injections, septic shock, acute psychosis, drugs and medication such as β-blockers, antipsychotic or cholesterol-lowering agents and, finally, after physical exercise (Table 1). With a latency of 24 to 72 h after exercise, a maximum of serum CK levels is noticed.

Table 2. Cumulative data of non-cardiac hyperCKemia studies.

Author		Cases	History		Neuro-Exam		EMG		Biopsy	
			norm.	abn.	norm.	abn.	norm.	abn.	norm.	abn.
Brewster	[3]	14	10	4	10	4	10	4	10	4
Joy	[10]	19	19	–	19	–	5	14	–	15
Kleppe	[11]	100	49	51	73	27	42	55	26	24
Prelle	[14]	3	2	1	2	1	3	–	1	2
Prelle	[15]	114	65	49	114	–	53	53	70	44
Reijneveld	[17]	31	2	29	31	–	22	9	7	24
Rowland	[18]	10	10	–	10	–	10	–	10	–
Sunohara	[21]	3	3	–	3	–	3	–	3	–
Weglinski	[22]	49	9	40	43	6	36	13	12	37
	Σ	343	169	174	305	38	184	148	139	150

CK-BB elevation is seen after stroke, subarachnoid hemorrhage, head trauma, epileptic seizures, and in meningoencephalitis and neoplasms of the brain.

CK-MB elevations are noticed after myocardial infarction, cardioversion or heart contusion.

In contrast, low CK levels are well known after prolonged bed-rest and muscle disuse, long-term corticoid treatment and hyperthyroidism.

The first report about the value of CK measurement in neuromuscular disorders appeared in 1959 [5]. In general, the level of CK reflects the severity of muscle membrane damage and necrosis, except in cases where loss of muscle fibers is so severe that CK begins to fall, as happens in the late stages of Duchenne dystrophy. Very high CK levels are seen only in a few conditions, notably rhabdomyolysis, regardless of the cause, inflammatory myopathies and X-linked muscular dystrophies. The term "idiopathic hyperCKemia" was created for cases without clinical and histopathological evidence of neuromuscular disorders [18].

Nowadays, a routine blood chemistry profile includes measurement of CK activity and an increasing number of patients with elevated levels are being identified. General practitioners, anesthesiologists, surgeons and neurologists might be confronted with such cases of unexplained non-cardiac hyperCKemia. Although this problem is not uncommon, it has been the subject of only a few studies, nine in all [3,10-11,14-15,17-18,21-22].

CUMULATIVE DATA ON NON-CARDIAC "IDIOPATHIC" HYPER-CK-EMIA

The available studies of hyperCKemia report on 343 patients aged 3-78 years, of which 89 were female (26%) and 254 male (74%) [3,10-11,14-15,17-18,21-22].

Table 3. Etiology of persistent non-cardiac "idiopathic" hyperCKemia.

Idiopathic hyperCKemia (of 343 patients tested)	$n = 144$
Myopathy (of 289 biopsies)	$n = 129$
Occupational and leisure activity (of 209 patients)	$n = 57$
Single episode of hyperCKemia (of 343 patients)	$n = 37$
Malignant hyperthermia susceptibility (of 64 in vivo contracture tests)	$n = 30$
Neurogenic changes on muscle biopsy (of 289 biopsies)	$n = 21$
Neuropathy (of 343 patients)	$n = 14$

General history including occupational and leisure activities in combination with family history revealed a possible predisposition to neuromuscular disorders in 174 patients (51%). In 57 of 209 patients (27%) manual work or leisure activities correlated with hyperCKemia.

Neurological examination disclosed evidence of a neuromuscular disorder in 38 patients (11%). Electromyographic examination (332 patients) was abnormal in 45% (myopathic pattern: 128 patients, neurogenic pattern: 14). Extensive histomorphological examination (including histochemistry with hematoxylin-eosin, trichrome Gomori, Oil red O, NADH-tetrazolium reductase, ATPase at different pH levels, PAS, myoadenylate deaminase, cytochrome oxidase, phosphorylase and immunohistology with cytoskeletal protein analysis, e.g. anti-dystrophin staining) revealed pathological findings in 150 patients (52%). A complete work-up failed to discover any abnormality in the history, clinical examination, electromyography and muscle biopsy of 42% (i.e. 144 cases with a diagnosis of "idiopathic hyperCKemia"). Thirty of 64 patients who were tested for susceptibility to malignant hyperthermia had positive halothane-caffeine contracture tests. None of the nine studies reviewed here found a significant correlation between the level of CK increase and the underlying neuromuscular disorder (Tables 2 and 3).

Persistent hyperCKemia in the absence of cardiac disorders is not an uncommon finding in routine laboratory tests. The magnitude of CK activity levels does not reflect underlying neuromuscular disorders but it should be noted that very high CK levels ($> 1000\,\mathrm{U/l}$) may indicate a myositis or, in younger subjects, myopathies related to dystrophin, desmin, caveolin, dysferlin, calpain, fukutin-related protein, sarcoglycan or spinal muscular atrophies [2-3,7,10-18, 21-22]. Among patients with slightly elevated CK activity (80-300 U/l), still in 50% there was a relation to subclinical neuromuscular disorders including motor neuron disease. Therefore, an effective differential diagnostic approach is needed.

Table 4. Diagnostic work-up of persistently elevated creatine kinase.

CK > 100 U/L
2× after 36 h rest; macroCK?

↙ ↘

CK-MB < 6% → **CK-MB > 6%**

CK-MB < 6%
↓
Neuromuscular history
drugs, occupation, physical activity, family history
↓
Neurological exam
myalgia, myotonia, cramps, fasciculation, paresis, muscle atrophy
↓
Laboratory assessments
LDH, CRP, calcium, phosphate, thyroid hormones, ANA, glucose, alcohol, acanthocytes
↓
Neurophysiological analysis
EMG, neurography
↘
Muscle biopsy
histology, immunohistochemistry, electron microscopy, biochemical analysis, genetics (FSHMD, myotonic dystrophy type 1 & 2)
in vitro contracture test for malignant hyperthemia

CK-MB > 6%
↓
Cardiac assessment

DIAGNOSTIC WORK-UP OF HYPER-CK-EMIA

History of Occupational and Leisure Activities, Family History

In the nine studies reviewed above, history of occupational and leisure activities in combination with a family history of neuromuscular disorders provided evidence as to the etiology of hyperCKemia in 51% of the patients. Isolated familial hyperCKemia is mostly related to specific neuromuscular disorders such as dystrophinopathies, caveolinopathies or McArdle's disease [1,12].

Manual work with repeated and sustained arm elevation or constrained postures, and leisure activities including excessive exercise, especially without prior training might lead to persistently elevated CK activity levels [4]. Therefore, as a first differential diagnostic step, a repeat measurement of CK activity should be done after at least 48 hours of rest. In 37 of 343 patients elevated CK activity was found only on one occasion [3,10-11,14-15,17-18,21-22]. Next, the presence of macro-CK should be excluded: a possible hint is provided by a very constant and persistent CK elevation with more than 25% CK-MB. In these patients, CK electrophoresis should be considered [8,20].

A history of muscle pain, cramps, muscle fatigue and weakness after exercise may help to identify patients with neuromuscular disorders. One should also be aware of the correlation between larger muscle mass and higher CK activity [9]. In addition, a brief search for underlying systemic disorders, e.g. hypothyroidism, hyperparathyroidism, collagen disorders, diabetes and chronic alcoholism, drugs and medication history should be performed and one may consider an investigation for blood acanthocytosis at this stage.

Neurological Exam and Neurophysiological Studies

The neurological examination in only 11% of the patients resulted in pathological findings, such as muscular weakness, myotonia, or diminished reflexes.

Next, needle electromyography (EMG) of at least two muscles in each of three extremities and neurography should be performed. Patients in whom EMG abnormalities are found, e.g. neurogenic or myopathic patterns, should be referred to a neuromuscular center with special capabilities for further assessment.

Muscle Biopsy

Part of the specialist assessment should be a muscle biopsy with electron microscopic investigations, cytoskeletal immunodiagnostics and Western blotting, and, in selected patients, with biochemical analysis. Routine muscle histochemistry that did not disclose an abnormality must be expanded to include histochemistry preparation for myoadenosine deaminase deficiency, phosphorylase, cytochrome oxidase and succinate dehydrogenase. In the end, immunohistological assessment of cytoskeletal proteins such as dystrophin, sarcoglycans, caveolin, dysferlin should be considered. In facioscapulohumeral muscular dystrophy (FSHMD), biopsy findings may be non-specific and genetic testing is appropriate in patients with suspected FSHMD.

In the studies mentioned, 144 patients were reported as suffering from "idiopathic hyperCKemia" but not all were examined with complete histochemistry and immunohistochemistry [3,10-11,14-15,17-18,21-22]. Some might therefore belong to the extensive group of skeletal muscle membrane disorders, e.g. the desminopathies, dystrophinopathies, caveolinopathies, dysferlinopathies, fukutin-related protein myopathies, sarcoglycanopathies or could suffer from McLeod myopathy [2,7,16].

In Vitro Contracture Test for Susceptibility to Malignant Hyperthemia

There was a considerable proportion (47%) of positive results in the 64 patients who had undergone the halothane-caffeine-contracture test for susceptibility to malignant hyperthermia (MH). This fact must alert anaesthesiologist to the possibility of MH susceptibility in patients with hyperCKemia. Reportedly, the risk of developing symptoms of autosomal-dominant MH is 1 in 25000 anesthesias [22]. The decision to perform the test in patients with elevated CK levels remains controversial [22], but testing for MH susceptibility should be considered in patients with hyperCKemia who are undergoing a diagnostic muscle biopsy. In general, in any patient with CK elevation the use of substances that are MH safe must be considered by the anesthesiologist (Table 4).

Interestingly, CK has no value as a screening test in patients without a personal or family history of MH, and even in patients at risk CK was little sensitive and non-specific.

CONCLUSION

In summary, persistent hyperCKemia indicates subclinical myopathy in up to half of the cases. Thus, an extensive assessment is advised in these patients. None of the reports reviewed mentioned additional screening for acanthocytosis. Given the presence of hyperCKemia in all patients with McLeod syndrome and in many with other types of neuroacanthocytosis, an appropriate screening procedure (see Chapter 9) must be included in the assessment. The rapid development of muscle molecular biology, in addition, will soon allow diagnosing patients with "idiopathic hyperCKemia" with greater precision than we can today.

REFERENCES

1. Afifi AK (1998) Idiopathic hyperCKemia revisited. *J Child Neurol* **13:** 251-252.
2. Blake DJ, Weir A, Newey SE, Davies KE (2002) Function and genetics of dystrophin and dystrophin-related proteins in muscle. *Physiol Rev* **82:** 291-329.
3. Brewster LM, de Visser M (1988) Persistent hyperCKemia: fourteen patients studied in retrospect. *Acta Neurol. Scand* **77:** 60-63.
4. Clarkson PM, Hubal MJ (2002) Exercise-induced muscle damage in humans. *Am J Phys Med Rehabil* **81:** S52-69.
5. Ebashi S, Toyokura Y, Momoi H, Sugita H (1959) High creatine phosphokinase activity of sera of progressive muscular dystrophy. *J. Biochem* **46:** 103-104.
6. ECCLS European committee for clinical laboratory standards (1988) Lund: *ECCLS central office,* document number 3-4
7. Emery AE (2002) The muscular dystrophies. *Lancet* **359:** 687-695.
8. Galasso PJ, Litin SC, Obrien JF (1993) The macroenzymes: a clinical review. *Mayo Clin Proc* **68:** 349-354.
9. Garcia W (1974) Elevated creatine kinase phosphokinase levels associated with large muscle mass. *J Am Med Assoc* **228:** 1395-1396.
10. Joy JL, Shin JOH (1989) Asymptomatic hyperCKemia: an electrophysiologic and histopathologic study. *Muscle Nerve* **12:** 206-209.
11. Kleppe B, Reimers CD, Altmann C, Pongratz DE (1995) Befunde bei 100 Patienten mit „ätiologisch ungeklärter" Erhöhung der Serumkreatinkinaseaktivität. *Med Klin* **90:** 623-627.

12. Merlini L, Carbone I, Capanni C, Sabatelli P, Tortorelli S, Sotgia F, Lisanti MP, Bruno C, Minetti C. (2002) Familial isolated hyperCKaemia associated with a new mutation in the caveolin-3 (*CAV*-3) gene. *J Neurol Neurosurg Psychiatry* **73:** 65-67.
13. Nevins MA, Saran M, Bright M, Lyon LJ (1973) Pitfalls in interpreting serum creatine phosphokinase activity. *J Am Med Assoc* **224:** 1382-1387.
14. Prelle A, Rigoletto C, Moggio M, Sciacco M, Comi GP, Ciscato P, Fagiolari G, Rapuzzi S, Bignotti V, Scarlato G (1996) Asymptomatic familiar hyperCKemia associated with desmin accumulation in skeletal muscle. *J Neurol Sci* **140:** 132-136.
15. Prelle A, Tancredi L, Sciacco M, Chiveri L, Comi GP, Battistel A, Bazzi P, Boneschi FM, Bagnardi V, Ciscato P, Bordoni A, Fortunato F, Strazzer S, Bresolin N, Scarlato G, Moggio M (2002) Retrospective study of a large population of patients with asymptomatic or minimally sympotomatic raised serum ceratine kinase levels. *J Neurol* **249:** 305-311.
16. Rando TA (2001) The dystrophin-glycoprotein complex, cellular signaling, and the regulation of cell survival in the muscular dystrophies. *Muscle Nerve* **24:** 1575-1594.
17. Reijneveld JC, Notermans NC, Linssen WHJP, Wokke JHJ (2000) Benign prognosis in idiopathic hyper-CK-emia. *Muscle Nerve* **23:** 575-579.
18. Rowland LP, Willner J, Cerri C, DiMauro S, Miranda A (1980) Approaches to the membrane theory of Duchenne muscular dystrophy. In: *Angelini C, Danieli GA, Fontanari D (eds.): Muscular Dystrophy – Advances and New Trends.* Amsterdam, Excerpta Medica, pp 3-13
19. Steeghs K, Benders A, Oelemans F, de Haan A, Heerschap A, Ruitenbeek W, Jost C, van Deursen J, Perryman B, Pette D, Bruckwilder M, Koudijs J, Jap P, Veerkamp J, Wieringa B (1997) Altered Ca (2+) responses in muscle with combined mitochondrial and cytosolic creatine kinase deficiencies. *Cell* **89:** 93-103.
20. Stein W (1988) *Laboratory diagnosis of acute myocardial infarction:* Darmstadt, Germany: GIT-Verlag
21. Sunohara N, Takagi A, Nonaka I, Sugita H, Satoyoshi E (1984) Idiopathic hyperCKemia. *Neurology* **34:** 544-547.
22. Weglinski MR, Wedel DJ, Engel AG (1997) Malignant hyperthermia testing in patients with persisting increased serum creatine kinase levels. *Anesth Analg* **84:** 1038-1041.

CHAPTER 11

PATHOLOGY OF NEUROACANTHOCYTOSIS AND OF HUNTINGTON'S DISEASE

Armando Martínez[1], María Angeles Mena[2], Zigmunt Jamrozik[3], and Justo García de Yébenes[4]

[1]Hospital Clínico, [2]Hospital Ramón y Cajal, [4]Fundación Jiménez Díaz; Madrid, Spain and [3]University of Warsaw Hospital, Poland

Abstract. The pathology of Huntington's disease is well characterized by atrophy, neuronal loss and gliosis of the striatum, with preferential damage of the small and medium spiny neurons. The surviving cells show nuclear and intracytoplasmic inclusions that stain with antibodies against ubiquitin and huntingtin. Involvement of other brain regions is common for the cerebral cortex (loss of the pyramidal glutamate bearing neurons, projecting to the striatum) and less frequent in other brain regions (globus pallidus, thalamus, subthalamic nucleus, substantia nigra, and cerebellum). In neuroacanthocytosis there is a more selective involvement of the striatum and globus pallidus, though the substantia nigra and medial thalamus may be affected in some cases. Involvement of the anterior horn cells and of the peripheral nerve is also described, and myopathic changes have been reported. No inclusions or specific immunocytochemical patterns have yet been reported in neuroacanthocytosis.

INTRODUCTION

The pathology of Huntington's disease (HD) is well known through hundreds of pathological studies in patients in different disease stages that had molecular confirmation of the diagnosis. More recently the initial changes in histology have been studied in transgenic models that allow for investigation of cellular changes before the presence of clinical abnormalities. The pathology of neuroacanthocytosis, however, presently is limited to a small number of cases without molecular confirmation and without complementary data provided by transgenic animals.

A. Danek (ed.), Neuroacanthocytosis Syndromes, 87–94.

Table 1. Grading of striatal pathology (according to *Vonsattel et al 1985* [17]).

- Grade 0 (1% of HD brains):
  No macroscopic changes, loss of 30-40% neurons without gliosis in the head of the caudate nucleus.
- Grade 1 (4% of HD brains):
  Atrophy of the tail of CN and 50% neuronal loss with gliosis in the head of the caudate nucleus.
- Grade 2 (16% of HD brains):
  Striatal atrophy is mild to moderate. Severe neuronal loss and gliosis.
- Grade 3 (54% of HD brains):
  Striatal atrophy is severe. Very severe neuronal loss and gliosis.
- Grade 4 (25% of HD brains):
  Loss of 95% of striatal neurons, preservation of accumbens nucleus in 50% of cases.

PATHOLOGY IN HD

Brain

There is great variability in the severity of changes in the brains of patients with HD, depending on disease duration and the size of the CAG expansion [16,17]. A grading system, proposed on the basis of the severity of striatal atrophy and the percentage of loss of striatal neurons, has been proposed (summarized in Table 1).

In general, gross striatal atrophy is considered to be present in around 95% of the patients with HD at the time of death (Figures 1a and b). Frontal atrophy is present in 80% of the patients. Morphometric studies reveal that the regional percentage of volume loss in different brain areas is the following: cerebral cortex 25%, thalamus 28%, caudate nucleus 57%, putamen 64%, globus pallidus pars medialis 27%, globus pallidus pars lateralis 50%, white matter 32% [17]. Cerebellar atrophy is often found in patients with juvenile onset or in those with advanced disease. Pathological changes in other brain areas, such as the limbic system or the hippocampus are rare, unless there is associated pathology, such as Alzheimer changes.

Microscopically, there is neuronal loss and gliosis of the striatum (Figures 2a and b). The severity of striatal changes has a regional gradient, increasing gradually along the antero-posterior, latero-medial, and ventro-dorsal axes. So, the neuropathological changes are more severe in the tail than in the body, and more severe in the body than in the head of the caudate nucleus, and the paraventricular portion is almost always more severely affected. Also, the dorsal portion of the putamen is much more involved than the ventral one. The nucleus accumbens is usually well preserved. The severity of gliosis in different brain areas parallels the loss of neurons [17].

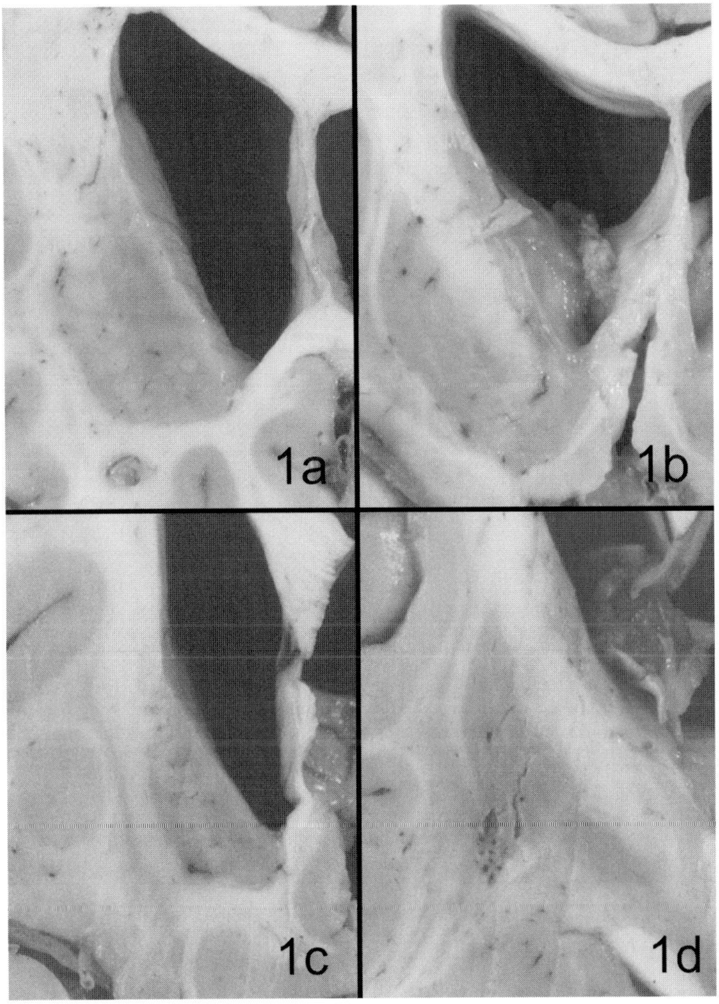

Figure 1. Coronal section of brains of patients with Huntington's disease (a and b) and neuroacanthocytosis (c and d), revealing atrophy of the striatum and, in the patient with neuroacanthoctytosis, of both striatum and pallidum.

Medium size spiny neurons are severely affected in the striatum of patients with HD. The enkephalin-containing spiny neurons projecting to the globus pallidus pars lateralis are more severely involved [1] than the spiny, substance P-containing neurons projecting to the globus pallidus pars medialis [11]. The striatal aspiny, NADPH-diaphorase positive neurons are preserved [4,5], as are the large aspiny cholinergic local-circuit neurons [3]. The difference in resistance of these neuronal subtypes is probably related to the types of

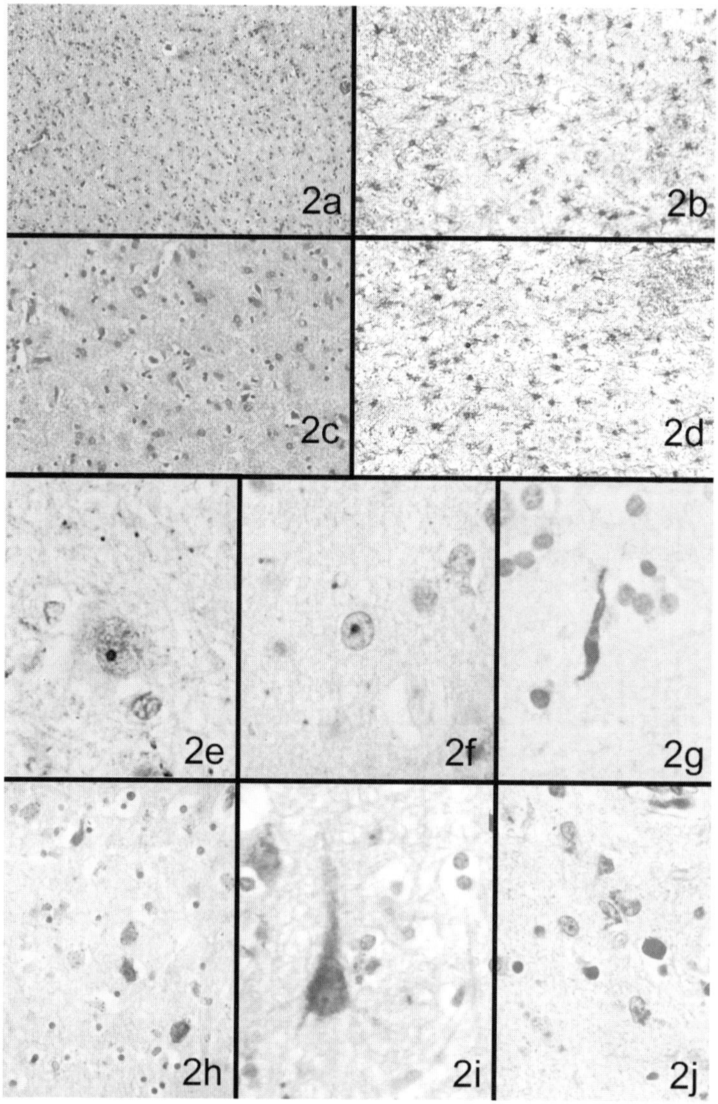

Figure 2. Microscopic sections of brains with HD (a-b; e-j) and neuroacanthocytosis (c-d). Magnification: ×100 (a), ×200 (b-d) and ×400 (e-j).
In the patient with HD, severe neuronal loss in the striatum (a: hematoxylin/eosine) and gliosis (b: GFAP immunocytochemistry) can be seen. Ubiquitin positive striatal intranuclear neuronal (e) and cortical intra-cellular inclusions (f), and dystrophic cortical neurites (g) are present in HD.
Polyglutamine immunoreactivity is observed in striatal (h) and cortical (i) neurons in HD (antibody generously provided by Drs. J.J. Lucas and J. Avila). Apoptosis of striatal neurons, shown by the TUNEL technique, is present in the striatum of the patient with HD (j).

Table 2. HD and its phenocopies ("Huntington's disease like", HDL).

• HDL1	(MIM 603218) Onset in the third decade, death after 15-20 years, absence of nuclear aggregates. Insertion in prion protein gene, resulting in familial prion disease [18].
• HDL2	(MIM 606438) Onset in the fourth decade, CTG/CAG trinucleotide repeat expansion in the gene for junctophilin-3 on chromosome 16 [23]. Nuclear ubiquitin positive aggregates [9] (see also Chapter 7).
• HDL3	Recessive, age at onset during the first decade, gene at 4p15.3. Saudi Arabian origin [8].
• HDL4	Dominant, onset in middle adulthood. Incoordination, dementia, depression. Gene unknown [12].

glutamate receptors that these neurons express and to their peculiar properties regarding energy production and metabolism.

The surviving striatal and cortical neurons of patients with HD show intracellular inclusions that stain with antibodies to ubiquitin (Figure 2e). Abnormal nuclear inclusions and neurites with dystrophic varicosities are numerous in the cerebral cortex (Figures 2f and 2g) which is revealed through their staining with antibodies to ubiquitin and antibodies raised against the NH2-terminal region of the expanded, mutated huntingtin. They do not show up with antibodies against other epitopes of the mutated protein, which reveals that these inclusions contain the abnormal peptide. Strong immunoreactivity against antibodies to polyglutamine expanded molecular N-terminals is found in striatal and cortical neurons (Figures 2h and 2i). Immunostaining for apoptotic cells in HD patients revealed nuclear labeling in some striatal neurons (Figure 2j).

Peripheral Pathology in Huntington's Disease

It is considered that HD does not involve tissues outside the brain. There is, however, at least one patient reported with a large CAG expansion and with histology-proven mitochondrial myopathy. That patient was a child with early onset HD, with epilepsy and severe mental deterioration, whose disease was unrecognized in his ancestors, raising the question of a recessive pattern of inheritance. The muscle biopsy was performed because the child had been misdiagnosed with neuroacanthocytosis. One of the authors (JGY) was personally informed by Dr. Michio Hirano (Columbia University, New York) of another case of histology-proven mitochondrial myopathy in a middle aged gentleman with a mid-size expansion. Abnormal complex I activity has been reported in patients with HD.

Pathology of Phenocopies of Huntington's Disease

Around 1% of the patients with clinical symptoms compatible with HD that undergo the molecular test for CAG expansions of the *IT15* gene do not

Table 3. Neuropathology of neuroacanthocytosis.

• Brain:	Gross atrophy restricted to striatum and pallidum. Loss of striatal dopamine, GABA and substance P and elevation of noradrenaline. Nigral atrophy in cases with parkinsonism.
• Myelon:	Rare loss of neurons in the anterior horn of the spinal cord.
• Nerve:	Peripheral neuropathy with demyelination.
• Muscle:	Rare changes suggestive of myopathy.

have mutations that produce an expanded polyglutamine residue of the huntingtin protein. In some of these patients there is information about their pathology. This, in general, is similar to that of HD. Clinical, genetic and pathological characteristics of these patients [8,9,12,18] are summarized in Table 2.

BRAIN PATHOLOGY IN NEUROACANTHOCYTOSIS

The number of neuropathological studies in patients with neuroacanthocytosis is very low. Other limitations of these studies include the lack of clinico-pathological correlations and lack of neuropathological studies in patients with different types of mutations of the gene that codes for chorein. The neuropathological information about the brain, therefore, is based on very few cases, such as those reported by Bird et al [19], Iwata et al [20], Galatioto et al [21], Burbaud et al [22], Sobue et al [15], de Yébenes et al [2] (2 cases reported and one additional more recent case not reported), Hardie et al [7] and Rinne et al [13,14] (3 patients: case 1 being a female manifesting McLeod syndrome).

With respect to HD, neuroacanthocytosis is characterized by a more selective atrophy of the basal ganglia, including caudate nucleus, putamen, and globus pallidus pars medialis and pars lateralis (Figures 1c and d). In these regions, there is neuronal loss and gliosis (Figures 2c and d) but there is no information about specific inclusions or immunocytochemical staining. There is no information about preferential involvement of striosomes or matrix or to the selective damage of specific neurotransmitter-containing neurons in the brain, though several neurotransmitters have been found abnormal in these patients. As in HD, the burden of striatal neuronal loss affects small and medium size striatal neurons. Only in one case is a gradient of severity of lesions mentioned, the caudate nucleus being affected mainly dorsally (case 1 from [13,14]).

The severity of involvement of the substantia nigra and the drop-out of nigrostriatal dopamine neurons is variable and correlates with the presence of parkinsonism [13]. Dopamine levels in the striatum are reduced [2]. Similarly, the levels of GABA and substance P are reduced in brains of these patients [2], suggesting damage of striatal neurons projecting to the pallidum and substantia nigra. That combination of nigral and striatal pathology could be mistaken with the pathological findings of multiple system atrophy. From the clinical point of view, these two disorders are quite different. The pathology is also very different, mostly because multiple system atrophy is characterized by glial cytoplasmic inclusions, which can be demonstrated with silver techniques or with immunoreactivity to synuclein [10].

The globus pallidus is almost as severely affected as the striatum, showing neuronal loss with concomitant gliosis [13-15]. Involvement of the thalamus is mild and, if any, affects mostly the medial thalamic nuclei. It is consistent with some neuronal loss and gliosis [14]. The subthalamic nucleus, the midbrain, pons, cerebellum, and cerebral cortex are normal in the majority of the reported cases.

The findings in the spinal cord are variable, from loss of anterior horn cells to absence of pathology, even in cases with clinical evidence of muscle weakness and wasting. Some authors suggest that the pattern of lesion is patchy and by no means uniform and, therefore, advocate extensive tissue sampling [15].

Peripheral nerve pathology, mostly consistent with patchy demyelination, has been found in more than half of the cases that had nerve biopsies. Some biopsies revealed hypertrophic Schwann cells and selective loss of heavily myelinated sensory fibers. Other cases showed evidence of axonal damage. EM studies have revealed giant axons, abnormal mitochondria and inclusions [7,15].

Though muscle weakness is usually attributed to anterior horn cell damage or peripheral neuropathy, there is evidence of clinical myopathy with nonspecific muscle atrophy in some cases [6].

REFERENCES

1. Albin RL, Reiner A, Anderson KD, Penney JB, Young AB (1990) Striatal and nigral neuron subpopulations in rigid Huntington's disease: Implications for the functional anatomy of chorea and rigidity-akinesia. *Ann Neurol* **27:** 357-65.
2. De Yébenes JG, Brin MF, Mena MA, De Felipe C, Martin del Río R, Bazan E, Martinez A, Fahn S, Del Río J, Vazquez A, Varela de Seijas E (1988) Neurochemical findings in neuroacanthocytosis. *Mov Dis* **3:** 300-312.
3. Ferrante RJ, Kowall NW, Beal MF, Martin ED, Richardson EP (1987) Morphologic and histochemical characteristics of a spared subset of striatal neurons in Huntington's disease. *J. Neuropathol Exp Neurol* **46:** 12-27.
4. Ferrante RJ, Beal MF, Kowall NW, Richardson EP, Martin JB (1987) Sparing of acetylcholinesterase-containing striatal neurons in Huntington's disease. *Brain Res* **411:** 162-166.
5. Ferriero DM, Arcavi LJ, Sagar SM, McIntosh TK, Simon RP (1988) Selective sparing of NADPH-diaphorase neurons in neonatal hypoxia-ischemia. *Ann Neurol* **24:** 670-676.
6. Gil-Nagel A, Morlan L, Balseiro J, de Yébenes JG, Cabello A, Martinez Martín P (1994) Neuroacanthocytosis with associated myopathy. *Neurologia* **9:** 165-168.
7. Hardie RJ, Pullon HWH, Harding AE, Owen JS, Pires M, Daniels GL, Imai Y, Misra VP, King RHM, Jacobs JM, Tippett P, Duchen LW, Thomas PK, Marsden CD (1991) Neuroacanthocytosis. A clinical, haematological and pathological study of 19 cases. *Brain* **114:** 13-51.
8. Kamburis M, Bohlega S, Al-Tahan A, Meyer BF (2000) Localization of the gene for a novel autosomal-recessive neurodegenerative Huntington-like disorder to 4p15.3 *Am J Hum Genet* **66:** 445-452.

9. Margolis RL, O'Hearn E, Rosenblatt A, Willour V, Holmes SE, Franz ML, Callahan C, Hwang BA, Troncoso JC, Ross CA (2001) A disorder similar to Huntington's disease is associated with a novel CAG repeat expansion. *Ann Neurol* **50:** 373-380.
10. Papp MI, Kahn JE, Lantos PL (1989) Glial cytoplasmic inclusions in the CNS of patients with multiple system atrophy. *J Neurol Sci* **94:** 79-100.
11. Richfield EK, Maguire Zeiss KA, Cox C, Gilmore J, Voorn P (1995) Reduced expression of preproenkephalin in striatal neurons from Huntington's disease patients. *Ann Neurol* **37:** 335-343.
12. Richfield EK, Vonsattel J-P, MacDonald ME, Sun ZQ, Deng Y-PP, Reiner A (2002) Selective loss of striatal preprotachykinin neurons in a phenocopy of Huntington's Disease. *Mov Dis* **17:** 327-332.
13. Rinne JO, Daniel SE, Scaravilli F, Harding AE, Marsden (1994a) Nigral degeneration in neuroacanthocytosis. *Neurology* **44:** 1629-1633.
14. Rinne JO, Daniel SE, Scaravilli F, Pires M, Harding AE, Marsden CD (1994b) The neuropathological features of neuroacanthocytosis. *Mov Dis* **9:** 297-304.
15. Sobue G, Mukai E, Fujii K, Mitsuma T, Takashi A (1986) Peripheral nerve involvement in familial chorea-acanthocytosis. *J Neurol Sci* **76:** 347-356.
16. Vonsattel J-P, DiFiglia M (1998) Huntington's disease. *J Neuropathol Exp Neurol* **57:** 369-384.
17. Vonsattel J-P, Myers RH, Stevens TJ, Ferrante RJ, Bird ED, Richardson EP (1985) Neuropathological classification of Huntington's disease. *J Neuropathol Exp Neurol* **44:** 559-77.
18. Moore RC, Xiang F, Monaghan J et al (2001). Huntington disease phenocopy is a familial prion disease. *Am J Hum Genet* **69:** 1385-1388.
19. Bird TD, Cederbaum S, Valpey RW, Stahl WL (1978) Familial degeneration of the basal ganglia with acanthocytosis: A clinical, neuropathological, and neurochemical study. *Ann Neurol* **3:** 253-258.
20. Iwata M, Fuse S, Sakuta M, Toyokura Y (1984) Neuropathological study of chorea-acanthocytosis. *Jap J Med* **23:** 118-122.
21. Galatioto S, Serra S, Batolo D, Marafioti T (1993) Amyotrophic choreo-acanthocytosis: a neuropathological and immunocytochemical study. *Ital J Neurol Sci* **14:** 49-54.
22. Burbaud P, Vital A, Rougier A et al (2002) Minimal tissue damage after stimulation of the motor thalamus in a case of chorea-acanthocytosis. *Neurology* **59:** 1982-1984.
23. Holmes SE, O'Hearn E, Rosenblatt A, Callahan C, Hwang HS, Ingersoll-Ashworth RG et al (2001) A repeat expansion in the gene encoding junctophilin-3 is associated with Huntington disease-like 2. *Nat Genet* **29:** 377-37.

CHAPTER 12

COGNITIVE AND NEUROPSYCHIATRIC FINDINGS IN MCLEOD SYNDROME AND IN CHOREA-ACANTHOCYTOSIS

Adrian Danek[1,2], Laura Sheesley[2], Michael Tierney[2], Ingo Uttner[1,3], and Jordan Grafman[2]

[1]Neurologische Klinik, Ludwig-Maximilians-Universität, München, Germany; [2]Cognitive Neuroscience Section, National Institute of Neurological Disorders and Stroke, NIH, Bethesda, MD, USA; [3]Neurologische Klinik, Universität Ulm, Germany

Abstract. Serial assessment of patient MT with McLeod syndrome (MLS) and a cross-sectional study in nine patients with genetically confirmed chorea-acanthocytosis (ChAc) are presented along with a review of more than 20 case reports each from the literature. The cognitive and neuropsychiatric findings may be summarized as a "fronto-subcortical syndrome" but are not unique to either condition since a similar syndrome is also shared by other basal ganglia degenerations, most notably Huntington's disease. Some distinction may be made on the basis of age of neuropsychiatric manifestations (MLS: fourth to fifth decade — ChAc: second to third decade) and the more pronounced features of ChAc. Correlation with the involvement of specific fronto-subcortical loops may be possible in the future.

INTRODUCTION

Impairment of higher brain function in neuroacanthocytosis, in addition to movement disorders and seizures was first observed in Levine's family [52,66] as well as in the cases of Critchley [18,19]. Subsequently, cognitive and neuropsychiatric changes have commonly been found (e.g. refs. 8,26,31,50,53,69,74, 81). A few cases of suicide have also been recorded (e.g. refs. 1,3,13), including the single patient yet treated with pallidotomy [34] (Y. Takiyama, personal communication 1999). Because of the greater detail provided, some case reports

A. Danek (ed.), Neuroacanthocytosis Syndromes, 95–115.

Table 1.

	age 53	age 55	age 58
WAIS, German analogue			
Full Scale IQ	116	126	117
Verbal IQ (VIQ)	128	130	128
Performance IQ (PIQ)	98	115	98
Subtests (scaled score)			
Information	15	14	14
Digit Span	10	14	10
Vocabulary	15	15	17
Arithmetic	17	17	15
Comprehension	17	14	17
Similarities	13	13	13
Picture Completion	10	15	12
Picture Arrangement	7	7	**5**
Block Design	10	8	**6**
Object Assembly	10	14	11
Digit Symbol	**5**	**5**	**4**
Stroop paradigm (T values)			
color word reading	44	46	**38**
color naming	42	51	**38**
interference	**28**	**39**	**28**
Verbal fluency (60 seconds each)			
total words (letters F, A, and S)	35	37	23

Serial cognitive assessment in patient MT, with McLeod syndrome, as performed in 2/1993, 3/1994 and 12/1997 using, among other tests, German adaptations of the Wechsler Adult Intelligence Scale [72] and of a Stroop paradigm [5]. The latter is rated using T values from a control group of clerical staff [5]. Abnormal results (lower than control mean minus one standard deviation) are typed in bold.

are notable: that of a boy with suggested autosomal-dominant inheritance [11], of two females with possible recessive chorea-acanthocytosis [27,57,80] and the mixed case series of Hardie and collaborators who characterized the changes as a "fronto-subcortical dementia" [39,47].

Since the cerebral cortex is usually not affected on post mortem examination (see preceding chapter), these observations could eventually shed light on the role of the basal ganglia in higher brain function. Yet an analysis of the published cases is only partially helpful since most authors lacked the techniques to properly distinguish among subtypes of neuroacanthocytosis.

Here, cognitive and neuropsychiatric observations as well as psychometric results from reports with a subsequent molecular diagnosis of either the McLeod syndrome or of chorea-acanthocytosis [21,29,45,63] are reviewed and our own, prospective findings in these two subtypes of neuroacanthocytosis are described [22,24].

MCLEOD SYNDROME

Our experience with MLS goes back to an individual patient, MT, who we followed from age 40 until his death at age 63 and who was the subject of several publications (case 1 of refs. 23,25,61,79; case 17 of ref. 21). Cerebral involvement in MLS was first clearly demonstrated when we reported MT's caudate atrophy, decrease of striatal D2 receptors and his slight cognitive impairment [24]. Unfortunately, neuropathological verification could not be performed.

During the course of his disease, MT was repeatedly assessed with psychometric methods (see Table 1). He had received a university education and had been successful in his job as an economist. At age 53, his full-scale IQ was 116 on a German version of the revised Wechsler Adult Intelligence Scale and there was a significant difference of 30 points between his verbal and performance IQ [72]. He scored consistently low on the "digit symbol" subtest.

Both encoding into verbal memory and retrieval from it were deficient: Long term storage and retrieval as well as list learning were found abnormal on the Selective Reminding Test [17] and encoding and recognition were at least one standard deviation below the mean of controls in a word list learning test [43] (Figure 1). Word paired-associates learning using eight distant pairs of nouns [76], the Benton Visual Retention Test [6] and the Rey-Osterrieth-Figure [62] were not affected. MT's result on Wisconsin Card Sorting [36] and FAS verbal fluency [7] were within normal limits, whereas his performance on a Stroop paradigm [5] showed increased susceptibility to interference.

Serial testing during five years of follow-up did not disclose a decrease in full scale IQ but a significant difference between verbal and performance IQ persisted (see Table 1). "Picture arrangement" and "digit symbol" scores decreased further and there was new, progressive impairment of three-dimensional visuoconstructive ability ("block design"). His results in the Stroop paradigm were already abnormal at age 53, with increased susceptibility to interference (T-value < 30: below two standard deviations of controls). His skills at color word reading and color naming deteriorated to values less than one standard deviation of the mean of controls as did verbal fluency, which had initially been above average. Impairment on word list learning and recall was progressive, whereas recognition memory appeared resistant (see Figure 1). Figural memory (Complex Figure drawing) remained within the normal range, too. The behavior of MT was always appropriate, although with disease progression, he showed increasing reactive depression.

The patient's brother, KT, presented at age 44 with seizures (case 2 of refs. 23,79; case 16 of ref. 21) and had noted increasing forgetfulness. On psychometric testing, memory, concentration and processing speed were found diminished. There were delusions of prosecution and of external control.

Reviewing 20 males with McLeod gene mutations [21], we noticed neuropsychiatric and cognitive changes in ten and seven cases, respectively (cases 1-10, Table 2). One patient has developed progressive dementia since [60].

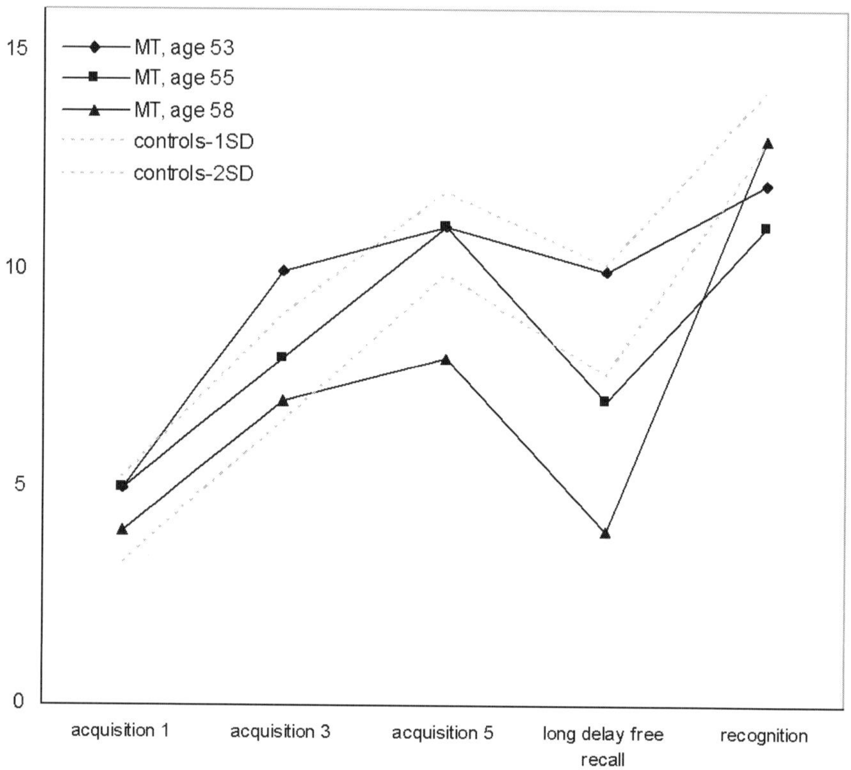

Figure 1. List learning of McLeod patient MT on a German analogue of the California Verbal Learning Test (forms A and B, 16 items), as compared to an age-matched control group [43]. The number of words named (from a 16-item list) is plotted in relation to the different stages of the learning procedure. Results in encoding and recall deteriorated with advancing age. When MT was tested at age 58, recall was most severely affected and he scored in the zero percentile range, at and below three standard deviations below the mean of the controls. On the other hand, his recognition of the list of words was little affected.

Of earlier case reports, a lady from Hardie's L family needs to be specifically mentioned since she is the only female carrier presently known who manifested a full clinical picture of a McLeod mutation (case 20 in Table 2). A few other female mutation carriers were included in the investigation that Jung et al [45] performed in a large Swiss-German family affected by a stop mutation of *XK* (cases 11-19 in Table 2). In Jung's male propositus (case 16 in Table 2), the disease had initially manifested at age 25 with a decrease in social competence. Psychiatric disorders were the predominant initial manifestations in five of seven affected males. On formal testing, Jung's index patient was disoriented for time and location and showed mild dementia, as did one maternal aunt (case 12 in Table 2). Among their relatives, two males and two females demonstrated moderate deficits in figural memory (cases 13,14,18, and 19 in Table 2), one with additional verbal executive dysfunction (case 18 in Table 2). One male was impaired on verbal as well as figural fluency tasks (case 14 in Table 2) and two males had normal test results yet displayed psychiatric changes (cases 12 and 15 in Table 2). Aphasia, apraxia or visuoperceptual deficits were not found in any family member.

Table 2 represents a collection of the neuropsychiatric and cognitive abnormalities that have been seen in McLeod syndrome, yet no prospective data will be available for some time. The glimpses these observations provide into the clinical spectrum include the fact that females may also be affected and that cognitive and neuropsychiatric findings need not occur together. Both domains may be affected independently (cases 5,14,17-19 and 3,7,15 of Table 2, respectively). MLS patients can also show complete sparing of higher functions such as the male with minor neuromuscular features (case 2 of ref. 61, case 6 of ref. 23, case 1 of ref. 21) who scored normally on an extensive psychometric investigation, including a full scale IQ of 116 (VIQ 114, PIQ 116). It can be argued that sparing of higher functions is owed to the young age of the patient (27 years). Many features of MLS such as areflexia and myopathy develop in an age-dependent manner and from the history of the patients reviewed in Table 2 this appears to hold true also for cognitive functions and personality.

In contrast to personality changes, alteration in cognition has not yet been described as a presenting symptom of MLS. Cognitive changes seem to manifest in older age groups, with a range from the mid thirties to the sixties (up to 20 years after onset as in case 10 of Table 2).

The psychiatric features that occurred most commonly were self-neglect and personality change. In single cases inappropriate conduct, including exhibitionism, was mentioned. Hoarding, as observed in case 4 who also had the habit of assembling the garbage collected into bizarre works of art, is an intriguing behavior that appears to be a caudate nucleus or frontal lobe sign [37]. It may be related to obsessive-compulsive features and such features have been observed in another patient (case 3). Emotional changes took the form of anxiety in two patients. There was also a description of emotional lability and of depression. Disorders of thought and of perception (paranoia, hallucinations) have been mentioned and one case of McLeod syndrome presented with a DSM-IV diagnosis of schizophrenia in the absence of neurological symptoms or signs (case 13 of Table 2 and ref. 82).

CHOREA-ACANTHOCYTOSIS

Information was collected on 23 cases from the literature with proven chorein gene mutations and was expanded by personal communication with the authors where possible.

Half of the patients with chorea-acanthocytosis showed neurobehavioral impairment already at presentation, with the earliest manifestations below the

Table 2.

patient (gender)	case identification	age at onset (years)	symptoms and signs at onset	neuropsychiatric findings
1 (m)	II.4 of ref. 71; III-7 of ref. 56; case of ref. 12; 8 of ref. 21	34	acanthocytosis, splenomegaly	exhibitionism
2 (m)	III-8 of ref. 55; 14 of ref. 21	37	md, depression	anxiety, delirium, hallucinations
3 (m)	ref. 30; 15 of ref. 21	42	md	anxiety, obsessive-compulsive features, depression
4 (m)	3 of ref. 23; 12 of ref. 21	51	antisocial behavior, nocturnal restlessness, md	self-neglect, hoarding, inappropriate and infantile conduct, depression
5 (m)	4 of ref. 23; 9 of ref. 21	42	muscle fatigue, md	unremarkable
6 (m)	11 of ref. 21	37	md	self-neglect
7 (m)	13 of ref. 21	21	splenomegaly	personality change
8 (m)	19 of ref. 21; 1 of ref. 60	58	md	self-neglect, inappropriate conduct
9 (m)	20 of ref. 21	44	seizure	apathy, paranoia, emotional lability
10 (m)	22 of ref. 21	49	seizure	none mentioned
11 (m)	II-5 of ref. 45	30	anxiety, depression	yes, but no details
12 (m)	IV-4 of ref. 45	26	anxiety, depression	yes, but no details
13 (m)	IV-5 of ref. 45	39	depression	schizophrenia
14 (m)	IV-6 of ref. 45	20	chorea	unremarkable
15 (m)	IV-7 of ref. 45	25	personality change	personality change
16 (m)	IV-13 of ref. 45	25	personality change	personality change
17 (f)	III-5 of ref. 45	75	dementia	unremarkable
18 (f)	IV-2 of ref. 45		family investigation	unremarkable
19 (f)	IV-11 of ref. 45		family investigation	unremarkable
20 (f)	5 of ref. 39; 1 of refs. 65, 64; see ref. 42	~ 50 s	seizures	distractible, self-neglect

age of 10 (siblings 5 and 6 of Table 3). One young man was diagnosed with schizophrenia in adolescence and neurological signs were noted only many years later ([13], case 22 of Table 3). Personality change was a presenting symptom in three patients. Cognitive complaints were much less common and rather vague (slowness, poor concentration).

All but two patients developed neuropsychiatric features by approximately the fourth decade with a wide range of abnormalities. Changes in social behavior and personality were commonly mentioned, such as disinhibition (6 instances) — including sexual disinhibition — decreased social skills (3 instances), and aggressive (3 instances) or immature, child-like behavior

Table 2 (cont.)

at age	years since onset	cognitive findings	at age	years since onset	patient
50	16	"deteriorating mental state"	53	19	1
49	12	none mentioned	—	—	2
54	12	normal (WAIS, WCST)	50	—	3
51	0	IQ 80 (estimated), probably reduced vs previous level, cognitive slowing	51	0	4
42	0	slightly impaired: working memory, verbal fluency, WCST, Stroop, Trail Making	42	0	5
37	0	perseveration	37	0	6
54	33	normal	48	—	7
61	3	dementia	61	3	8
60	16	memory complaints	60	16	9
—	—	working memory reduced	71	22	10
52	0	none mentioned	52	—	11
49	0	normal: vm, fm, vef, fef	49	—	12
48	0	impaired: fm; normal: vm, vef, fef	48	9	13
40	—	impaired: fm, vef, fef; normal: vm	40	20	14
35	0	normal: vm, fm, vef, fef	35		15
49	0	demented, impaired: vm, fm, vef, fef	49	24	16
76	—	demented, impaired: vm, fm, vef, fef	76	0	17
43		impaired: fm, vef; normal: vm, fef	43	0	18
51	—	impaired: fm; normal: vm, vef, fef	51	0	19
57	7	impaired: shifting of attention, working memory, planning	57	7	20

Neuropsychiatric and cognitive findings in twenty patients with mutations in the McLeod gene *XK*. Cases 1 to 10 are from our earlier review [21], cases 11-20 have been reported by others. Please, note the presence of findings in female gene carriers of this X-linked mutation.
Abbreviations: md movement disorder, vm verbal memory, fm figural memory, vef verbal executive functions, fef figural executive functions.

(2 instances). These patients often neglected their own affairs (3 instances), showed poor hygiene or even inflicted wounds on themselves. Apathy was noted in 3 cases, but hyperactivity, impulsivity, and even agitation were more frequent (5 instances). As a peculiar feature, many reports comment about the patients' distractibility. Obsessions and compulsions are also commonly

Table 3.

patient (gender)	case identification	age at onset (years)	symptoms and signs at onset	neuropsychiatric findings
1 (m)	1 of ref. 39; family 1 of refs. 63,67	37	unsteady gait	personality change, emotional lability, antisocial behavior
2 (f)	2 of refs. 39,47; 3 of refs. 64,65; family 1 of refs. 63,67	39	perioral dyskinesia, dysarthria, dysphagia	increasing agitation, emotional lability
3 (f)	3 of refs. 39,47; family 1 of refs. 63,67	40	micrographia, dysarthria, chorea, gait abnormality, intellectual impairment	none mentioned
4 (f)	4 of refs. 39,47; family 1 of refs. 63,67	44	depression, obsessional features, checking behavior	emotional lability, perseveration, distractible, disinhibited
5 (f)	11 of refs. 39,47; family 9 of refs. 63,67	5	temper tantrums, school phobia, refusal	infantile, immature behavior, obsessive-compulsive disorder; limited spontaneous conversation; emotional lability
6 (m)	12 of ref. 39; family 9 of refs. 63,67	8	poor school performance, dyslexia	distractible, disinhibited, infantile behavior; emotional lability
7 (f)	16 of refs. 39,47; sister of case 36 of ref. 29	29	chorea	anxiety, impulsive, distractible, self-neglect, less sociable, "frontal affect"
8 (m)	18 of refs. 39,47; 32 of ref. 29	33	dysarthria, unsteady gait	none mentioned
9 (m)	case of refs. 14,15, 32,75; family 2 of refs. 63,67; pc Ferrer 1995	32	seizures, cognitive slowing, memory complaints, personality change	self-neglect, disinhibition
10 (m)	2 of ref. 54; family 3 of refs. 63,67	23	perioral dyskinesia, chorea, unsteady gait	anxiety, psychomotor agitation apathy, distractible, slow cognition, hyperactivity, aggressive reactions
11 (m)	1 of refs. 46, 74; family 5 of refs. 63,67	32	dysarthria, perioral dyskinesia, restlessness	depression
12 (m)	1 of refs. 29,41; pc Hiersemenzel 1996	33	motor restlessness, poor concentration	self-neglect, poor hygiene, self-inflicted wounds, sexual acts in public
13 (f)	5 of ref. 29; AH from family 1 of ref. 9	42	perioral dyskinesia	agitation, emotional lability

Table 3 (cont.)

at age	years since onset	cognitive findings	at age	years since onset	patient
42	5	mildly demented	60	23	1
42	3	full scale IQ 86 severely impaired vs premorbid IQ	42	3	2
		impaired: ef, mem; intact: la, vp	54	15	
	—	VIQ 94, PIQ 96	44	4	3
		cognitive deterioration (VIQ& PIQ 88), impaired: ef; intact: mem, la, vp	49	9	
44	0	VIQ 87, PIQ 89 (moderate decrease vs premorbid level); intact: ef, mem, la, vp	47	3	4
24	0	VIQ 81, PIQ 106 (mild decrease vs premorbid level) impaired: ef, intact: mem, la, vp	24	19	5
23	0	poor visuospatial skills, writing, reading; poor concentration	16	0	6
29	0	moderate IQ reduction vs premorbid, impaired: mem, ef; intact: la, vp	36	7	7
—	—	mild IQ reduction vs premorbid; impaired: ef; intact: mem, la, vp	36	3	8
32	0	IQ 96, slow cognition	32	0	9
		IQ 84	37	5	
36	13	normal intelligence	36	—	10
38	15				
32	0	IQ 137	38	—	11
∼ 40	7	IQ and speed of processing low average, visuomotor tasks and task shifting subnormal	39		12
		lack of concentration, fluctuating attention, mental inertia	42	9	
∼ 50	8	none mentioned	—	—	13

Table 3 (cont.)

patient (gender)	case identification	age at onset (years)	symptoms and signs at onset	neuropsychiatric findings
14 (m)	family 1 of ref. 9	30	perioral dyskinesia, vocal tics, dysphagia, chorea, dystonia	apathy
15 (m)	family 1 of ref. 9	20	dystonia	apathy
16 (f)	family 1 of ref. 9	20	akinesia, obsessional features	diagnosis of schizophrenia
17 (f)	SAH from refs. 9,10; 15 of ref. 29; pc Bohlega 1998	28	seizures	antisocial and aggressive behavior, depression
18 (m)	JG from ref. 9; 17 of ref. 29	21	dysarthria, unsteady gait	distractible, obsessional features
19 (m)	case 16 of refs. 29,78; pc Wihl and Saft 2001; see Chapter 28	17	vocal tics	personality change, obsessive-compulsive features, distractible, reactive depression
20 (m)	18 of refs. 29,73; Troiano, pc 2000	6	seizures	personality change
21 (m)	41 of refs. 29,70; pc Sorrentino 1998	20	personality change; erioral dyskinesia	sexual disinhibition, aggressive behavior
22 (m)	case 2 of ref. 13, brother of case 42 of ref. 29	17	schizophrenia (note: movement disorder, gait instability, dysphagia, tongue dyskinesia were noted only 17 years later)	paranoia, agitation, hallucinations (visual, auditory), incoherence; diagnosis: paranoid schizophrenia; paranoia, gambling, suicide at age 41
23 (f)	42 of ref. 29; pc Chouinard 2003	34	movement disorder, personality change	disinhibition, impulsivity, emotional lability

observed (5 instances). Mood disorders often take the form of "emotional lability" (7 instances), but depression occurs, too (3 instances, one reactive). Anxiety and paranoia are less often noted.

Chorea-acanthocytosis is a dementing disease: a discrepancy between predicted general intelligence (as estimated from the patients' skills at properly pronouncing uncommon words) and their actual performance in an intelligence test is frequently seen. Only three subjects were regarded as normal on psychometric examination (cases 10, 11, 21 of Table 3), whereas the majority had developed cognitive impairment by the fifth decade. Again, the two siblings 5 and 6 of Table 2 were affected already at a very young age: the girl had devel-

Table 3 (cont.)

at age	years since onset	cognitive findings	at age	years since onset	patient
30	0	intelligence below average for age	30	0	14
22	2	none mentioned	—	—	15
20	0	none mentioned	—	—	16
30	2	VIQ 69, PIQ 78	30	2	17
26	5	none mentioned	—	—	18
38	21	impaired: general intelligence, vm, fm, fef, selective attention	39	22	19
35	29	impaired cognition, slowing, MMSE 21/30	35	29	20
20	0	psychometry normal	37	—	21
35	0	impaired: ef, mem, attention, lack of judgment	39	22	22
39 34	0	"frontal subcortical deficit"	43	9	23

Neuropsychiatric and cognitive findings in 23 patients collected from the literature with proven mutations in the chorein gene.
Abbreviations: md movement disorder, mem memory, ef executive functions, vp visuoperceptual functions, la language, pc personal communication, VIQ verbal intelligence quotient, PIQ performance IQ, MMSE mini mental status examination.

oped behavioral problems in school, including ritualistic touching and kissing of objects. At age 24 her IQ was low [39]. Her brother was diagnosed as dyslexic, with poor visuospatial skills, writing and spelling. His educational attainments were below average when he left school aged 16 years [39]. Language and visuoperceptual skills were not found impaired in other cases.

In spite of the wealth of information that can be gained from retrospective analyses and case reports a systematic study has so far been lacking.

We therefore prospectively examined a group of eight patients in whom chorein gene mutations had been found with a battery of cognitive tests (Table 4). Six of the patients in our study were male, two were female. Their ages ranged from 26 to 51 years and they had shown onset of motor symptoms at ages 20 to 34; 6 to 8 years before our investigation. Three patients (cases 4-6) were siblings. Two patients were severely dysarthric and had lost vocal communication, which is not uncommon in later stages of the disease. They nevertheless communicated efficiently with the help of an electronic keyboard which synthesizes speech sounds but did not take the National Adult Reading Test [59] or the Stroop procedure that require pronunciation skills and fluent naming under time constraints.

A ninth patient with a chorein mutation (37 of ref. 29) was seen clinically but could not be formally tested due to the combined effects of dysarthria and a language barrier. Nevertheless his case illustrates some behavioral aspects of ChAc. This 36 year old man had been divorced in the course of his disease and now lived with his parents. Disease onset was about ten years earlier, with unsteady gait and leg dystonia, followed by personality change and deterioration of manners. Episodes of binge eating and drinking were reported. He would pass gas or belch in public and would neither excuse nor correct his conduct. He had also become negligent with respect to his clothing and to religious traditions, in spite of a strict orthodox religious background. Neither the knowledge from his earlier rabbinic studies nor his theological discussion skills were felt altered by his father, but in everyday life he had become unable to make up his mind. On examination there was limb and perioral hyperkinesia, including vocal tics, and his behavior often was driven by external stimuli: he was easily distracted, showed echopraxia, and repeatedly touched objects in his view.

Habitual object touching was noted in another patient of our series, too (case 7). Repetitive behaviors associated with eating, drinking and drugs were also observed in patient 2, a 35 year old business administration graduate with an onset of motor symptoms nine years previously. He would continuously smoke or handle cigarettes and had changed his eating manners, piling large amounts of food on his plate and excessively consuming select items such as peanut butter and milk. He also had changed his social manners and, for example, would not wait for instructions during testing. He did not keep a socially appropriate distance, asking the examiner questions about her private life, or would abruptly stop a conversation to smoke a cigarette. Like patients 8 and 9, he belched without excuse and repeatedly passed gas in the presence of others.

One patient (case 1, onset with echolalia and vocal tics at the age of 27, parkinsonian and aphonic at the age of 44) was remarkable for an increase in sexual behavior that had to be controlled with estradiol medication. His wife in addition mentioned memory problems that became obvious as he failed to manage the successive steps necessary to run computer programs. Patient 3, a 30 year old history graduate, for the past five years had also developed

Table 4.

Patient	1	2	3	4	5	6	7	8
age (gender)	44 (m)	35 (m)	30 (f)	51 (m)	42 (m)	40 (f)	36 (m)	26 (m)
case identification	24 of ref. 29	14 of ref. 29	23 of ref. 29	22 of ref. 29	brother of 4	sister of 4	21 of ref. 29	this report
age of onset	27	27	24	35	24	22	29	20
first sign	md	md	md	md	md	seizure	md	md
NART								
estimated IQ	—	106	118	—	111	107	114	105
WAIS-III								
Full IQ	85	95	113	99	87	84	99	95
VIQ	91	100	118	101	86	88	97	94
PIQ	78	89	106	95	89	80	102	97
Subtests (scaled score)								
Digit Span	7	8	10	9	8	9	11	**6**
Arithmetic	**5**	11	15	9	**6**	**6**	10	10
Comprehension	10	11	14	11	**6**	**6**	7	10
Picture Arrangement	9	10	12	13	8	**6**	11	10
Block Design	**6**	**5**	12	**6**	8	7	10	7
Object Assembly	**5**	7	11	**6**	4	9	10	9
Digit Symbol	**2**	**6**	**5**	**5**	**6**	**5**	**5**	**2**
Matrix Reasoning	**5**	9	15	11	8	10	11	11
Symbol Search	**5**	**6**	8	**5**	7	**5**	7	**4**

Results of prospective cognitive assessment in eight patients with chorea-acanthocytosis using WAIS-III [77] complemented by NART [59] to estimate premorbid intelligence. Those WAIS subtests, in which each patient scored at least 7 (corresponding to the mean minus one standard deviation) are not detailed. Patients 1 and 4, with severe dysarthria, did not participate in NART. Scaled scores lower than 7 are typed in bold.
Abbreviations: WAIS-III Wechsler Adult Intelligence Scale, 3rd revision, NART National Adult Reading Test, md movement disorder.

memory problems (mainly affecting short-term memory and new learning) as had patient 7. The latter had also shown obsessive and compulsive features.

Except for case 2, all patients suffered from epilepsy. Antiepileptic drugs provided good seizure control but might also have impaired performance. While they underwent cognitive testing, the patients took phenytoin (cases 1,5), phenobarbital (cases 5,6), clonazepam (case 6), carbamazepine (cases 1,4,6,7), oxcarbazepine (case 1), valproate (case 8), and levetiracetam (case 1). Further clinical details of these patients, including video documentation of their movement disorders, are being reported elsewhere (Danek et al, in preparation).

Intelligence testing confirmed the hypothesis of a decrease from premorbid levels in ChAc (Table 4). Mean Full IQ on the WAIS-III was 95, corresponding to a decrease of more than one standard deviation from the mean estimated premorbid IQ of 110. The character of ChAc as a progressive dementing disease is further supported by looking at the influence of time: The gap between premorbid and current IQ widens with disease duration (Table 4).

With respect to the test profile, the mean results for Performance IQ (92) were slightly inferior to the mean Verbal IQ (97). Of the single subtests, the mean age scaled scores of the patients were average (score 10) or above average for the subtests picture completion (12,3), letter-number-sequencing (11,6), information (10,5), similarities (10,1), vocabulary (10,0), matrix reasoning (10,0), and picture arrangement (9,9). Somewhat lower but within the normal range were the patient mean scores on the subtests for comprehension (9,4), arithmetic (9,0), digit span (8,5), block design (7,6), and object assembly (7,6).

The symbol search results (5,9) and, in particular, the digit symbol subtest scores (4,5) stood out. They were between one and two standard deviations below the mean of healthy control subjects (corresponding to scores of 7 and of 4, respectively). Interestingly in Huntington's disease, digit symbol is affected early in HD gene carriers [33,48,49], is linked to HD gene triplet repeat numbers [38], and is selectively correlated with tissue loss in the putamen (density of dopamine D2- and dopamine transporter receptors and putamen size) [4,68].

Looking at individual digit symbol results, two patients were markedly impaired and no patient scored within the normal range. In one patient, the memory component was clearly outweighed by impaired performance in the control task of symbol copying which indicates decreased visuomotor skill. Ocular motor scanning in particular plays a role in the digit symbol subtest [44] and impaired saccadic eye movements in chorea-acanthocytosis, such as documented in three of our cases [35] could have possibly contributed to the patients' abnormal results. Additional data that were collected in our patients to check incidental learning and the motor aspects of the test suggest that involvement of both components, visuomotor skill as well as short-term memory, underlie the impairment on the digit symbol subtest in chorea-acanthocytosis.

Memory itself can be severely impaired in ChAc. On the California Verbal Learning Test the group means were slightly below average, but there was a high degree of individual variation (Figure 2). Interestingly, patients that were severely affected (cases 4 and 1: mute, long disease duration; case 8: severe motor impairment) performed above average in the memory measures. The patient who performed worst (case 7) was most afflicted with epileptic seizures and had been treated with various antiepileptic drugs. Nevertheless, memory performance in ChAc appears unrelated to epileptic brain damage

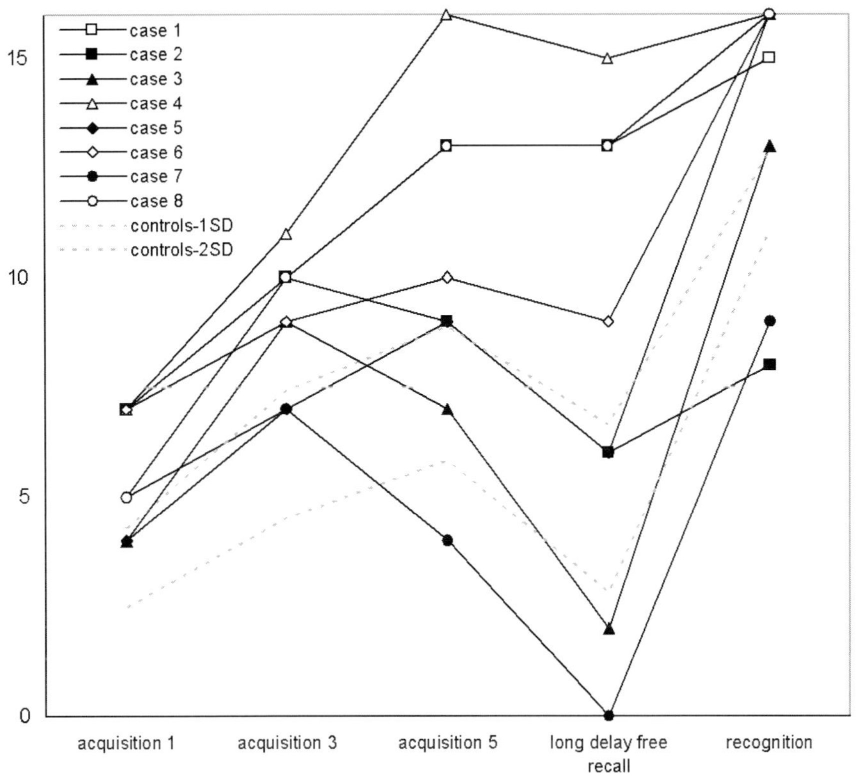

Figure 2. List learning of eight patients with chorea-acanthocytosis on the California Verbal Learning Test in comparison to a control group of 288 adults with a mean age of 48 years [28]. There is a wide range of individual performance (words named out of a 16-item list) which was unrelated to disease duration or verbal IQ.

or drug treatment since patient 3 who performed almost as badly had shown memory impairment before suffering three rare epileptic seizures and was not on medication while tested.

Among "executive tests", verbal fluency and resistance to interference were subnormal in some of the patients. In the Stroop procedure (Figure 3), the result of less vulnerability by interference as the disease progresses appears paradoxical, but might indicate increasing mental rigidity.

Verbal fluency was abnormal in more than half of the group and showed no relationship with disease duration. On the Behavioral Assessment of the Dysexecutive Syndrome task there was mild impairment and the results on Wisconsin Card Sorting and the Tower of London were heterogeneous. Aphasia, apraxia, neglect or perceptual disorders were not found in any ChAc patient.

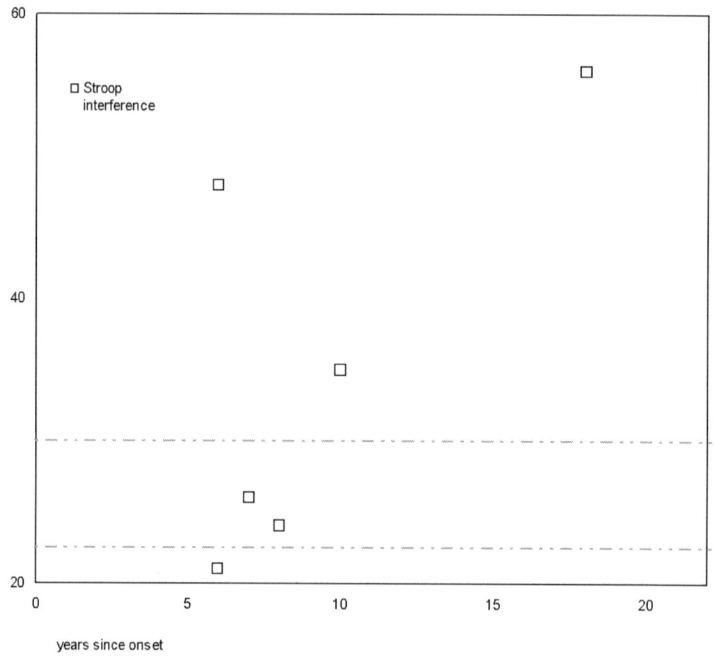

Figure 3. Resistance to interference (Stroop procedure) in six patients with chorea-acanthocytosis as related to time since first manifestation of the disease. The horizontal lines correspond to the values of a control group (number of correctly named colors in the interference condition, during 45s; group mean minus one and two standard deviations according to the norms of Daigneault for the Golden version of the Stroop procedure as contained in the Unified Huntington's Disease Rating Scale; age group 45-65; for details see reference 58).

SUMMARY

Although the pattern of neurobehavioral impairment (which may be summarized as a fronto-subcortical type of dementia) is similar in McLeod syndrome and in chorea-acanthocytosis, abnormal findings are more common and more severe in the latter and occur at an earlier age. Thus, clinical differences between the two conditions could be due to a slower progression in MLS of an essentially identical disease process. Additionally, there might be differences in structural pathology, i.e. specific lesion patterns, and thus specific neuropsychiatric features for each disease. The motor symptomatology supports such an assumption (based on the principle of distributed representation) since dyskinesias of orofacial structures and, particularly, of the tongue are very typical for ChAc and uncommon in MLS. The personality changes that are so pronounced in ChAc might then be related to damage of specific parts of the basal ganglia that usually are spared in MLS or affected later.

It should be noted that the neuropsychiatric and cognitive observations described above are not specific for ChAc or MLS but may occur in a variety of basal ganglia diseases. Of these, Huntington's disease shows the most overlap of neurobehavioral features [40].

To explain individual symptomatology, an influential concept suggests that the manifestation of neuropsychiatric features in basal ganglia disorders is mirrored in the involvement of specific frontal-subcortical connections [20]. These frontal-striatal-pallidal-thalamic loops are organized in parallel and serve different aspects of motor, cognitive and emotional functions. In spite of the good evidence for this concept from animal studies [2], detailed knowledge as to the basic anatomy and to the clinical correlation is lacking in humans. So far, the pertinent human anatomical data have rested on analogy [16] and more direct methods have only just become available [16a]. One to one correlations between the proposed loops and clinical symptoms or syndromes have been difficult to find. Huntington's disease has often been studied as a model for these assumptions and it has been proposed that the personality changes in HD may relate to dysfunction of ventrobasal and orbital parts of the frontal lobe, psychosis to medial caudate pathology and depression to dorsomedial caudate pathology [51]. The study of neuroacanthocytosis syndromes (that lack structural damage to cortical structures) could supplement this approach.

In conclusion, neuropsychiatric and cognitive features are apparent in almost all patients with McLeod syndrome or chorea-acanthocytosis at some time during their disease progression. MLS patients appear to be affected much later and neuropsychiatric features seem to be more common in patients with ChAc. Memory, visuomotor and executive functions are particularly impaired and cortical signs such as aphasia, apraxia or agnosia are lacking. So far, it has not been possible to reliably distinguish among McLeod syndrome, chorea-acanthocytosis, Huntington's disease and other basal ganglia disorders just on the basis of their neuropsychiatric or cognitive features.

Acknowledgements

The collection of many of the data listed above was possible only with the help of Colvin Redman, Arthur Hays, Alessandro Malandrini, Maria-Teresa Dotti, Wolfgang Kalckreuth, François Tison, Ruth Walker, Andrew Marshall, Ian Sutton, David Nicholl, Alastair Corbett, Burton Scott, Andrea Németh, Mitchell Brin, W.A. Symmans†, L. Marsh†, Xavier Ferrer, Lutz-Peter Hiersemenzel, Saeed Bohlega, Günther Wihl, Carsten Saft, André Troiano, Guiseppe Sorrentino, Sylvain Chouinard, Saidi Mohiddin, Lameh Fananapazir, Barbara Karp, and Mark Hallett. The molecular data have been provided by the group of Tony Monaco, Oxford, by Carol Dobson-Stone in particular. Gisela Stenglein-Krapf and Mimi Courtney Arnold assisted in the neuropsychological testing.

REFERENCES

1. Aguilar i Bascompte JL, Berga L, Merino A, Domenech JM, Vives-Corrons JL (1988) A further case of choreo-acanthocytosis. *Acta Haematol Basel* **80:** 175-176.
2. Alexander GE, Crutcher MD, DeLong MR (1990) Basal ganglia-thalamocortical circuits: parallel substrates for motor, oculomotor, "prefrontal" and "limbic" functions. *Prog Brain Res* **85:** 119-146.
3. Alonso ME, Teixeira F, Jimenez G, Escobar A (1989) Chorea-acanthocytosis: Report of a family and neuropathological study of two cases. *Can J Neurol Sci* **16:** 426-431.
4. Bäckman L, Robins-Wahlin TB, Lundin A, Ginovart N, Farde L (1997) Cognitive deficits in Huntington's disease are predicted by dopaminergic PET markers and brain volumes. *Brain* **120:** 2207-2217.
5. Bäumler G (1984) *Farbe-Wort-Interferenztest (FWIT) nach J.R.Stroop.* Verlag für Psychologie Dr. C.J. Hogrefe, Göttingen
6. Benton AL (1990) *Der Benton-Test.* Huber, Bern
7. Benton AL, Hamsher KS, Varney NR, Spreen O (1983) *Contributions to neuropsychological assessment.* Oxford University Press, New York
8. Bird TD, Cederbaum S, Valpey RW, Stahl WL (1978) Familial degeneration of the basal ganglia with acanthocytosis: A clinical, neuropathological, and neurochemical study. *Ann Neurol* **3:** 253-258.
9. Bohlega S, Al Jishi A, Dobson-Stone C et al (2003) Chorea-acanthocytosis: Clinical and genetic findings in three families from the Arabian peninsula. *Mov Dis* **18:** 403-407.
10. Bohlega S, Riley W, Powe J, Baynton R, Roberts G (1998) Neuroacanthocytosis and aprebetalipoproteinemia. *Neurology* **50:** 1912-1914.
11. Borchardt CM, Jensen C, Dean CE, Tori J (2000) Case study: Childhood-onset tardive dyskinesia versus choreoacanthocytosis. *J Am Acad Child Adolesc Psychiatry* **39:** 1055-1058.
12. Brin MF, Hays A, Symmans WA, Marsh WL, Rowland LP (1993) Neuropathology of McLeod phenotype is like choreaacanthocytosis (CA). *Can J Neurol Sci* **20 (Suppl.):** S234.
13. Bruneau MA, Lesperance P, Chouinard S (2003) Schizophrenia-like presentation of neuroacanthocytosis. *J Neuropsychiatry Clin Neurosci* **15:** 378-380.
14. Burbaud P, Rougier A, Ferrer X et al (2002) Improvement of severe trunk spasms by bilateral high-frequency stimulation of the motor thalamus in a patient with chorea-acanthocytosis. *Mov Dis* **17:** 204-207.
15. Burbaud P, Vital A, Rougier A et al (2002) Minimal tissue damage after stimulation of the motor thalamus in a case of chorea-acanthocytosis. *Neurology* **59:** 1982-1984.
16. Burruss JW, Hurley RA, Taber KH et al (2000) Functional neuroanatomy of the frontal lobe circuits. *Radiology* **214:** 227-230.
16a. Lehéricy S, Ducros M, Van de Moortele PF et al (2004) Diffusion tensor fiber tracking shows distinct corticostriatal circuits in humans. *Ann Neurol* **55:** 522-529.
17. Buschke H, Fuld PA (1974) Evaluating storage, retention, and retrieval in disordered memory and learning. *Neurology* **24:** 1019-1025.
18. Critchley EMR, Betts JJ, Nicholson JT, Weatherall DJ (1970) Acanthocytosis, normolipoproteinemia and multiple tics. *Postgrad Med J* **46:** 698-701.
19. Critchley EMR, Clark DB, Wikler A (1967) An adult form of acanthocytosis. *Trans Am Neurol Assoc* **92:** 132-137.
20. Cummings JL (1993) Frontal-subcortical circuits and human behavior. *Arch Neurol* **50:** 873-880.
21. Danek A, Rubio JP, Rampoldi L et al (2001) McLeod neuroacanthocytosis: Genotype and phenotype. *Ann Neurol* **50:** 755-764.

22. Danek A, Sheesley L, Tierney M, Grafman J (2003) Cognitive findings in patients with chorea-acanthocytosis. *J Int Neuropsychol Soc* **9:** 528.
23. Danek A, Tison F, Rubio J et al (2001) The chorea of McLeod syndrome. *Mov Dis* **16:** 882-889.
24. Danek A, Uttner I, Vogl T, Tatsch K, Witt TN (1994) Cerebral involvement in McLeod syndrome. *Neurology* **44:** 117-120.
25. Danek A, Witt TN, Stockmann HBAC et al (1990) Normal dystrophin in McLeod myopathy. *Ann Neurol* **28:** 720-722.
26. de Yébenes JG, Brin MF, Mena MA et al (1988) Neurochemical findings in neuroacanthocytosis. *Mov Dis* **3:** 300-312.
27. Delecluse F, Deleval J, Gérard J-M, Michotte A, Zegers de Beyl D (1991) Frontal impairment and hypoperfusion in neuroacanthocytosis. *Arch Neurol* **48:** 232-234.
28. Delis DC, Kramer JH, Kaplan E, Ober BA (2000) *CVLT-II, California Verbal Learning Test, Second Edition – Adult Version.* The Psychological Corporation, San Antonio, TX
29. Dobson-Stone C, Danek A, Rampoldi L et al (2002) Mutational spectrum of the *CHAC* gene in patients with chorea-acanthocytosis. *Eur J Hum Genet* **10:** 773-781.
30. Dotti MT, Battisti C, Malandrini A et al (2000) McLeod syndrome and neuroacanthocytosis with a novel mutation in the XK gene. *Mov Dis* **15:** 1282-1284.
31. Dubinsky RM, Hallett M, Levey R, DiChiro G (1989) Regional brain glucose metabolism in neuroacanthocytosis. *Neurology* **39:** 1253-1255.
32. Ferrer X, Julien J, Vital C, Lagueny A, Tison F (1990) La Chorée-acanthocytose. *Rev Neurol (Paris)* **146:** 739-745.
33. Foroud T, Siemers E, Kleindorfer D, Bill DJ (1995) Cognitive scores in carriers of Huntington's disease gene compared to noncarriers. *Ann Neurol* **37:** 657-664.
34. Fujimoto Y, Isozaki E, Yokochi F et al (1997) A case of chorea-acanthocytosis successfully treated with posteroventral pallidotomy [Japanese]. *Clin Neurol (Tokyo)* **37:** 891-894.
35. Gradstein L, Danek A, Grafman J, Fitzgibbon EJ (2003) Eye movements in chorea-acanthocytosis. *Invest Ophthalmol Vis Sci* **44:** 1934.
36. Grant DA and Berg EA (1948) *The Wisconsin Card Sorting Test.* Psychological Assessment Resources, Inc., Odessa, Florida
37. Hahm DS, Kang Y, Cheong SS, Na DL (2001) A compulsive collecting behavior following an A-com aneurysmal rapture. *Neurology* **56:** 398-400.
38. Hahn-Barma V, Deweer B, Durr A et al (1998) Are cognitive changes the first symptoms of Huntington's disease? A study of gene carriers. *J Neurol Neurosurg Psychiatry* **64:** 172-177.
39. Hardie RJ, Pullon HWH, Harding AE et al (1991) Neuroacanthocytosis: A clinical, haematological and pathological study of 19 cases. *Brain* **114:** 13-49.
40. Harper PS (1996) *Huntington's Disease.* W.B. Saunders, London, Philadelphia, Toronto, Sydney, Tokyo
41. Hiersemenzel L-P, Johannes S, Themann P, Hofferberth B (1996) Die Choreoakanthozytose, ein neurologisch-hämatologisches Syndrom. *Nervenarzt* **67:** 490-495.
42. Ho MF, Chalmers RM, Davis MB, Harding AE, Monaco AP (1996) A novel point mutation in the McLeod syndrome gene in neuroacanthocytosis. *Ann Neurol* **39:** 672-675.
43. Ilmberger J (1988) *Münchner Verbaler Gedächtnistest.* Institut für Medizinische Psychologie der Universität München
44. Joy S, Fein D, Kaplan E (2003) Decoding digit symbol: speed, memory, and visual scanning. *Assessment* **10:** 56-65.
45. Jung HH, Hergersberg M, Kneifel S et al (2001) McLeod syndrome: A novel mutation, predominant psychiatric manifestations, and distinct striatal imaging findings. *Ann Neurol* **49:** 384-392.
46. Kaplan PW, Erwin CE, Bowman MH, Massey EW (1986) Evoked potentials in choreoacanthocytosis. *Electroenceph clin Neurophys* **63:** 349-352.

47. Kartsounis LD, Hardie RJ (1996) The pattern of cognitive impairment in neuroacanthocytosis. A frontosubcortical dementia. *Arch Neurol* **53:** 77-80.
48. Kirkwood SC, Siemers E, Hodes ME et al (2000) Subtle changes among presymptomatic carriers of the Huntington's disease gene. *J Neurol Neurosurg Psychiatry* **69:** 773-779.
49. Kirkwood SC, Siemers E, Stout JC et al (1999) Longitudinal cognitive and motor changes among presymptomatic Huntington disease gene carriers. *Arch Neurol* **56:** 563-568.
50. Kito S, Itoga E, Hiroshige Y, Matsumoto N, Miwa S (1980) A pedigree of amyotrophic chorea with acanthocytosis. *Arch Neurol* **37:** 514-517.
51. Lauterbach EC, Cummings JL, Duffy J et al (1998) Neuropsychiatric correlates and treatment of lenticulostriatal diseases: A review of the literature and overview of research opportunities in Huntington's, Wilson's, and Fahr's diseases. *J Neuropsychiatry Clin Neurosci* **10:** 249-266.
52. Levine IM, Estes JW, Looney JB (1968) Hereditary neurological disease with acanthocytosis, a new syndrome. *Arch Neurol* **19:** 403-409.
53. Limos LC, Ohnishi A, Sakai T et al (1982) "Myopathic" changes in chorea-acanthocytosis. *J Neurol Sci* **55:** 49-58.
54. Malandrini A, Fabrizi GM, Palmeri S et al (1993) Choreo-acanthocytosis like phenotype without acanthocytes: Clinicopathological case report. *Acta Neuropathol (Berl)* **86:** 651-658.
55. Malandrini A, Fabrizi GM, Truschi F et al (1994) Atypical McLeod syndrome manifested as X-linked chorea-acanthocytosis, neuromyopathy and dilated cardiomyopathy: Report of a family. *J Neurol Sci* **124:** 89-94.
56. Marsh WL, Marsh NJ, Moore A et al (1981) Elevated serum creatine phosphokinase in subjects with McLeod syndrome. *Vox Sang* **40:** 403-411.
57. Medalia A, Merriam A, Sandberg M (1989) Neuropsychological deficits in choreoacanthocytosis. *Arch Neurol* **46:** 573-575.
58. Mitrushina MN, Boone KB, D'Elia LF (1999) *Handbook of Normative Data for Neuropsychological Assessment.* Oxford University Press, New York,Oxford
59. Nelson HE, Willison JR (1991) *The National Adult Reading Test (NART).* NFER-Nelson, Windsor
60. Nicholl DJ, Sutton I, Dotti MT et al (2004) White matter abnormalities on MRI in neuroacanthocytosis. *J Neurol Neurosurg Psychiatry* **75:** 1200-1201.
61. Oechsner M, Buchert R, Beyer W, Danek A (2001) Reduction of striatal glucose metabolism in McLeod choreoacanthocytosis. *J Neurol Neurosurg Psychiatry* **70:** 517-520.
62. Osterrieth PA (1944) Le test de copie d'une figure complexe. *Arch Psychol* **30:** 206-356.
63. Rampoldi L, Dobson-Stone C, Rubio JP et al (2001) A conserved sorting-associated protein is mutant in chorea-acanthocytosis. *Nat Genet* **28:** 119-120.
64. Rinne JO, Daniel SE, Scaravilli F, Harding AE, Marsden CD (1994) Nigral degeneration in neuroacanthocytosis. *Neurology* **44:** 1629-1632.
65. Rinne JO, Daniel SE, Scaravilli F et al (1994) The neuropathological features of neuroacanthocytosis. *Mov Dis* **9:** 297-304.
66. Rovito DA, Pirone FJ (1963) Acanthrocytosis associated with schizophrenia. *Am J Psychiatry* **120:** 182-185.
67. Rubio JP, Danek A, Stone C et al (1997) Chorea-acanthocytosis: Genetic linkage to chromosome 9q21. *Am J Hum Genet* **61:** 899-908.
68. Sánchez-Pernaute R, Künig G, del Barrio Alba A et al (2000) Bradykinesia in early Huntington's disease. *Neurology* **54:** 119-125.
69. Serra S, Arena A, Xerra A, Gugliotta AM, Galatioto S (1986) Amyotrophic choreoacanthocytosis: Is it really a very rare disease? *Ital J Neurol Sci* **7:** 521-524.
70. Sorrentino G, De Renzo A, Miniello S, Nori O, Bonvita V (1999) Late appearance of acanthocytes during the course of chorea-acanthocytosis. *J Neurol Sci* **163:** 175-178.

71. Symmans WA, Shepherd CS, Marsh WL et al (1979) Hereditary acanthocytosis associated with the McLeod phenotype of the Kell blood group system. *Br J Haematol* **42:** 575-583.
72. Tewes U (1991) *HAWIE-R: Hamburg-Wechsler Intelligenztest für Erwachsene.* HAWIE-R. Bern, Huber.
73. Troiano AR, Trevisol-Bittencourt PC (1999) Neuroacantocitose, relato de caso. *Arq Neuropsiquiatr* **57:** 489-494.
74. Vance JM, Pericak-Vance MA, Bowman MH et al (1987) Chorea-Acanthocytosis: A report of three new families and implications for genetic counselling. *Am J Med Genet* **28:** 403-410.
75. Vital A, Bouillot S, Burbaud P, Ferrer X, Vital C (2002) Chorea-acanthocytosis: Neuropathology of brain and peripheral nerve. *Clin Neuropathol* **21:** 77-81.
76. von Cramon DY, Hebel N, Schuri U (1988) Verbal memory and learning in unilateral posterior cerebral artery infarction. *Brain* **111:** 1061-1077.
77. Wechsler, D. (1997) *WAIS-III, Wechsler Adult Intelligence Scale, Third Edition.* The Psychological Corporation, San Antonio, TX
78. Wihl G, Volkmann J, Allert N et al (2001) Deep brain stimulation of the internal pallidum did not improve chorea in a patient with neuro-acanthocytosis. *Mov Dis* **16:** 572-575.
79. Witt TN, Danek A, Reiter M et al (1992) McLeod syndrome: A distinct form of neuroacanthocytosis. *J Neurol* **239:** 302-306.
80. Wyszynski B, Merriam A, Medalia A, Lawrence C (1989) Choreoacanthocytosis: Report of a case with psychiatric features. *Neuropsychiatry, Neuropsychol Behav Neurol* **2:** 137-144.
81. Yamamoto T, Hirose G, Shimazaki K et al (1982) Movement disorders of familial neuroacanthocytosis syndrome. *Arch Neurol* **39:** 298-301.
82. Jung HH, Haker H (2004) Schizophrenia as a manifestation of X-linked McLeod-neuroacanthocytosis syndrome. *J Clin Psychiatry* **65:** 722-723.

CHAPTER 13

EPILEPSY IN NEUROACANTHOCYTOSIS

HARTMUT MEIERKORD
Neurologische Klinik und Poliklinik, Universitätsklinikum Charité, Berlin, Germany

Abstract. Epileptic seizures are not an uncommon complication of neuroacanthocytosis. Their pathophysiology is not known, yet possible mechanisms are reviewed here. Antiepileptic drug treatment should be started already after the occurrence of the first seizure of a patient with neuroacanthocytosis. Not only general side-effects must be considered when deciding about a specific drug but also effects upon the movement disorder should be taken into account. Due to the rarity of the condition therapeutic recommendations are necessarily empirical yet carbamazepine and oxcarbazepine appear particularly well-suited.

INTRODUCTION

It is well known that in some neurodegenerative conditions epileptic seizures are exceedingly rare whereas in others they represent common additional features. The occurrence of epilepsy in a degenerative brain disorder is of theoretical interest, since this may allow unique insights into the process of epileptogenesis. Furthermore, there are important practical implications since the clinician needs to choose the appropriate treatment for the individual patient. This article focuses on the particularly common association of epilepsy and neuroacanthocytosis.

The term neuroacanthocytosis with or without the McLeod phenotype is used in this chapter to denote all affected cases with normal lipoproteins and vitamin E levels [21]. Not included are conditions such as abetalipoproteinemia (Bassen-Kornzweig), HARP (hypoprebetalipoproteinemia, acanthocytosis, retinitis pigmentosa, pallidal degeneration) and other variants of late infantile neuroaxonal dystrophy (Hallervorden-Spatz, PKAN) which are also sometimes summarized under the term neuroacanthocytosis.

A. Danek (ed.), Neuroacanthocytosis Syndromes, 117–122.

FEATURES OF EPILEPSY

Frequency

In the first systematic review of the literature regarding epilepsy in neuroacanthocytosis we pointed out that in 25% of the 48 patients described in 1990 epileptic seizures had been reported at some stage in the course of the condition [16]. In the series of Hardie et al [9] more than one third of the 19 cases developed epileptic seizures. In the two siblings with neuroacanthocytosis reported by Schwartz et al [19] epileptic seizures occurred in both. Also, in one of the two siblings described by Aasly et al [1] seizures were particularly striking in one case. Kazis et al [12] and Troiano and Trevisol-Bittencourt [22] also reported cases of neuroacanthocytosis and epileptic seizures as a prominent feature. In the McLeod syndrome, too, epileptic seizures have been reported but appear to be slightly less common, occurring in 5 of the 22 cases reported by Danek et al [5]. According to an approximate estimate one should expect the occurrence of epileptic seizures in at least every fourth case of neuroacanthocytosis and slightly less often in the McLeod syndrome.

Seizure Type

Regarding seizure type in the vast majority generalized tonic-clonic seizures have been described and there are only few cases in which a preceding focal onset was clinically identifiable [23]. In most cases epileptic seizures have been reported to occur at some stage in the course of the condition but it is important to note that in a substantial number of cases seizures may also occur as the presenting feature long before the development of the movement disorder [1,9,12,19,22]. In rare cases the epilepsy may even take the form of tonic-clonic status epilepticus as in the case of Meierkord and Shorvon [16], in case 16 of Hardie's series [9] and in the case of Kazis et al [12]. Response to treatment has been reported favourable in some cases but others have not become seizure free. There is some evidence that a number of patients died during a seizure or during status epilepticus.

CLASSIFICATION OF SEIZURES AND EPILEPSY

Because of its implications for a rational antiepileptic drug therapy the correct classification of epileptic seizures occurring in neuroacanthocytosis is crucial. Are the tonic-clonic seizures that occur in neuroacanthocytosis primary or secondarily generalized? To answer this question one has to recall that in cases in which the underlying cause of the epilepsy is either unknown or irreversible, treatment and prognosis usually depend upon characteristics of the ictal events. However, in some patients treatment and prognosis may also be determined by diagnosis of a specific epileptic syndrome, based on seizure type and further additional clinical and para-clinical information. Consequently, it has been useful to devise a classification of epileptic seizures and also a classification of the epilepsies and epileptic syndromes [3,4]. The seizure classification is based entirely on distinctive behavioral and electrophysiological features of the epileptic events. Seizures are termed partial when behavioral or EEG evidence indicates an onset in a part of the brain limited to one hemisphere, they are termed generalized when they appear to begin bilaterally. When partial seizures evolve into generalized tonic-clonic seizures, they are called secondarily generalized seizures. Consequently, generalized tonic-clonic seizures may occur in both primary and secondary epilepsies. In the former, the basic defect is undetermined and in general there is no detectable structural brain lesion, whereas in the latter the structural lesion of the brain is presumably diffuse or multifocal so that epileptic seizures appear to begin in a generalized fashion. Clearly, the generalized tonic-clonic seizures that occur in neuroacanthocytosis have to be classified as secondarily generalized tonic-clonic seizures.

POSSIBLE NEUROBIOLOGICAL BASIS OF EPILEPSY

Neuropathology

In post-mortem studies the most consistent neuropathological finding of neuroacanthocytosis is extensive neuronal loss and gliosis in the caudate and putamen, pallidum and substantia nigra. The thalamus may be mildly affected but the subthalamic nucleus, cerebral cortex, pons, medulla and cerebellum appear to be spared [9,17]. Given the finding that the cerebral cortex is relatively spared how can the frequent occurrence of epileptic seizures be explained?

Electrophysiological studies have indicated a crucial role of the caudate nucleus in the control of epileptic seizures. Stimulation of the caudate nucleus in cats and patients with epilepsy led to a significant decrease of epileptiform discharges [20,24]. In a recent case report, a patient with left striatal atrophy and hemidystonia showed periodic lateralized epileptiform discharges, confined to the affected hemisphere [7]. Thus, a well-functioning striatum including the caudate, putamen or both should attenuate seizure spread [6]. Consequently, cell loss, and inherently a possible deficient functioning, might be related to insufficient endogenous seizure control as seen in patients with neuroacanthocytosis.

Acid-Base Balance

It is well known from clinical and experimental studies that intra- and/or extracellular alteration in acid-base balance may profoundly affect the seizure threshold [8]. In this context it is interesting to note the changes that Kay et al [11] described in the major anion transport protein (band 3 protein) in erythrocytes from patients with neuroacanthocytosis. The protein exhibited an

Table 1. Non-CNS side effects of antiepileptic drugs (*Leppik 2001* [13]).

organ system	side effect	antiepileptic drugs (AEDs)
liver	enzyme induction	CBZ, FBM, LTG, PB, PHT, TPM
	enzyme inhibition	FBM, PHT, TPM, VPA
	hepatitis	All AEDs metabolized by the liver, but especially FBM and VPA
heart	conduction disturbances, exacerbation of congestive heart failure	CBZ (infrequent)
pancreas	pancreatitis	VPA
	inhibited insulin release	PHT
	increased insulin effect	TPM
kidney	renal calculi	TPM
	hyponatremia	CBZ, OCBZ
skin	minor rash	CBZ, PHT, PB
	Stevens-Johnson syndrome	LTG, all other AEDs with reactive metabolites
connective tissue	gingival hyperplasia	PHT
	lupus erythematosus	PHT
gastrointestinal system	nausea, vomiting	CBZ, PHT, VPA
	diarrhea	VPA
systemic	weight gain	GBP, VPA
	weight loss	FBM, TPM

AED: antiepileptic drug, CBZ: carbamazepine, FBM: felbamate, GBP: gabapentin, GVB: vigabatrin, LTG: lamotrigine, OCBZ: oxcarbazepine, PB: phenobarbital, PHT: phenytoin, TGB: tiagabine, TPM: topiramate, VPA: valproate.

increased molecular weight causing changes in anion transport with subsequent alterations in acid-base equilibrium. The protein is important in the maintenance of acid-base balance. Since this membrane protein is present in erythrocytes and in central nervous system neurons the finding may also contribute to the increased seizure susceptibility in patients with neuroacanthocytosis.

TREATMENT WITH ANTIEPILEPTIC DRUGS

Although a diagnosis of epilepsy formally can only be made in the case of recurrent seizures, in patients with neuroacanthocytosis antiepileptic drug therapy should be started already after a first seizure. The rationale behind this is that the risk of further seizures is extremely high if a drug treatment has not been started. Given the complications that may be associated with tonic-clonic seizures (from minor tongue biting to traffic accidents and even sudden death) the antiepileptic drug treatment should be efficient and lack systemic and neurological toxicity. Since epileptic seizures in neuroacanthocytosis have a focal onset (see above) a substance appropriate for this context should be selected.

Table 2. Antiepileptic drug-induced extrapyramidal signs and symptoms.

	CBZ	FBM	GBP	GVB	LTG	PB	PHT	TGB	VPA
chorea	×	×	×		×	×	×		
tremor	×		×	×	×		×	×	×
dystonia		×	×		×	×	×		
orofacial dyskinesia	×					×	×		
parkinsonism							×		×
ballism							×		
Tic/Tourette	×						×		

AED: antiepileptic drug, CBZ: carbamazepine, FBM: felbamate, GBP: gabapentin, GVB: vigabatrin, LTG: lamotrigine, PB: phenobarbital, PHT: phenytoin, TGB: tiagabine, VPA: valproate.

Table 1 summarizes some important non-CNS side effects of antiepileptic drugs. Since cardiac and other organ involvement is frequently seen in neuroacanthocytosis, this should be taken into account. In addition possible CNS side effects – in particular those affecting the movement disorder – must be taken into account (Table 2).

In cases with high seizure frequency or in which status epilepticus has occurred carbamazepine or oxcarbazepine should be chosen. In less severe epilepsy, as indicated by low seizure frequency, gabapentin should be tried. In each single case, however, the antiepileptic medication needs to be tailored to the situation of the individual patient.

REFERENCES

1. Aasly J, Skandsen T, Ro M (1999) Neuroacanthocytosis – the variability of presenting symptoms in two siblings. *Acta Neurol Scand* **100:** 322-325.
2. Alvarez-Gomez MJ, Vaamonde J, Narbona J et al (1993) Parkinsonian syndrome in childhood after sodium valproat administration. *Clin Neuropharmacol* **16:** 451-455.
3. Commission on Classification and Terminology of the International League Against Epilepsy (1981): Proposal for revised clinical and electroencephalographic classification of epileptic seizures. *Epilepsia* **22:** 489-501.
4. Commission on Classification and Terminology of the International League Against Epilepsy (1989): Proposal for revised classification of epilepsies and epileptic syndromes. *Epilepsia* **30:** 389-399.
5. Danek A, Rubio JP, Rampoldi L, Ho M, Dobson-Stone C, Tison F, Symmans WA, Oechsner M, Kalckreuth W, Watt JM, Corbett AJ, Hamdalla HH, Marshall AG, Sutton I, Dotti MT, Malandrini A, Walker RH, Daniels G, Monaco AP (2001) McLeod neuroacanthocytosis: genotype and phenotype. *Ann Neurol* **50:** 755-764.
6. Dreifuss S, Vingerhoets FJ, Lazeyras F, Andino SG, Spinelli L, Delavelle J, Seeck M (2001) Volumetric measurements of subcortical nuclei in patients with temporal lobe epilepsy. *Neurology* **57:** 1636-1641.
7. Gross DW, Quesney LF, Sadikot AF (1998) Chronic periodic lateralized epileptiform discharges during sleep in a patient with caudate nucleus atrophy: insights into the anatomical circuitry of PLEDs. *EEG Clin Neurophysiol* **107:** 434-438.

8. Gutschmidt KU, Stenkamp K, Buchheim K, Heinemann U, Meierkord H (1999) Anticonvulsant actions of furosemide in vitro. *Neuroscience* **91:** 1471-1481.
9. Hardie RJ, Pullon HWH, Harding AE, Owen JS, Pires M, Daniels GL, Imai Y, Misra VP, King RHM, Jacobs JM, Tippett P, Duchen LW, Thomas PK, Marsden CD (1991) Neuroacanthocytosis. *Brain* **114:** 13-49.
10. Karas BJ, Wilder BJ, Hammond et al (1983) Treatment of valproate tremors. *Neurology* **33:** 1380-1382.
11. Kay MM (1991) Band 3 in aging and neurological disease. *Ann NY Acad Sci* **621:** 179-204.
12. Kazis A, Kimiskidis V, Georgiadis G, Voloudaki E (1994) Neuroacanthocytosis presenting with epilepsy. **242:** 415-417.
13. Leppik IE (2001) Issues in the treatment of epilepsy. *Epilepsia* **42(suppl. 4):** 1-6.
14. Leppik IE, Gram L, Deaton R et al (1999) Safety of tiagabine: summary of 53 trials. *Epilepsy Research* **33:** 235-277.
15. Lombroso CT (1999) Lamotrigine-induced tourettism. *Neurology* **52:** 1191-1194.
16. Meierkord H, Shorvon S (1990) Epilepsie bei Neuroakanthozytose. *Nervenarzt* **61:** 692-694.
17. Rinne JO, Daniel SE, Scaravilli F, Pires M, Harding AE, Marsden CD (1994) The neuropathological features of neuroacanthocytosis. *Mov Dis* **9:** 297-304.
18. Robertson PL, Garofalo EA, Silverstein FS (1993) Carbamazepine-induced tics. *Epilepsia* **34:** 965-968.
19. Schwartz MS, Monro PS, Leigh PN (1992) Epilepsy as the presenting feature of neuroacanthocytosis in siblings. *J Neurol* **239:** 261.
20. Sramka M, Chkhenkeli SA (1990) Clinical experience in intraoperational determination of brain inhibitory structures and application of implanted neurostimulators in epilepsy. *Stereotact Funct Neurosurg* **54-55:** 56-59.
21. Stevenson VL, Hardie RJ (2001) Acanthocytosis and neurological disorders. *J Neurol* **248:** 87-94.
22. Troiano AR, Trevisol-Bittencourt PC (1999) Neuroacanthocytosis. A case report. *Arq Neuropsiquiatr* **57:** 489-494.
23. Vance JM, Pericak-Vance MA, Bowman MH, Payne CS, Fredane L, Diddique T, Roses AD, Massey EW (1987) Choreo-acanthocytosis: a report of three new families and implications for genetic counselling. *Am J Med Genet* **28:** 403-410.
24. Vella N, Ferraro G, Caravaglios G, et al (1991) A feature of caudate control of focal hippocampal epilepsy: evidence for an anterograde pathway. *Exp Brain Res* **85:** 240-242.

CHAPTER 14

SLEEP FEATURES IN CHOREA-ACANTHOCYTOSIS

LEJA DOLENC-GROŠELJ[1], JANEZ JAZBEC[2], AND JAN KOBAL[1]

[1]Division of Neurology, University Medical Centre, and [2]University Children's Hospital, Ljubljana, Slovenia

Abstract. Sleep pattern was studied in two patients with chorea-acanthocytosis. Both patients originated from a large Slovenian family. A whole night polysomnographic recording (PSG) was performed. Sleep pattern disturbances similar to Huntington's disease patients were detected in both patients. However, we found an increased duration of slow wave sleep and longer latency of rapid eye movement sleep, in comparison to the previous studies done in ChAc patients. An important decrease in involuntary movements during sleep was detected in both patients.

INTRODUCTION

Chorea-acanthocytosis (ChAc) is a rare autosomal-recessive disorder characterized by involuntary movements, muscle wasting and mild intellectual decline. The disorder may be phenotypically confused with Huntington's disease (HD). Specifically, however, acanthocytes are detected in peripheral blood films. Degeneration of the basal ganglia, closely resembling that seen in HD, results in atrophy of the putamen and caudate nucleus [1].

Sleep studies in patients with HD [6] showed a disturbed sleep pattern. Increased sleep onset latency was combined with reduced sleep efficiency, frequent nocturnal awakenings, decreased slow wave sleep (SWS) and increased density of sleep spindles. Sleep abnormalities were correlated to the atrophy of caudate nucleus. Only a few sleep studies have been performed in ChAc patients revealing variable results [4,5].

A. Danek (ed.), Neuroacanthocytosis Syndromes, 123–125.

Table 1. Sleep structure, as found on night polysomnography recording (PSG) in two patients with ChAc.

	Patient 1	Patient 2
Recording time in bed	5 h 42 min	5 h 41 min
Total sleep time (TST)	1 h 54 min	2 h 34 min
Number of awakenings	15	6
Wake time during sleep	3 h 2 min	2 h 22 min
Sleep efficiency	33%	46%
Sleep latency (min)	3 h 7 min	27 min
Percentage of stage 1 (of TST)	7,5%	1,9%
Percentage of stage 2 (of TST)	21,1%	15,5%
Percentage of stage 3 (of TST)	1,3%	5,3%
Percentage of stage 4 (of TST)	63,2%	40%
Percentage of stage 5 (of TST)	7%	37,2%
REM latency (min)	3 h 49 min	3 h 59 min

REM: Rapid Eye Movement Sleep

PATIENTS AND METHODS

Our study was performed in two symptomatic ChAc patients (brother and sister) from a large Slovenian family, using whole night polysomnographic recording (PSG). EEG (using the Grey-Walter montage), EOG, EMG of submental muscles along with respiratory parameters, ECG and EMG of pretibial muscles were performed. Involuntary movements during sleep were observed by infrared camera. Both patients were free of other neurological and/or somatic disorders except ChAc and were free of any therapy.

RESULTS

PSG recording in our ChAc patients did not detect any sleep apnea, restless legs syndrome or periodic limb movement. Infrared camera observation revealed an important decrease of involuntary movements in both patients during sleep. Increased sleep latency was detected in one patient. Arousals were frequent during the observed sleep period and the patients' sleep efficiency was very low (33%-46%). Slow wave sleep (SWS) was normal but rapid eye movement (REM) sleep was decreased, and REM latency was prolonged in both patients (Table 1).

CONCLUSIONS

Our study confirms that sleep pattern is disturbed in patients with ChAc. Increased sleep onset latency and reduced sleep efficiency in ChAc patients resemble the sleep disturbances found in HD patients. Similar to findings in HD [2], we observed an important decrease in involuntary movements during sleep in our two patients with ChAc. However, important differences were also detected as we compared their sleep patterns to HD patients. The time spent in stage III and IV (SWS) was proportionally longer in ChAc as compared to HD [3,5], and an increased latency of REM sleep was detected in our ChAc patients. No increase in sleep spindle density during stage II was detected.

In summary, the disturbances in sleep pattern in ChAc are similar to HD. In comparison to HD, however, we observed an increase of time spent in SWS, an increased latency of REM sleep and absence of sleep spindle changes during stage II sleep in our patients with ChAc.

REFERENCES

1. Bruyn GW. Chorea-acanthocytosis (1986) In: *Handbook of Clinical Neurology. North Holland, Amsterdam;* **49:** 327-334.
2. Fish DR, Sawyers D, Allen P et al (1990) The effect of sleep on the dyskinetic movements of Parkinson's disease, Gilles de la Tourette's syndrome, Huntington's disease and torsion dystonia. *Arch Neurol* **47:** 216-218.
3. Hansotia P, Wall R, and Berendes J (1985) Sleep disturbances and severity of Huntington's disease. *Neurology* **35:** 1672-1674.
4. Hori A, Kazukawa S, Nakamura I, Endo M (1985) Electroencephalographic findings in neuroacanthocytosis. *Electroencephalogr Clin Neurophysiol* **61:** 352-358.
5. Silvestri R, Raffaele M, De Domenico P et al (1995) Sleep features in Tourette's syndrome, neuroacanthocytosis and Huntington's chorea. *Neurophysiol Clin* **25:** 66-77.
6. Weigand M, Moller AA, Lauer CJ et al (1991) Nocturnal sleep in Huntington's disease. *J Neurol* **283:** 203-208.

CHAPTER 15

NEUROMUSCULAR FINDINGS IN EIGHT ITALIAN FAMILIES WITH NEUROACANTHOCYTOSIS

MARIA TERESA DOTTI, ALESSANDRO MALANDRINI, AND ANTONIO FEDERICO

Department of Neurological and Behavioral Sciences, University of Siena, Italy

Abstract. We report the muscle and peripheral nerve biopsy findings in a series of Italian patients with McLeod syndrome and chorea-acanthocytosis (ChAc). Myopathy was the prominent feature in McLeod syndrome, with central nucleation and necrotic fibers as the more constant findings. Occasionally mitochondrial changes, as a secondary phenomenon, were observed. Nerve pathology mostly consisted in myelinic lesions, such as thin myelin sheaths and onion bulb formations. In ChAc patients, peripheral nerve biopsy showed loss of large myelinated fibers, axonal degeneration and axoplasmic accumulation of membrano-lamellar profiles. These findings were indicative of a distal axonopathy. Muscle pathology in ChAc mainly showed neurogenic atrophy. In both diseases, the pathological process underlying the neuromuscular involvement remains to be determined.

INTRODUCTION

Neuromuscular involvement is a constant and often subclinical sign in McLeod syndrome and chorea-acanthocytosis (ChAc), two different neurogenetic disorders sharing common clinical and neuropathological features. Reduced or absent deep tendon reflexes, hypotonia, muscle atrophy and increased serum creatine kinase level are hallmarks of both syndromes and are distinguishing features with respect to Huntington's disease. Heterogeneous muscle and nerve biopsy findings have been reported in single cases of neuroacanthocytosis syndromes. Here we provide a comparative description of the clinical and neuromuscular features of 11 patients, three with McLeod syndrome and eight with ChAc, from eight Italian families.

A. Danek (ed.), Neuroacanthocytosis Syndromes, 127–138.

Table 1. Clinical findings in McLeod patients.

Case/gender (age at death)	Age of onset, symptoms	Movement Disorder	Psychiatric changes	Neuromuscular signs
FA/M (†55y)	37y, psychiatric changes/ movement disorder	Orofacial dyskinesias, limb chorea, dysarthria, dystonia	Depression, deliriousness, hallucinations, insomnia	CK 1200 U/l areflexia
FN/M (†59y)	43y, psychiatric changes	Facial dyskinesias, limb chorea	Anxiety, hallucinations, insomnia	areflexia
BF/M (†59y)	42y, movement disorder/ psychiatric changes	Limb chorea, dystonia	Anxiety, depression, obsessive/ compulsive behavior, insomnia	CK 3260 U/l areflexia

MCLEOD SYNDROME

Clinical Cases

The clinical and molecular genetic characteristics of the three patients with McLeod syndrome have been reported [14,6,4] and their main features are summarized in Table 1. All had short stature (156 to 164 cm) and showed psychiatric changes which were the presenting feature in the two siblings, FA and FN. Movement disorders, present in all three cases, included limb chorea (3/3), orofacial dyskinesias (OFD) (2/3), dysarthria (1/3), dysphagia (1/3) and dystonia (2/3). In one case extrapyramidal hypertonia was present. The siblings had severe cardiac involvement (dilated cardiomyopathy). Neuromuscular signs included areflexia, marked serum CK increase, and EMG/NCV alterations (Table 2). No patient had clear signs of mental deterioration or seizures. Peculiar clinical features included severe insomnia and sudden death in all cases. Death was probably related to cardiac pathology in two cases (FA and FN). The other case (BF) who also had symptoms suggesting autonomic dysfunction, died during sleep. Brain MRI of this patient showed unusual mild white matter signal abnormalities and caudate signal hyperintensity.

Table 1 (cont.)

Case	Systemic signs	MRI	Kell antigens	*XK* gene mutation
FA	7% acanthocytes, dilated cardiomyopathy, short stature	Caudate atrophy	Weak expression	W36X
FN	? acanthocytes, arrhythmia, dilated cardiomyopathy, short stature	not performed	not tested	W36X
BF	8% acanthocytes hypersalivation, sweating, short stature	Cerebral atrophy, caudate and white matter hyperintensity	Weak expression	R133X

Nerve and Muscle Biopsy Findings

The main findings of McLeod patients are summarized in Table 2.

Nerve biopsy was performed in two cases (FA and BF). In both cases, it showed mild neuropathic changes with predominantly myelinic involvement consisting of fibers with thin myelin sheath and onion bulb formations (Figure 1). Rare axonal degenerations were also observed in case BF.

Muscle biopsy showed primary myopathic changes with a high percentage of central nucleation and necrotic fibers in both cases (Figure 2). Some COX-negative fibers and morphological abnormalities of mitochondria were observed in FA. Concomitant mild neurogenic atrophy was also present in both patients. In order to detect deficient expression of muscle proteins, we performed an immunocytochemical study using antibodies against proteins of the sarcolemma (dystrophin, alpha, beta, gamma, and delta sarcoglycans, caveolin), myofibrillar network (desmin), extracellular matrix (α2-laminin, collagen IV, β1-laminin) and nuclear envelope (emerin, laminin A/C). No alteration was detected with respect to controls.

Discussion

The muscle biopsy findings of these McLeod patients were similar to most other reported cases [6,8,11,14,20,21,23]. In this syndrome, skeletal muscle involvement, ranging from asymptomatic to mild/severe myopathy, invariably occurs. Since the McLeod locus was found in the same region as that of dystrophin,

Table 2. Nerve and muscle biopsy findings in McLeod patients.

Case	EMG/NCVs	Nerve biopsy	Muscle biopsy
FA	Slight denervation. Collateral reinnervation. Normal NCVs.	Normal density of myelinated fibers. Several onion bulb formations. Some axons with thin myelin sheaths.	18% central nucleation. Mild neurogenic atrophy. Variation of fiber size.
BF	Mild chronic neurogenic changes. Mild sensory-motor neuropathy.	Normal density myelinated fibers. Some axons with thin myelin sheaths. Rare axonal degeneration. Rare onion bulb formations.	12% central nucleation. Several basophilic fibers and clusters of necrotic fibers. Mild neurogenic atrophy.
CONCLUSION		Mild neuropathic changes with prevalent myelin damage.	Myopathic changes (necrotic fibers, central nucleation). Mild neurogenic atrophy.

a possible interaction between both genes to determine the skeletal muscle damage was initially suggested, but later was shown to be incorrect when the

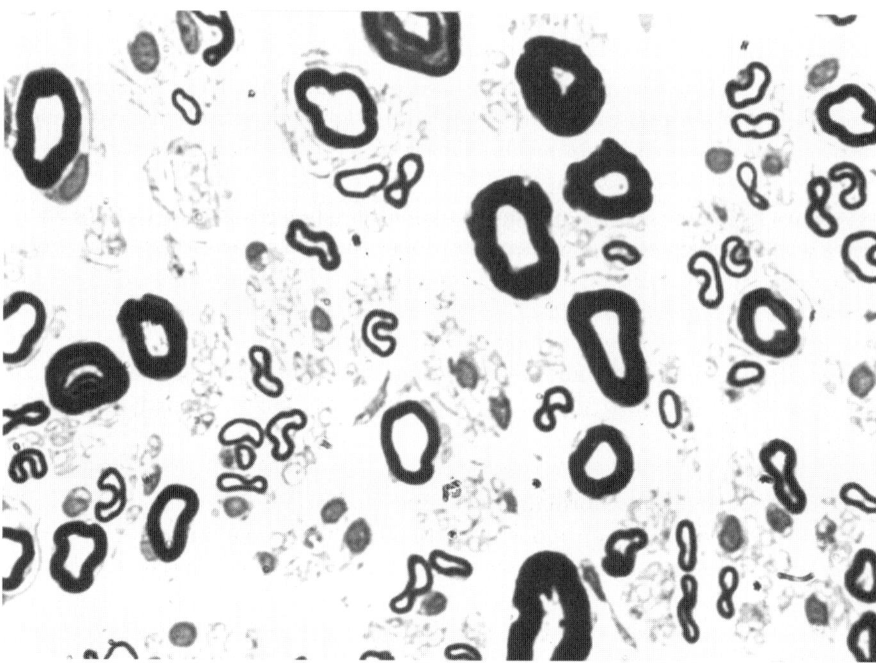

Figure 1. Nerve biopsy from a McLeod patient (semi-thin section × 800). Slight reduction in density of myelinated fibers and some onion bulb formations.

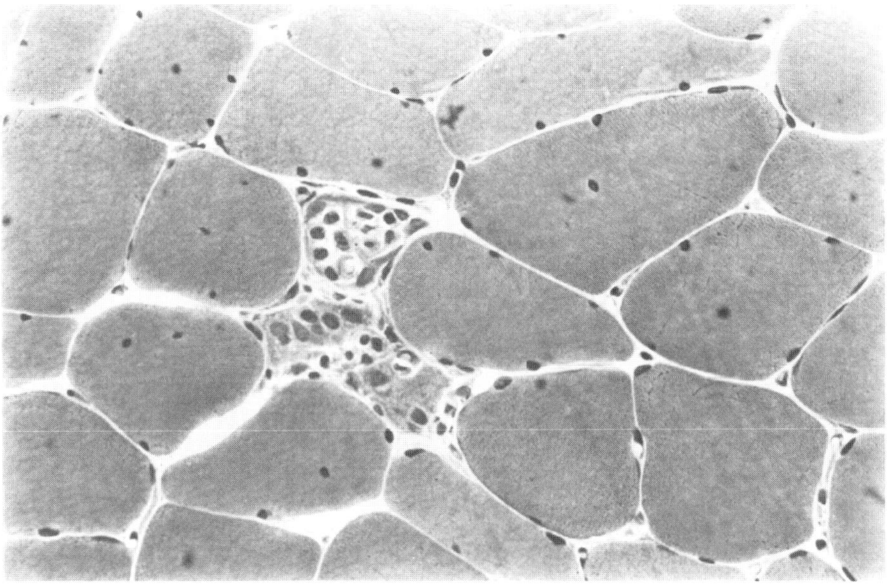

Figure 2. Muscle biopsy from a McLeod patient (H&E $\times 300$). Small cluster of necrotic fibers and central nucleation.

McLeod locus was mapped several thousands of kilobases from the DMD gene. The myopathy in McLeod syndrome must therefore be regarded as a separate phenomenon. Muscle biopsy usually shows slight/mild variation in fiber size with necrotic fibers and central nucleation. Serum CK levels are usually elevated. In a recent study [1], large macrophage infiltrates were observed. A particularly severe myopathy has been described by Kawakami et al [11]. In addition to myopathic findings, type 2 fiber atrophy was recently reported as a distinctive feature [10]. Antibodies against XK and Kell antigens showed strong expression of the protein in type 2 fibers of muscles of normal subjects but significantly reduced expression in those of McLeod patients [10]. In the present cases (see Table 2) we confirmed the presence of myopathic changes in muscle biopsy specimens. Central nucleation seemed to be the most prominent feature. We also observed COX-negative fibers in one case (FA) associated with slight morphological abnormalities of mitochondria. Mitochondrial changes have never previously been reported in McLeod syndrome; in the present case, they may be a secondary phenomenon of unknown origin. We stress that the immunocytochemical results were normal.

Besides myopathic involvement, neurogenic atrophy and grouping have frequently been reported in muscle biopsies of McLeod patients.

Much work has been done on muscle pathology but nerve biopsy has only been performed in a few cases. Patients usually do not show clear clinical

Table 3. Clinical findings in ChAc patients.

Case/sex/age	Age of onset, symptoms	Psychiatric changes	Seizures	Movement Disorder
1/M/†43y	32y, seizures	Depression, insomnia	+	OFD, B, LC, D, Dt
2/ F/ 42y	33y, seizures	Anxiety, insomnia	+	Dt, OFD
3/M/ 43y	31y, involuntary movements	Obsessive-compulsive behavior, insomnia	+	OFD, Dt, B, LC, D
4/ F/ 31y	25y, involuntary movements	Anxiety	+	OFD, Dt, B, Dp, LC
5/ F/ 37y	15y, psychiatric changes	Agitation, hallucinations, suicide attempt	–	Dt, Dp, LC, OFD
6/M/†31y	17y, seizures	Depression, aggressivity, disinhibition	+	OFD, B, Dp, Dt, LC, D
7/M/†38y	23y, involuntary movements	Agitation, anxiety	+	OFD, B, LC, D
8/M/†29y	23y, involuntary movements	Not reported	–	OFD, LC

signs of peripheral neuropathy, though an unexplained absence of reflexes is reported in most. Clinical, neurophysiological and biopsy studies have not yet led to a unifying hypothesis on the type of peripheral neuropathy. Nerve biopsies usually show the co-existence of axonal damage and demyelination. In the present cases, demyelinating features seem to predominate. It would be of interest to perform immunocytochemical analysis of the expression of XK protein in the peripheral nerve of McLeod patients and healthy subjects.

CHOREA-ACANTHOCYTOSIS

Clinical Cases

Only three of the eight ChAc patients (which includes two pairs of siblings: cases 1, 2 and 7, 8 respectively) have already been reported (cases 6-8 [13]). Case 7 and his family were included in the linkage analysis of Rubio et al [17] and subsequently in the mutation analysis of Rampoldi et al [16]. In cases 1 to 5, mutation analysis was also carried out and different mutations of the *CHAC* gene were found [5].

Table 3 summarizes the main clinical features of the eight ChAc patients. Five cases were males and three females. Age at onset ranged from 15 to 33 years. Three patients died suddenly during sleep after duration of illness < 16 years. The presenting symptoms were involuntary movements (4/8), seizures

Table 3 (cont.)

Case	Neuromuscular signs	Systemic signs	MRI
1	↑CK, ↑LDH, areflexia	6% acanthocytes, muscle wasting	Caudate/putamen atrophy and signal alteration
2	↑↑CK, ↑LDH, hyporeflexia	6% acanthocytes, muscle atrophy	Normal
3	↑↑CK, ↑LDH, areflexia	60% acanthocytes, hypotension	Cortical atrophy
4	↑CK, ↑LDH, hyporeflexia	60% acanthocytes, mitral valve prolapse	Normal
5	↑CK, ↑LDH, hyporeflexia	↑acanthocytes, muscle atrophy	Caudate atrophy
6	↑↑CK, areflexia	No acanthocytes, muscle wasting	Caudate atrophy
7	↑↑CK, areflexia	11% acanthocytes, muscle atrophy	Caudate/putamen atrophy and signal alteration
8	np	np	np

+: present; -: absent; OFD: orofacial dyskinesias; B: lip/tongue biting; LC: limb chorea; D: dystonia; Dt: dysarthria; Dp: dysphonia; np: not performed

(3/8) and psychic manifestations (1/8). Involuntary movements were detected in all patients with the following characteristic: orofacial dyskinesia (8/8), limb chorea (7/8), lip/tongue biting (5/8), and dystonia (4/8). Dysphagia and/or dysarthria were found in several cases (5/8). Psychiatric changes were present in seven patients, being particularly severe in three of them. In one case coprolalia was observed. All patients showed slight/severe mental deterioration. A high frequency of seizures (6/8) was found. Signs of neuromuscular involvement included hypo-/areflexia (7/7), muscle atrophy or wasting (5/7), increased serum CK levels (7/7) and EMG alterations (5/7).

Nerve and Muscle Biopsies

Nerve biopsy was performed in six cases (Table 4). Common findings included: a) slight reduction in myelinated fibers, with particular involvement of those of large diameter; b) degenerating axons, and in some cases, c) atrophic axons (cases 1 and 2). In some cases involvement of unmyelinated fibers was also observed. Regeneration clusters were found in case LM only. The most striking electron microscopy (EM) feature was accumulation of membrano-lamellar profiles in the axoplasm (Figures 3 and 4). All these findings were indicative of axonal neuropathy.

Table 4. EMG/NCVs and nerve biopsy findings in ChAc patients.

Case	EMG/NCVs	Nerve biopsy	CONCLUSION
1	Severe sensory motor neuropathy	Slight ↓ in myelinated fiber density. Rare axonal degenerations. Early onion bulbs, axons with thin myelin sheath. TEM: axonal membrano-lamellar profiles. Rare degenerating unmyelinated fibers. Atrophic axons.	Axonal neuropathy
2	Predominant sensory axonal neuropathy	Slight ↓ in myelinated fiber density. Rare axonal degeneration. Axonal atrophy. TEM: axonal membrano-lamellar profiles. Collagen pockets.	Axonal neuropathy
4	Normal	Slight ↓ in myelinated fiber density. Rare axonal degeneration.	Axonal neuropathy
5	Normal	Slight ↓ of myelinated fibers. Some regeneration cluster. TEM: axonal accumulation of dense bodies and membrano-lamellar profiles.	Axonal neuropathy
6	Severe axonal sensory motor neuropathy	Severe loss of myelinated fibers (large diameter fibers absent). TEM: axonal filaments and lamellar profiles accumulation.	Axonal neuropathy
7	Slight axonal neuropathy	Slight ↓ of myelinated fibers. Rare axonal degeneration. TEM: axonal accumulation of filaments.	Axonal neuropathy

TEM: transmission electron microscopy

Table 5. Muscle biopsy findings in ChAc.

Case	Muscle biopsy	CONCLUSION
1	Almost complete substitution with fibroadipose tissue. Grouping atrophy with necrotic fibers and clumps of vesicular nuclei.	Neurogenic atrophy
4	Rare necrotic cells. Nuclear centralization (6%). Mild tendency to grouping.	

Muscle biopsy was only performed in two patients (1 and 4) and the findings suggested neurogenic atrophy (Table 5).

Discussion

Peripheral neuropathy may be a prominent feature of ChAc [7,8,12,13,15,18,22]. Indeed in the past the term "amyotrophic chorea" was used to describe severe atrophy of the limbs. Most studies concerned with peripheral neuropathy in ChAc were performed before molecular diagnostic tests were available. Find-

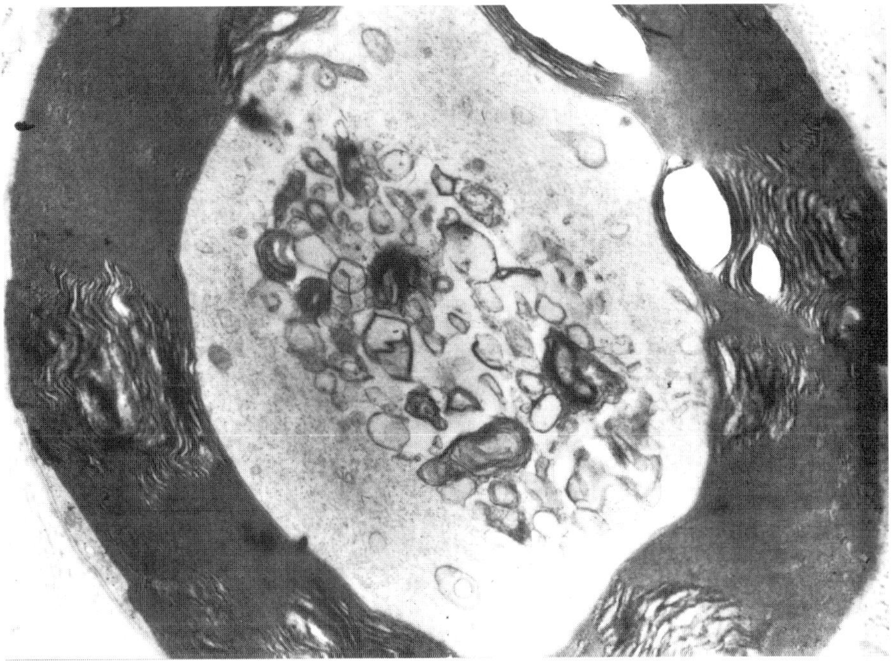

Figure 3. Nerve biopsy from a ChAc patient (TEM × 9000). Membrano-lamellar profiles in the axoplasma of a myelinated axon.

ings indicative of axonal damage, such as loss of myelinated fibers, especially those of large diameter, have usually been reported. In the present cases (see Table 4), significant findings were: 1) slight/mild loss of myelinated axons, 2) some axonal degeneration, 3) slight remyelination signs and 4) cytoskeletal abnormalities. This indicates that the pathological process underlying the peripheral neuropathy is distal axonopathy. The presence of some early onion bulbs and probably remyelinated fibers may be due to remyelination after demyelination, the latter following axonal atrophy. We stress the scarcity of regeneration clusters; it is possible that the pathological process affects the cell body and its ability to regenerate. The most striking feature were cytoskeletal changes such as intra-axonal accumulation of membrano-lamellar material, filaments and dense bodies, and signs of axonal atrophy (disappearance of neurofilaments with relative increase in neurotubules, and relative thickening of the myelin sheath). Cytoskeletal accumulation of neurofilaments has already been reported. This finding, recalling neuroaxonal dystrophy, suggests impairment of axonal transport as pathogenetic mechanism. Neurons probably fail to maintain adequate effective metabolism in the distal part of the axon. The loss of motoneurons in the anterior horn of the spinal cord and of spinal ganglion cells, observed in autopsy cases [13,18], supports this hypothesis. The rela-

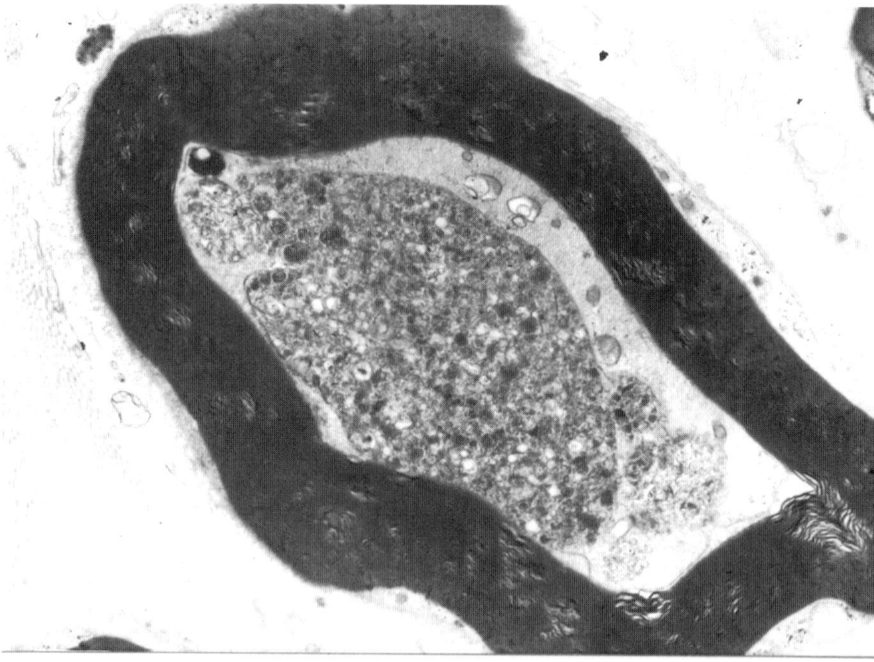

Figure 4. Nerve biopsy from a ChAc patient (TEM × 9000). Dense bodies and granulo-lamellar material within a myelinated axon.

tion between peripheral neuropathy and genetic lesion is still unclear; however, the deficit of *CHAC* protein may influence axonal transport. When antibodies against this protein become available, it will be interesting to look for its expression in peripheral nerve structures (e.g. Schwann cells) or the neuron cell body (dorsal root ganglia or motoneurons).

In ChAc patients, skeletal muscle involvement has usually been regarded as secondary to peripheral neuropathy (neurogenic atrophy). Besides chronic denervation, Limos et al [12] reported myopathic findings (10% central nucleation and fiber splitting) in four ChAc patients. In one of the present cases, we also found chronic denervation with rare necrotic fibers and 7% central nucleation. It is noteworthy that slight/mild central nucleation and rare necrotic fibers may also indicate neurogenic atrophy. However, when there is mild/severe neurogenic atrophy, concomitant primary myopathic findings may be obscured and can be misdiagnosed.

REFERENCES

1. Barnett MH, Yang F, Iland H, Pollard JD (2000) Unusual muscle pathology in McLeod syndrome. *J Neurol Neurosurg Psychiatry* **69:** 655-657.

2. Burbaud P, Rougier A, Ferrer X, Guehl D, Cuny E, Arne P, Gross Ch, Bioulac B (2002) Improvement of severe trunk spasms by bilateral high-frequency stimulation of the motor thalamus in a patient with chorea-acanthocytosis. *Mov Dis* **17:** 204-207.
3. Danek A, Tison F, Rubio JP, Oechsner M, Kalckreuth W, Monaco AP (2001) The chorea of McLeod syndrome. *Mov Dis* **16:** 882-889.
4. Danek A, Rubio JP, Rampoldi L, Ho M, Dobson-Stone C, Tison F et al (2001) McLeod neuroacanthocytosis: genotype and phenotype. *Ann Neurol* **50:** 755-764.
5. Dobson-Stone C, Danek A, Rampoldi L, Hardie R et al (2002) Mutational spectrum of the *CHAC* gene in patients with chorea-acanthocytosis. *Eur J Hum Genet* **10:** 773-781.
6. Dotti MT, Battisti C, Malandrini A, Federico A, Rubio JP, Curciarello G, Monaco AP (2000) McLeod syndrome and neuroacanthocytosis with a novel mutation in the XK gene. *Mov Dis* **15:** 1282-1284.
7. Ferrer X, Julien J, Vital C, Lagueny A, Tison F (1990) Chorea-acanthocytosis. *Rev Neurol* **146:** 739-745.
8. Hardie RJ, Pullon HWH, Harding AE, Owen JS, Pires M, Daniels GL, Imai Y, Misra VP, King RHM, Jacobs JM, Tippett P, Duchen LW, Thomas PK, Marsden CD (1991) Neuroacanthocytosis. A clinical, haematological and pathological study of 19 cases. *Brain* **114:** 13-49.
9. Hirayama M, Hamano T, Shiratori M, Mutoh T, Kumano T, Aita T, Kuriyama M (1997) Chorea-acanthocytosis with polyclonal antibodies to ganglioside GM1. *J Neurol Sci* **151:** 23-24.
10. Jung HH, Russo D, Redman C, Brandner S (2001) Kell and XK immunohistochemistry in McLeod myopathy. *Muscle Nerve* **24:** 1346-1351.
11. Kawakami T, Takiyama Y, Sakoe K et al (1999) A case of McLeod syndrome with unusually severe myopathy. *J Neurol Sci* **166:** 36-39.
12. Limos LC, Ohnishi A, Sakai T, Fujii N, Goto I, Kuroiwa Y (1982) "Myopathic" changes in chorea-acanthocytosis. Clinical and histopathological studies. *J Neurol Sci* **55:** 49-58.
13. Malandrini A, Fabrizi GM, Palmeri S, Ciacci G, Salvatori C, Berti G, Bucalossi A, Federico A, Guazzi GC (1993) Chorea-acanthocytosis like phenotype without acanthocytes: clinicopathological case report. A contribution to the knowledge of the functional pathology of the caudate nucleus. *Acta Neuropathol* **86:** 651-658.
14. Malandrini A, Fabrizi GM, Truschi F, di Pietro G, Moschini F, Bartalucci P, Berti G, Salvatori C, Bucalossi A, Guazzi GC (1994) Atypical McLeod syndrome manifested as X-linked chorea-acanthocytosis, neuromyopathy and dilated cardiomyopathy: report of a family. *J Neurol Sci* **124:** 89-94.
15. Ohnishi A, Sato Y, Nagara H, Sakai T, Iwashita H, Kuroiwa Y, Nakamura T, Shida K (1981) Neurogenic muscular atrophy and low density of large fibres of sural nerve in chorea-acanthocytosis. *J Neurol Neurosurg Psychiatry* **44:** 645-648.
16. Rampoldi L, Dobson-Stone C, Rubio JP, Danek A et al (2001) A conserved sorting-associated protein is mutant in chorea-acanthocytosis. *Nat Genet* **28:** 119-120.
17. Rubio JP, Danek A, Dobson-Stone C, Chalmers R et al (1997) Chorea-acanthocytosis: genetic linkage to chromosome 9q21. *Am J Hum Genet* **61:** 899-908.
18. Sobue G, Mukai E, Fujii K, Mitsuma T, Takahashi A (1986) Peripheral nerve involvement in familial chorea-acanthocytosis. *J Neurol Sci* **76:** 347-356.
19. Sorrentino G, De Renzo A, Miniello S, Nori O, Bonavita V (1999) Late appearance of acanthocytes during the course of chorea-acanthocytosis. *J Neurol Sci* **163:** 175-178.
20. Swash M, Schwartz MS, Carter ND, Heath R, Leak M, Rogers KL (1983) Benign X-linked myopathy with acanthocytes (McLeod syndrome), its relationship to X-linked muscular dystrophy. *Brain* **106:** 717-733.
21. Takashima H, Sakai T, Iwashita H, Matsuda Y, Tanaka K, Oda K, Okubo Y, Reid ME (1994) A family of McLeod syndrome, masquerading as chorea-acanthocytosis. *J Neurol* **124:** 56-60.

22. Vita G, Serra S, Dattola, R, Santoro M, Toscano A, Venuto C, Carrozza G, Baradello A (1989) Peripheral neuropathy in amyotrophic chorea-acanthocytosis. *Ann Neurol* **26:** 583-587.
23. Witt TN, Danek A, Reiter M, Heim MU, Dirschinger J, Olsen EGJ (1992) McLeod syndrome: a distinct form of neuroacanthocytosis. Report of two cases and literature review with emphasis on neuromuscular manifestations. *J Neurol* **239:** 302-306.

CHAPTER 16

CARDIAC INVOLVEMENT IN THE NEUROACANTHOCYTOSIS SYNDROMES

SAIDI A. MOHIDDIN AND LAMEH FANANAPAZIR
Cardiovascular Branch, National Heart Lung and Blood Institute, NIH, Bethesda, MD, USA

Abstract. The profound and progressive neurological features largely dominate the clinical course of neuroacanthocytosis. However, the development of a cardiomyopathy is reported in up to two thirds of cases, and sudden death from cardiac arrest is not infrequent. This review summarizes current understanding of familial hypertrophic and dilated cardiomyopathies and the cardiac findings in the neuroacanthocytosis syndromes. Unraveling the molecular basis of neuroacanthocytosis may provide us with a novel cardiomyopathic mechanism.

INTRODUCTION

The neuroacanthocytoses (NA) are rare, closely-related clinical syndromes that are characterized by neurological abnormalities and erythrocyte acanthocytosis. There is often a skeletal myopathy as demonstrated by persistently elevated plasma creatine kinase. Many of the patients are reported to have an unusual cardiomyopathy that may be associated with increased risk of sudden death. However, cardiac involvement is poorly defined and has been limited to mostly case history descriptions. The cardiac phenotype may resemble familial hypertrophic cardiomyopathy (FHC), a genetic disease that results in primary cardiac hypertrophy and is associated with hypercontractile left ventricular (LV) contraction. The molecular defect has been defined in about half of the cases of FHC as mutations affecting genes that encode components of the sarcomere, the repetitive contractile unit of the myocardium. In other cases of NA, there is a dilated cardiomyopathy (DCM) with LV dilatation and impaired systolic function. Allelic and non-allelic sarcomeric gene mutations may also

A. Danek (ed.), Neuroacanthocytosis Syndromes, 139–152.

cause familial DCM (FDC), a heart disease associated with dilatation and impaired LV contraction. Progressive damage and dilatation of the LV is also a feature of several neuromuscular, metabolic and other syndromes. Increasingly, syndromes that include an FHC phenotype are also being recognized, and their molecular causes that may have non-sarcomeric consequences. Syndromic cardiomyopathies challenge sarcomere-based theories of the etiology of FHC and FDC, and their characterization will increase our understanding of cardiac cellular biology.

Areas of uncertainty concerning neuroacanthocytosis include (1) cardiac phenotype; (2) mechanism(s) of sudden death; (3) cellular consequences of causative mutations in the heart and other organs; and (4) the phenotype, if any, associated with the heterozygous state. Accurate characterization of the cardiac manifestations in affected patients may aid their management, especially with reference to therapeutic strategies that improve heart failure, and prevent disease progression or sudden death. In this chapter we first describe clinical features of the cardiomyopathies, then review cardiac involvement in NA, and speculate on potential cardiomyopathic mechanisms.

CARDIOMYOPATHY

The cardiomyopathies are broadly defined as diseases of the myocardium associated with cardiac dysfunction [51]. Further classification on the basis of morphologic and hemodynamic criteria divides the cardiomyopathies into (1) hypertrophic cardiomyopathy; (2) dilated cardiomyopathy (DCM) which includes FDC; (3) arrhythmogenic right ventricular cardiomyopathy (ARVC); and (4) restrictive cardiomyopathy. This classification does not assume etiology nor does it incorporate other diagnostic modalities such as genetic testing. Advances in understanding of the molecular basis of the cardiomyopathies has revealed similarities between types of cardiomyopathy ("different" cardiomyopathies can be allelic) and has, at the same time, shown an unexpected diversity in their causes (there is a high degree of allelic and non-allelic heterogeneity). Additionally, other disorders of heart muscle with a genetic basis (such as familial catecholaminergic ventricular tachycardia and familial Wolff-Parkinson-White syndrome) do not easily fit into this scheme [22,39]. However, for the present, this classification is useful for making patient management decisions. This chapter is restricted to discussions of FHC and FDC, cardiomyopathies that have been associated with NA.

FAMILIAL HYPERTROPHIC CARDIOMYOPATHY (FHC)

Phenotype and Clinical Features

FHC is the single most common cause of sudden death in otherwise healthy young people. FHC is a genetic disease occurring in approximately 1 in

$500 - 1,000$ of the general population [15,17,65]. The myopathy is often undiagnosed and asymptomatic cases are often unrecognized. Usually, FHC develops with progressive asymmetric LV hypertrophy (LVH) during the period of rapid body growth of adolescence, but it may be present in childhood or even before birth. Progressive hypertrophy after age 20 is uncommon, but initial diagnosis in old age is not. The hypertrophy predominantly involves the LV, and is often more marked in FHC than in any other cardiac disease. It represents hypertrophy and hyperplasia of several cell types, including cardiac myocytes, fibroblasts, and smooth muscle cells, along with excessive collagen and matrix deposition [20]. The normal parallel arrangement of myocytes is often disturbed (fiber disarray). Some sarcomeric mutations also result in subclinical and non-progressive skeletal myopathies characterized by central core disease and ragged red fibers [48].

Increased wall thickness and cardiac remodeling, resulting in a narrow LV outflow tract, and high left ventricular ejection fractions, may cause dynamic intra-ventricular obstruction. LV diastolic abnormalities, myocardial ischemia and atrial or ventricular arrhythmias are also common features. Patients may complain of dyspnea, chest discomfort, light-headedness, presyncope, syncope, tiredness, or palpitations. Symptoms are often exacerbated by exertion, dehydration, postural changes, or following meals.

At physical exam, a bifid arterial pulse with a sharp upstroke and a systolic murmur (often intensified by the Valsalva maneuver) suggest dynamic interventricular obstruction and associated mitral regurgitation. A third or fourth heart sound may be heard and reflect increased LV stiffness during diastolic filling. However, physical findings are often unremarkable and their absence does not exclude cardiomyopathy. Electrocardiographic abnormalities are common and include increased voltages, conduction abnormalities and alterations in ventricular repolarization (QT interval). Diagnosis is made by echocardiography, following identification of a hypertrophied LV that has developed in the absence of another cause of LVH such as aortic stenosis or hypertension. Morphological variants are readily distinguishable at echocardiography and allow division into non-obstructive and obstructive forms of disease (Figure 1). While this allows selection of patients likely to have symptomatic benefit from interventions that relieve outflow obstruction, the pattern of LVH has a poor relationship with other abnormalities (such as diastolic dysfunction, ischemia, risk of sudden death) and to the molecular causes of FHC. In a subset of patients, there is progressive thinning of the LV walls with chamber enlargement and deterioration of systolic function that may result in a DCM like phenotype and death from congestive heart failure. Progressive left atrial enlargement is frequently seen, and predisposes to atrial fibrillation with its adverse hemodynamic and thrombo-embolic consequences. Sudden death occurs in 1-4% of patients per year, and is most commonly attributed to ventricular tachycardia and fibrillation, but may also result from brady-arrhythmias, such as heart block [18,19]. Other investigations document functional impairment

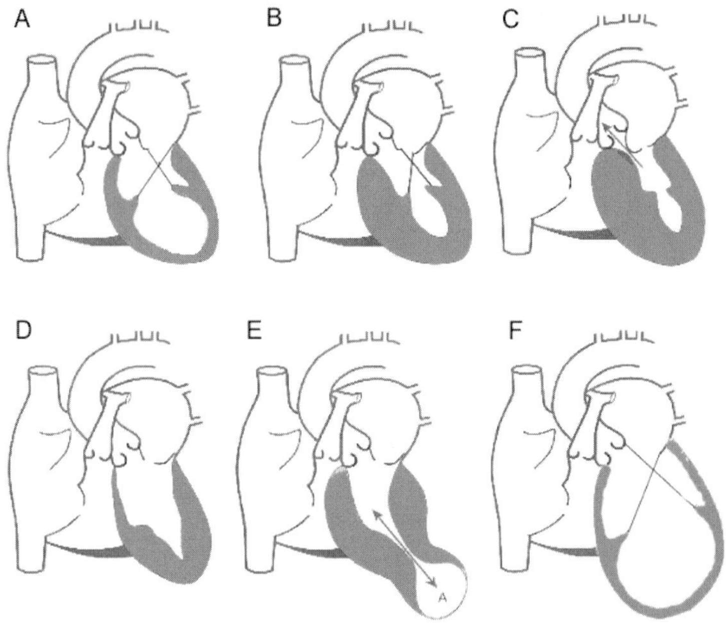

Figure 1. Morphologic variants of familial hypertrophic cardiomyopathy (FHC) and dilated cardiomyopathy (DCM). (A) Normal morphology, (B) Asymmetric hypertrophy of the septum is the commonest phenotype. (C) Sub-aortic stenosis results when increased ejection velocities pull the anterior mitral valve leaflet across a narrow outflow tract towards the septum, as a result of the Bernoulli Principle. (D) Apical FHC is commonest in Japanese populations. (E) Mid-cavity obstructive FHC due to excessive hypertrophy of the papillary muscles appose in systole to divide the LV cavity. An apical aneurysm may result. (F) DCM with thinned LV walls and a more spherical, less efficient shape.

and estimate the risk of sudden death. These include exercise stress testing, Holter monitoring, myocardial thallium scintigraphy and cardiac catheterization with hemodynamic, angiographic and electrophysiologic assessment.

In all phenotypic respects, even among patients who share the same genetic cause, FHC can be extremely varied. A current focus of our research is to determine what the genetic basis of FHC implies for diagnosis and management.

Genetic Basis of FHC

Since the earliest description of FHC, its genetic basis has been appreciated. Autosomal-dominant inheritance is most frequently observed, but recessive inheritance is also reported. In common with many other hereditary diseases, research in the last decade has brought significant advances in the understanding of its molecular basis, and has revealed an unexpected degree of heterogeneity. **Sarcomeric FHC**. Up to half of the cases result from the dominant-negative effects of a mutant protein component of the sarcomere, the highly organized contractile unit of cardiac and skeletal muscle [2,21,38,40,44,58,59]. At least nine different genes are implicated, including those encoding beta myosin heavy chain, cardiac actin, myosin binding protein C, alpha tropomyosin, the troponins, titin and the light chains of myosin. These were the first detected causes, and FHC has been considered a disease of the sarcomere with the hypothesis that cardiac remodeling develops in order to compensate for abnormal sarcomeric contractility. Thus, increasing contractile unit number, and reducing cavity sizes and wall tension accommodate impaired contractility [2,59]. At present, more than 100 mutations in these nine genes alone have been causally associated with FHC [21].

While there is little doubt that the sarcomeric (molecular) lesion is responsible for FHC in many affected individuals, it is increasingly apparent that the development of the cardiac phenotype occurs as a result of processes more complex than suggested by the "compensatory theory". First, there is dramatic phenotypic variability and incomplete penetrance in unrelated individuals and pedigrees that have an identical sarcomeric mutation; sarcomeric dysfunction may not alone be sufficient to cause FHC. Alternatively, more subtle forms of compensation account for incomplete penetrance and phenotypic variability [4,16]. Second, as many as half the FHC patients do not have a sarcomeric mutation despite an apparently similar cardiac phenotype; primary sarcomeric dysfunction is not necessary for the development of FHC. Third, there is no consistent association between in-vitro assays of function at the molecular level and the resulting phenotype. Compared to wild type, some sarcomeric mutant proteins lead to an apparent gain of function, some in no detectable difference and others in a loss of function [10,45]. Fourth, syndromic variants of FHC are difficult to attribute to sarcomeric lesions; for example in Leopard syndrome, Friedreich's ataxia and in NA [9,30]. Finally, transgenic studies indicate that non-sarcomeric molecular lesions can cause cardiac hypertrophy and research has identified similar defects in human subjects [22]. Of the several categories of candidate causes of FHC, genes encoding proteins involved in the production of ATP and proteins regulating intracellular calcium are of current interest.

Non-Sarcomeric FHC. Mutations affecting energy production in the myocyte are an increasingly recognized molecular etiology of FHC. Mutations affecting *PRKAG2*, a gene that encodes a regulatory subunit of cyclic AMP dependent kinase (AMPK), are associated with a cardiomyopathy syndrome characterized by FHC that progresses to LV dysfunction, atrio-ventricular bypass tracts, atrial fibrillation, and bradycardia from heart block and sinus node disease [22]. AMPK is activated by increases in AMP relative to ATP and is thought to act as an energy charge sensor [27]. Activated AMPK increases ATP production, inhibits non-essential cellular processes, and alters gene expression.

Mutations in nuclear and mitochondrial genes encoding mitochondrial proteins have also been associated with FHC (Table 1). Expanded trinucleotide repeats and missense (mainly null) mutations in the *FRDA* gene encoding frataxin, a nuclear encoded mitochondrial protein, results in Friedreich's ataxia [30]. Up to 75% of patients have an FHC phenotype that frequently progresses to DCM [13]. The absence of frataxin may lead to abnormal iron metabolism and oxidative damage to mitochondria [35]. Mutations affecting nuclear and mitochondrion-encoded components of the electron transport system and beta-oxidation have been associated with a variety of cardiomyopathic states [25].

The skeletal myopathy associated with sarcomeric FHC is characterized by central core disease and ragged red fibers, histopathological features similar to mitochondrial myopathies [48]. An alternative to the compensatory hypothesis is that sarcomeric mutations, perhaps by reducing the efficiency of transduction of ATP to mechanical energy, result in abnormal energy metabolism and FHC.

Myocellular Ca^{2+} concentrations are tightly regulated. Increases in intracellular Ca^{2+} initiate sarcomeric contraction, and its rapid removal leads to relaxation. Ca^{2+} dependent signaling in myocytes also mediates the hypertrophic gene response to a variety of hypertrophic stimuli [8,41]. To date, no abnormalities in Ca^{2+} regulatory genes have been reported as causes of human FHC. However, a multitude of transgenic models have shown that abnormalities in several of these genes are sufficient to produce hypertrophic cardiomyopathy [32,33,56].

Characterization of the components of the "*hypertrophic circuitry*" (Figure 2) will help identify new causes of FHC, help predict outcomes in affected individuals and may suggest novel therapies.

FAMILIAL DILATED CARDIOMYOPATHY

Phenotype and Clinical Features

DCM is characterized by left ventricular dilatation and systolic dysfunction [34]. Patients are at risk of death from congestive cardiac failure, cardiac arrhythmias and thromboembolism. Pathologic changes include myocyte atrophy and death and fibrosis.

Presenting symptoms include fatigue, weakness, effort intolerance, pulmonary congestion, pre-syncope and syncope, chest pain and stroke. Screening of family members and other groups increasingly detects presymptomatic cases. Electrocardiographic abnormalities are common, though nonspecific, and include conduction defects and diminished or increased voltages. Echocardiography typically shows LV enlargement, global LV systolic impairment, diastolic filling abnormalities and reduced LV wall thickness.

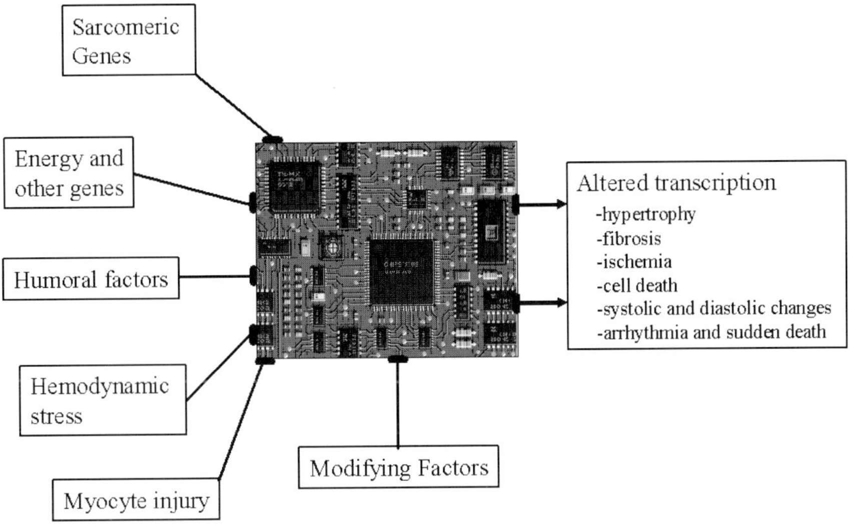

Figure 2. Hypertrophic circuitry. Similar cardiac phenotypes develop as a result of a variety of inputs. Evidence suggests the interaction of several pathways leads to altered gene expression, myocyte hypertrophy and a variety of associated changes such as myocardial fibrosis and arrhythmia. The variability in 'output' phenotype that often results from identical inputs is likely due to a variety of additional modifying inputs.

Early, preferably pre-clinical diagnosis may allow prognostically beneficial therapy to prevent progressive systolic impairment and death [34]. Investigations are similar to those for FHC, and aim to determine the severity of systolic impairment, conduction abnormalities and risk of arrhythmias. A careful history, blood tests, myocardial biopsy and screening of family members may indicate the likely etiology.

Genetic Basis of DCM

In many cases, the cause of DCM is never determined. It has been estimated that DCM is familial (FDC) in approximately one-third of cases and transmission may be autosomal-dominant or recessive, or X-linked [1,3,54,57,64]. Several genetic abnormalities associated with FDC have been described and mutations in several other genes have been implicated in transgenic animal studies. DCM may be an important and life threatening complication of several neuromuscular (and metabolic) diseases such as the muscular dystrophies, mitochondrial syndromes and myotonic dystrophy (Table 1). Cardiac complications of metabolic and neuromuscular syndromes have been reviewed recently [25,54].

Products of genes associated with FDC include those localized to the cytoskeletal, nuclear envelope, sarcomere and mitochondrion; still others are

Table 1. Cardiac involvement in selected familial neurological syndromes.

Syndrome (cardiac phenotype)	Gene Affected	Gene Product
Friedreich's ataxia (LVH, LVD, HB)	*FRDA*	frataxin (nuclear encoded mitochondrial protein)
Kearns-Sayre (LVH, LVD,HB)	mtDNA deletions	mitochondrial proteins
MELAS (LVH, LVD,HB)	mtDNA deletions	mitochondrial proteins
MERRF (LVH, LVD,HB)	mtDNA deletions	mitochondrial proteins
Myotonic dystrophy 1 (HB, VT)	*DMPK*	myotonic dystrophy protein kinase
Myotonic dystrophy 2 (HB, VT)	*ZNF9*	zinc finger protein
Duchenne muscular dystrophy (LVD)	*DMD*	dystrophin (cytoskeletal protein)
Becker muscular dystrophy (LVD*)	*DMD*	allelic with Duchenne
Limb-girdle muscular dystrophy** (HB, LVD)	several	at least 15 different AD and AR variants
Emery-Dreifuss – X-linked (HB, LVD)	*emerin*	emerin (nuclear membrane protein)
Emery-Dreifuss – AD & AR (HB, LVD)	*LMNA****	lamins A and C (nuclear membrane proteins)
Desmin related myopathies (LVH, LVD, HB, RCM)	several	desmin, alpha-B crystallin, and others

* less severe than Duchenne muscular dystrophy,
** some variants, such as those associated with mutations in the sarcoglycan genes, are associated with LV dilatation and dysfunction.
*** *LMNA* mutations also associated with limb-girdle muscular dystrophy type 1B, Charcot Marie Tooth disease type 2B-1, and can also result in a dilated cardiomyopathy, often with conduction abnormalities in the absence of skeletal muscle weakness.
mtDNA: mitochondrial DNA; LVH: LV hypertrophy; LVD: LV dilatation and systolic dysfunction; HB: heart block; RCM: restrictive cardiomyopathy; AD: autosomal-dominant; AR: autosomal-recessive.

implicated by transgenic studies. It has been suggested that FDC results from abnormalities in propagation of contractile force generated by the sarcomere to the extracellular matrix [59]. Non-familial DCM, such as resulting from ischemic heart disease, viral myocarditis, alcohol abuse, vitamin deficiency, chemotherapy and hemochromatosis, are characterized by diffuse myocardial damage and cell death [31,46,55,60]. In FDC, it is likely that cell damage and death result from a variety of insults such as from sarcomeric contraction inadequately restrained by the cytoskeleton and cellular "ischemia" resulting from energy deficit.

NEUROACANTHOCYTOSIS

Cardiac Findings

Progressive neurological impairment and acanthocytosis are the typical and diagnostic clinical features of the NA syndromes [11,28,66]. In both X-linked McLeod syndrome and autosomal-recessive chorea-acanthocytosis (ChAc), chorea, parkinsonism, psychiatric disturbance and peripheral neuropathy increase in severity over several decades and dominate patient concerns and clinical care. A mild skeletal myopathy is usually present, characterized by persistently raised plasma creatine kinase and mixed neurogenic-myopathic features on microscopy. The rate of disease progression and the phenotype is variable, but the similarities between McLeod's and ChAc are striking.

Cardiomyopathy has been reported in several reports of apparently solitary and familial cases of NA (Table 2). Cardiomyopathy is a feature of McLeod syndrome and ChAc, though the association is better described in the former where cardiac abnormalities are present in at least two-thirds of cases [5,7,11,14,24,26,28,36,37,61-63,66]. The dominance of neurological features and their ability to obscure cardiac symptoms (such as exertion intolerance, episodic impairment of consciousness) may have led to under-detection of cardiac involvement. Cardiac findings in NA patients include LVH, LV dilatation and systolic impairment, conduction abnormalities and sudden death from cardiac arrest. In several of the reports describing DCM, the ECG demonstrated increased voltages and the LV walls were often thickened, suggesting progression from a hypertrophic to a dilated phase. Sudden death from cardiac arrest is a common mode of death in NA patients.

Elucidating the role of the respective gene products offers a novel perspective on the genesis of cardiac hypertrophy and LV dysfunction.

Molecular Basis of the Neuroacanthocytosis Syndromes

Most NA patients have mutations in one of two genes: (1) McLeod syndrome is caused by missense and deletion mutations of *XK* [29], and (2) autosomal-recessive NA is linked to mutations in *CHAC* [49]. Further analysis may reveal other molecular defects as reports indicate autosomal-dominant and sporadic forms of the disease.

The role of the *XK* gene product (XK) is not known. XK and the red cell Kell antigen co-precipitate in protein purification and are covalently associated transmembrane proteins [6,50]. Kell is expressed in red cells, the testes, cardiac and skeletal muscle and in the brain; distribution of XK is less well described but appears to be similar [50,52]. Stable expression of Kell is markedly reduced in patients with the McLeod syndrome. Sequence homology and in vitro assays

Table 2. Reported cardiac findings in NA patients.

Reference		NA Phenotype	Echo/CC	ECG	SD
Spitz et al	[62]	ChAc	n/a	LAFB	–
Gross et al	[24]	ChAc	normal	VH	–
Hardie et al	[28]*	ChAc and McLeod	n/a	n/a	yes
Cavalli et al	[7]	ChAc	LVH	normal	–
		ChAc	LVH	normal	–
Sorrentino et al	[61]	ChAc	normal	normal	–
Caballero et al	[5]	ChAc	n/a	normal	–
Malandrini et al	[36]	uncertain	normal	RBBB	31 years
		uncertain	normal	normal	–
		uncertain	n/a	n/a	29 years
Faillace et al	[14]	McLeod [67]	LVH, LVD	1° HB, LVH	–
Witt et al	[66]	McLeod	LVH, DCM	RBBB, AF	–
		McLeod	LVH, DCM	AF	–
Malandrini et al	[37]	McLeod	DCM	1° HB, LBBB	–
		McLeod	n/a	LVH	58 years
		McLeod	n/a	n/a	68 years
Takashima et al	[63]	McLeod	normal	LAFB	–
Hanaoka et al	[26]	McLeod	n/a	normal	–
Danek et al	[11]**	McLeod	varied	varied	yes

* 19 patients are reported, no cardiac evaluations are reported, but two brothers who died suddenly at age 31 had marked LVH at autopsy.

** Cardiac assessments performed in 17 patients, of whom 11 had abnormal results. The nature of these is not fully reported.

Echo/CC: echocardiogram or cardiac catheterization findings; ECG: electrocardiogram; n/a: not available; SD: sudden death; LVH: criteria for left ventricular hypertrophy met at either echocardiography or ECG; LVD: left ventricular systolic dysfunction; 1° HB: first degree heart block; LBBB, RBBB: left and right bundle branch block respectively; LAFB: left anterior fasicular block or left axis deviation; AF: atrial fibrillation.

identify Kell as a zinc endopeptidase showing specificity for the precursor peptides the Big Endothelins (Big ET, particularly Big ET-3) and releases active peptides [6,29,50]. Endothelins promote myocyte hypertrophy and inotropy, and have a role in vasomotor regulation, neurotransmission and organogenesis [23,42,47,53]. An *XK* knockout mouse has recently been developed and has been shown to have diminished ET-3 levels (M. Ho, personal communication). The cardiac phenotype in these mice is not reported.

However, a physiological role for Kell in endothelin metabolism has not been shown. Additionally, the (normal) null Kell genotype also results in the absence of Kell antigens and a reduction in the amount of XK protein in red cell membranes, but is not associated with features of NA [12]. This finding suggests that it is the absence of normal XK rather than of Kell that results in the McLeod phenotype. The absence of XK may lead to loss of a modulatory effect on Kell or to another, as yet undetermined, loss of function.

Hydropathy analysis of XK's amino acid sequence predicts ten transmembrane segments and it is hypothesized that it acts as a membrane transporter which has structural similarities to glutamate transporters [49]. As deletions are the commonest mutations associated with McLeod, it seems likely that the phenotype is due to haploinsufficiency.

The striking similarities between ChAc and McLeod's suggest XK and chorein have intimate functional relationships. *CHAC*'s product, chorein, is also expressed in a variety of cell types that include neurons, red cells and cardiac myocytes (C. Dobson-Stone, personal communication). Chorein's function is entirely unknown, but sequence analyses indicate homology to yeast proteins implicated in trafficking of transmembrane proteins from the Golgi. Although this offers a hypothetical association with transmembrane XK, abnormalities affecting other transmembrane systems might be expected and XK itself is expressed normally in ChAc.

Red cells are subject to persistent low grade hemolysis and there are persistently high levels of creatine kinase and hepatic transaminases, suggesting the molecular lesion leads to a loss of membrane integrity, leakiness and perhaps cell death. It is not established if these features result from a specific membranopathic effect or instead reflect cell damage and death from other processes. The associated cardiomyopathy and consideration of its known causes also suggests XK and chorein may have roles in energy production and its regulation, maintenance of the cytoskeleton and Ca^{2+} homeostasis.

SUMMARY

The NA syndromes are diseases with a multi-system phenotype resulting from abnormalities in proteins with functions unknown. Cardiac involvement is common and manifestations include dilated and hypertrophic cardiomyopathy, conduction disease and increased risks from sudden death. The prevalence, features and prognosis of NA associated cardiomyopathy are poorly described. At the National Institutes of Health, we were conducting a study that seeked to (1) describe the cardiac and skeletal phenotype associated with both McLeod syndrome and ChAc and determine its prevalence and prognosis, (2) determine the phenotype, if any, associated with the heterozygous state, (3) provide insights into potential mechanisms of sudden death, and (4) investigate the function of XK and chorein [43].

Cardiologic evaluation is indicated in NA patients who may gain symptomatic and prognostic benefit from appropriate treatment. The molecular basis of NA may represent a novel cardiomyopathic mechanism, will improve our understanding of the myocellular hypertrophic circuitry and may offer novel therapeutic targets.

REFERENCES

1. Arbustini E, Morbini P, Pilotto A, Gavazzi A, Tavazzi L (2000) Familial dilated cardiomyopathy: from clinical presentation to molecular genetics. *Eur Heart J* **21:** 1825-1832.
2. Bonne G, Carrier L, Richard P, Hainque B, Schwartz K (1998) Familial hypertrophic cardiomyopathy: from mutations to functional defects. *Circ Res* **83:** 580-593.
3. Bowles NE, Bowles KR, Towbin JA (2000) The "final common pathway" hypothesis and inherited cardiovascular disease. The role of cytoskeletal proteins in dilated cardiomyopathy. *Herz* **25:** 168-175.
4. Brugada R et al (1997) Role of candidate modifier genes on the phenotypic expression of hypertrophy in patients with hypertrophic cardiomyopathy. *J Investig Med* **45:** 542-551.
5. Caballero IR et al (2000) Autosomal recessive chorea-acanthocytosis linked to 9q21. *Neurologia* **15:** 132-135.
6. Cartron JP et al (1998) Insights into the structure and function of membrane polypeptides carrying blood group antigens. *Vox Sang* **74(Suppl 2):** 29-64.
7. Cavalli G, de Gregorio C, Nicosia S, Melluso C, Serra S (1995) Cardiac involvement in familial amytrophic chorea with acanthocytosis: description of two new clinical cases. *Ann Ital Med Int* **10:** 249-252.
8. Chien KR (2000) Meeting Koch's postulates for calcium signaling in cardiac hypertrophy. *J Clin Invest* **105:** 1339-1342.
9. Coppin BD, Temple IK (1997) Multiple lentigines syndrome (LEOPARD syndrome or progressive cardiomyopathic lentiginosis). *J Med Genet* **34:** 582-586.
10. Cuda G, Fananapazir L, Epstein ND, Sellers JR (1997) The in vitro motility activity of beta-cardiac myosin depends on the nature of the beta-myosin heavy chain gene mutation in hypertrophic cardiomyopathy. *J Muscle Res Cell Motil* **18:** 275-283.
11. Danek A, et al (2001) McLeod neuroacanthocytosis: genotype and phenotype. *Ann Neurol* **50:** 755-764.
12. Daniels GL, et al (1996) A combination of the effects of rare genotypes at the XK and KEL blood group loci results in absence of Kell system antigens from the red blood cells. *Blood* **88:** 4045-4050.
13. Dutka DP, Donnelly JE, Nihoyannopoulos P, Oakley CM, Nunez DJ (1999) Marked variation in the cardiomyopathy associated with Friedreich's ataxia. *Heart* **81:** 141-147.
14. Faillace RT, Kingston WJ, Nanda NC, Griggs RC (1982) Cardiomyopathy associated with the syndrome of amyotrophic chorea and acanthocytosis. *Ann Intern Med* **96:** 616-617.
15. Fananapazir L (1999) Advances in molecular genetics and management of hypertrophic cardiomyopathy. *JAMA* **281:** 1746-1752.
16. Fananapazir L, Epstein ND (1994) Genotype-phenotype correlations in hypertrophic cardiomyopathy. Insights provided by comparisons of kindreds with distinct and identical beta-myosin heavy chain gene mutations. *Circulation* **89:** 22-32.
17. Fananapazir L, Epstein ND (1995) Prevalence of hypertrophic cardiomyopathy and limitations of screening methods. *Circulation* **92:** 700-704.
18. Fananapazir L, Chang AC, Epstein SE, McAreavey D (1992) Prognostic determinants in hypertrophic cardiomyopathy. Prospective evaluation of a therapeutic strategy based on clinical, Holter, hemodynamic, and electrophysiological findings. *Circulation* **86:** 730-740.
19. Fananapazir L, McAreavey D (1997) Hypertrophic cardiomyopathy: evaluation and treatment of patients at high risk for sudden death. *Pacing Clin Electrophysiol* **2:** 478-501.
20. Ferrans VJ, Rodriguez ER (1983) Specificity of light and electron microscopic features of hypertrophic obstructive and nonobstructive cardiomyopathy. Qualitative, quantitative and etiologic aspects. *Eur Heart J* **4(Suppl F):** 9-22.
21. FHC Mutation Database. http://www.angis.org.au/Databases/Heart/dbsearch.html
22. Gollob MH, et al (2001) Identification of a gene responsible for familial Wolff-Parkinson-White syndrome. *N Engl J Med* **344:** 1823-1831.

23. Gray GA, Battistini B, Webb DJ (2000) Endothelins are potent vasoconstrictors, and much more besides. *Trends Pharmacol Sci* **21:** 38-40.
24. Gross KB, Skrivanek JA, Carlson KC, Kaufman DM (1985) Familial amyotrophic chorea with acanthocytosis. New clinical and laboratory investigations. *Arch Neurol* **42:** 753-756.
25. Guertl B, Noehammer C, Hoefler G (2000) Metabolic cardiomyopathies. *Int J Exp Pathol* **81:** 349-372.
26. Hanaoka N, et al (1999) A novel frameshift mutation in the McLeod syndrome gene in a Japanese family. *J Neurol Sci* **165:** 6-9.
27. Hardie DG, Hawley SA (2001) AMP-activated protein kinase: the energy charge hypothesis revisited. *Bioessays* **23:** 1112-1119.
28. Hardie RJ et al (1991) Neuroacanthocytosis. A clinical, haematological and pathological study of 19 cases. *Brain* **114:** 13-49.
29. Ho M et al (1994) Isolation of the gene for McLeod syndrome that encodes a novel membrane transport protein. *Cell* **77:** 869-880.
30. Johnson WG (1995) Friedreich ataxia. *Clin Neurosci* **3:** 33-38.
31. Kearney MT, Cotton JM, Richardson PJ, Shah AM (2001) Viral myocarditis and dilated cardiomyopathy: mechanisms, manifestations, and management. *Postgrad Med J* **77:** 4-10.
32. Kirchhefer U et al (2001) Cardiac hypertrophy and impaired relaxation in transgenic mice overexpressing triadin-1. *J Biol Chem* **276:** 4142-4149.
33. Kiriazis H, Kranias EG (2000) Genetically engineered models with alterations in cardiac membrane calcium-handling proteins. *Annu Rev Physiol* **62:** 321-351.
34. Leier CV (2001) Dilated cardiomyopathy. *Curr Treat Op Cardiovasc Med* **3:** 451-462.
35. Lodi R, Taylor DJ, Schapira AH (2001) Mitochondrial dysfunction in Friedreich's ataxia. *Biol Signals Recept* **10:** 263-270.
36. Malandrini A et al (1993) Choreo-acanthocytosis like phenotype without acanthocytes: clinicopathological case report. A contribution to the knowledge of the functional pathology of the caudate nucleus. *Acta Neuropathol (Berl)* **86:** 651-658.
37. Malandrini A et al (1994) Atypical McLeod syndrome manifested as X-linked chorea-acanthocytosis, neuromyopathy and dilated cardiomyopathy: report of a family. *J Neurol Sci* **124:** 89-94.
38. Marian AJ, Roberts R (2001) The molecular genetic basis for hypertrophic cardiomyopathy. *J Mol Cell Cardiol* **33:** 655-670.
39. Marks AR, Priori S, Memmi M, Kontula K, Laitinen PJ (2002) Involvement of the cardiac ryanodine receptor/calcium release channel in catecholaminergic polymorphic ventricular tachycardia. *J Cell Physiol* **190:** 1-6.
40. Mohiddin SA et al (2003) Utility of genetic screening in hypertrophic cardiomyopathy: Prevalence and significance of novel and double (homozygous and heterozygous) β-Myosin mutations. *Genetic Testing* **7:** 21-27.
41. Molkentin JD et al (1998) A calcineurin-dependent transcriptional pathway for cardiac hypertrophy. *Cell* **93:** 215-228.
42. Mortensen LH (1999) Endothelin and the central and peripheral nervous systems: a decade of endothelin research. *Clin Exp Pharmacol Physiol* **26:** 980-984.
43. National Institutes of Health Clinical Studies: http://clinicalstudies.info.nih.gov/index.html. Keyword Neuroacanthocytosis.
44. Olson TM et al (2002) Myosin light chain mutation causes autosomal-recessive cardiomyopathy with mid-cavitary hypertrophy and restrictive physiology. *Circulation* **21;105:** 2337-2340.
45. Palmiter KA et al (2000) R403Q and L908V mutant beta-cardiac myosin from patients with familial hypertrophic cardiomyopathy exhibit enhanced mechanical performance at the single molecule level. *J Muscle Res Cell Motil* **21:** 609-620.

46. Piano MR (2002) Alcoholic cardiomyopathy: incidence, clinical characteristics, and pathophysiology. *Chest* **121:** 1638-1650.
47. Pieske B et al (1999) Functional effects of endothelin and regulation of endothelin receptors in isolated human nonfailing and failing myocardium. *Circulation* **99:** 1802-1809.
48. Poetter K et al (1996) Mutations in either the essential or regulatory light chains of myosin are associated with a rare myopathy in human heart and skeletal muscle. *Nat Genet* **13:** 63-69.
49. Rampoldi L et al (2001) A conserved sorting-associated protein is mutant in chorea-acanthocytosis. *Nat Genet* **28:** 119-120.
50. Redman CM, Russo D, Lee S (1999) Kell, Kx and the McLeod syndrome. *Baillieres Best Pract Res Clin Haematol* **12:** 621-635.
51. Richardson P et al (1996) Report of the 1995 World Health Organization/International Society and Federation of Cardiology Task Force on the Definition and Classification of cardiomyopathies. *Circulation* **93:** 841-842.
52. Russo D, Wu X, Redman CM, Lee S (2000) Expression of Kell blood group protein in nonerythroid tissues. *Blood* **96:** 340-346.
53. Ruwhof C, van der Laarse A (2000) Mechanical stress-induced cardiac hypertrophy: mechanisms and signal transduction pathways. *Cardiovasc Res* **47:** 23-37.
54. Sachdev B, Elliott PM, McKenna WJ (2002) Cardiovascular complications of neuromuscular disorders. *Curr Treat Op Cardiovasc Med* **4:** 171-179.
55. Saltzberg MT (2000) Secondary and infiltrative cardiomyopathies. *Curr Treat Op Cardiovasc Med* **2:** 373-384.
56. Sato Y et al (1998) Cardiac-specific overexpression of mouse cardiac calsequestrin is associated with depressed cardiovascular function and hypertrophy in transgenic mice. *J Biol Chem* **273:** 28470-28477.
57. Schonberger J, Seidman CE (2001) Many roads lead to a broken heart: the genetics of dilated cardiomyopathy. *Am J Hum Genet.* **69:** 249-260.
58. Seidman CE, Seidman JG (1998) Molecular genetic studies of familial hypertrophic cardiomyopathy. *Basic Res Cardiol* **93(Suppl 3):** 13-16.
59. Seidman JG, Seidman C (2001) The genetic basis for cardiomyopathy: from mutation identification to mechanistic paradigms. *Cell* **104:** 557-567.
60. Singal PK, Iliskovic N (1998) Doxorubicin-induced cardiomyopathy. *N Engl J Med* **339:** 900-905.
61. Sorrentino G, De Renzo A, Miniello S, Nori O, Bonavita V (1999) Late appearance of acanthocytes during the course of chorea-acanthocytosis. *J Neurol Sci* **163:** 175-178.
62. Spitz MC, Jankovic J, Killian JM (1985) Familial tic disorder, parkinsonism, motor neuron disease, and acanthocytosis: a new syndrome. *Neurology* **35:** 366-370.
63. Takashima H et al (1994) A family of McLeod syndrome, masquerading as chorea-acanthocytosis. *J Neurol Sci* **124:** 56-60.
64. Towbin JA, Bowles NE (2001) Molecular genetics of left ventricular dysfunction. *Curr Mol Med* **1:** 81-90.
65. Wigle ED (2001) Cardiomyopathy: The diagnosis of hypertrophic cardiomyopathy. *Heart* **86:** 709-714.
66. Witt TN et al (1992) McLeod syndrome: a distinct form of neuroacanthocytosis. Report of two cases and literature review with emphasis on neuromuscular manifestations. *J Neurol* **239:** 302-306.
67. Marsh WL (1983) Deleted antigens of the Rhesus and Kell blood groups: Association with cell membrane defects. In: *Garraty G (ed.) Blood group antigens and disease.* Arlington, Virginia: American Association of Blood Banks 165-185

CHAPTER 17

ERYTHROCYTE MEMBRANE ABNORMALITIES IN NEUROACANTHOCYTOSIS: EVIDENCE FOR A NEURON-ERYTHROCYTE AXIS?

G.J.C.G.M. Bosman[1], M.W.I.M. Horstink[2], and W.J. De Grip[1]

[1]Department of Biochemistry and [2]Department of Neurology; Nijmegen Center for Molecular Life, Sciences, Nijmegen, The Netherlands

Abstract. There are no indications for the involvement of one specific erythrocyte membrane protein in the formation of acanthocytes in neuroacanthocytosis, with two notable exceptions: the XK protein and the anion exchanger band 3. Changes in structure and/or function of erythrocyte band 3 are consistent findings in patients with neuroacanthocytosis. These changes are probably secondary to other processes, which may include changes in the lipid composition of the erythrocyte membrane. As band 3 is associated with a number of membrane and cytosolic proteins, changes in band 3 not only affect cell shape, but also total cell homeostasis. Elucidation of the disease-related changes in band 3 may be instrumental in the elucidation of the molecular causes of neuroacanthocytosis.

INTRODUCTION

Clinical diagnosis of neurodegenerative movement disorders is complicated by the mixed clinical symptoms and by the common absence of accompanying specific laboratory data [24]. Certain movement disorders, however, are associated with the presence of abnormally shaped erythrocytes. In such patients, affected by one of the neuroacanthocytosis syndromes, the red cell abnormality in association with a nervous system disorder indicates that a component common to neurons and erythrocytes may be specifically involved.

A. Danek (ed.), Neuroacanthocytosis Syndromes, 153–159.

ERYTHROCYTE MEMBRANE PROTEINS AND CELL SHAPE

Two theories aim to explain the regulation of erythrocyte shape: the coupled lipid bilayer theory and the protein network scaffold theory, with most explanatory power for the latter. In this theory, the scaffold for cell shape is the cytoskeleton, consisting of spectrin tetramers that are cross-linked at junction points by short filaments of actin, in a complex with 4.1 protein, tropomodulin, tropomyosin, and adducin. At their central region, the spectrin tetramers are bound to the membrane by ankyrin/band 3 complexes. The interaction between these proteins may be regulated by phosphorylation [3]. Changes in the cytoskeleton network itself as well as changes in ankyrin and band 3, that affect the interaction between the cytoskeleton and the plasma membrane, result in abnormal cell shapes, varying from acanthocytes to spherocytes to stomatocytes [12].

ERYTHROCYTE MEMBRANE PROTEINS AND NEURO-ACANTHOCYTOSIS

For practically all the proteins of the erythrocyte membrane there is no known connection between the occurrence of mutations, the presence of acanthocytes, and the neurological symptoms of the neuroacanthocytosis syndromes. Mutations in the spectrins, 4.1 protein, glycophorin or in the ankyrins may be associated with abnormal erythrocyte morphologies such as spherocytosis and ovalocytosis, but not with the presence of acanthocytes [12]. While most of these proteins are also expressed in neurons, at present there are no data that link mutations in these proteins with neuroacanthocytosis. There is, however, one important caveat, in view of the difficulty to recognize acanthocytes in blood smears [24], and exemplified by the serendipity of the discovery of one acanthocytosis-related mutation in band 3 [17]: it remains possible that a neuroacanthocytosis-related mutation in one of the erythrocyte membrane proteins has simply not been noticed yet. One exception is clear: mutations in the *XK* gene that result in the absence of the XK protein in the erythrocyte membrane are associated with the McLeod syndrome [15]. However, the scarcity in the knowledge on the characteristics of the XK protein (integral membrane protein, associated with the Kell protein, expressed in erythroid and non-erythroid tissues [15,24]; see also Chapter 22) make it difficult even to speculate on its involvement in erythrocyte shape and neurodegeneration.

ERYTHROCYTE ANION EXCHANGER BAND 3

Although the exact nature of the neuron-erythrocyte axis remains to be elucidated, in all neuroacanthocytosis and neuroacanthocytosis-like syndromes this connection seems to be visible as changes in the structure and/or function of the band 3 protein. Band 3 or AE1 is an integral membrane protein that belongs to the anion exchanger (AE) gene family. Members of the AE family are expressed in all cells examined so far, including neurons [1,16]. Band 3 is a multifunctional protein, that is involved in chloride/bicarbonate exchange and regulation of intracellular pH, in the interaction between plasma membrane and cytoskeleton, in binding of glycolytic enzymes, in transport of NO through the circulation, in binding of denatured hemoglobin, and in the recognition and removal of damaged and old cells by the immune system [1,16,22]. It is suggested that band 3 forms the scaffold for a protein assembly that includes the Rh protein complex and CD47 [25], and that may include and affect the activity of the glucose transporter as well [7]. Changes in band 3 structure do not only lead to alterations in erythrocyte shape, but also to altered anion transport characteristics and increased autoimmunoreactivity [16]. Immunochemical and functional studies in neuronal cells and in brains of patients with Alzheimer's disease indicate that brain band 3 has the same function as erythrocyte band 3, and show that senescent erythrocyte-like expression of senescent cell antigens occurs in degenerating neurons [9,18].

ERYTHROCYTE ANION EXCHANGER BAND 3 AND NEUROACANTHOCYTOSIS

Thus band 3, with its central function in the maintenance of erythrocyte shape and homeostasis, is a likely instrument in the elucidation of the common link between neurodegeneration and erythrocyte abnormalities in neuroacanthocytosis: (1) Most mutations in band 3 either have no detectable effect on erythrocyte morphology, or are associated with spherocytosis or ovalocytosis but not with acanthocytosis. However, the one known mutation in an erythrocyte membrane protein – with the exception of the XK protein – that is associated with acanthocytosis, is a point mutation in band 3 [13,17]; (2) In all cases of (neuro)acanthocytosis where the erythrocyte membrane proteins have been examined with anti-band 3 antisera, indications for changes in band 3 structure could be detected (Table 1); (3) The latter changes are associated with changes in band 3 function (anion transport and/or ankyrin binding; Table 1). It is not clear if other changes such as increased phosphorylation of band 3 [20] and anti-band 3 immunoreactivity [18] are cause or effect, and if they are part of a more systemic phenomenon. It is tempting to speculate that the observed decrease in high-affinity ankyrin binding sites [17] results in loss of anchorage of the cytoskeleton to band 3 and permits evagination of the membrane, leading to acanthocyte formation. However, in acanthocytes from a patient with abetalipoproteinemia the number or ankyrin binding was increased, not decreased [17]. These observations may be caused by the differences in dissociation constants between those of binary complexes *in vitro* and those of higher-order complexes *in situ*, and illustrate that acanthocyte formation may have multiple causes.

Table 1. Summary of band 3-related changes in erythrocytes from patients with neuroacanthocytosis and neuroacanthocytosis-like syndromes.

Reference	neurological symptoms	acanthocytes	band 3 structure	anion transport	other changes
[17]	–	+	increased Mw	↑	ankyrin binding ↓
[17] Abeta	–	+	breakdown	↓	ankyrin binding ↑
[18]	+	+	altered	↑	autoimmune reactivity
[10] proband	+	+	breakdown	↓	autoimmune reactivity
[10] siblings	+	+/–	breakdown	↓	autoimmune reactivity
[20]	+	+	ND	↑	phosphorylation ↑
[11]	*	–	breakdown*	ND	autoimmune reactivity

Abeta, erythrocytes from a patient with abetalipoproteinemia; +, present; –, absent; ↑, increased; ↓, decreased; ND, not determined; *, various extrapyramidal movement disorders (for further explanation see the text); altered, exposure of senescent cell antigens; Mw, apparent molecular weight.

ERYTHROCYTE BAND 3 IN THE DIAGNOSIS OF EXTRAPYRAMIDAL MOVEMENT DISORDERS

These observations led us to the hypothesis that changes in band 3 structure and function could be helpful in differential diagnosis of amyotrophic extrapyramidal movement disorders with axonal neuropathy in general [10,11]. For this purpose, we investigated the specificity of the findings in patients with neuroacanthocytosis by immunoblot analysis of the erythrocyte membranes of 14 patients with various other extrapyramidal disorders such as dystonia, Parkinson's disease, parkinsonism, idiopathic myoclonic syndromes, Huntington's disease, and hereditary motor and sensory neuropathy (HMSN1a). Analysis of erythrocyte membrane proteins with anti-band 3 antiserum showed for most patients immunoblot patterns that are identical to those obtained for control donors (Figure 1). These patterns consist of a major immunoreactive band of 95kDa (native band 3), together with smaller, naturally occurring breakdown products [8,16]. Patterns that indicate increased breakdown are seen in the erythrocytes from one patient (out of four) with Parkinson's disease (Figure 1, lane 1), a patient with myoclonic dystonia (Figure 1, lane 4), a patient with Huntington's disease (Figure 1, lane 5), and one patient (out of two) with spasmodic torticollis (not shown). The concentration of erythrocyte-bound IgG, the physiological aging parameter [16], was in the low to normal range for all patients (0.1 - 2.1 fgm IgG/1000 erythrocytes; normal range 0.7 - 2.2 fgm/1000

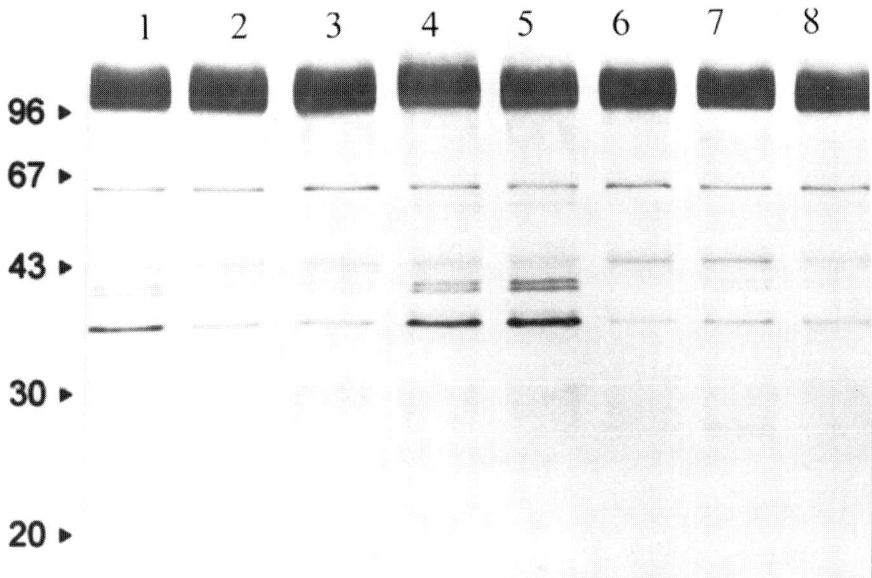

Figure 1. Immunoblots of erythrocyte membranes comparing band 3 patterns of erythrocytes from patients with various extrapyramidal disorders, and controls.
Erythrocyte membrane proteins were analyzed by SDS-polyacrylamide gel electrophoresis, blotting, and antibodies against erythrocyte band 3 as described before [10]. 1, Parkinson's disease; 2, healthy control; 3, essential myoclonus; 4, myoclonic dystonia; 5, Huntington's disease; 6, HMSN Ia; 7, Meige syndrome; 8, Dystonia musculorum deformans. Numbers at arrowheads represent molecular weight calibration.

RBC), with the exception of that of one patient with mitochondrial encephalomyopathy with parkinsonism (6.2 fgm/1000 RBC). It remains to be established whether these patterns are identical to those observed for neuroacanthocytosis patients [10,18]. Interestingly, the IgG contents of the erythrocytes with an apparently increased band 3 break-down were exceptionally low, i.e. below the detection limit of 0.1 fgm/1000 RBC. Together with the conspicuous presence of high numbers of young erythrocytes (unpublished observations), this low IgG content indicates an increased erythrocyte turnover and an abnormal erythrocyte aging process.

MEMBRANE LIPIDS AND NEUROACANTHOCYTOSIS

Some data indicate that changes in phospholipid distribution and/or fatty acid composition are associated with acanthocytosis: (1) abnormalities in the fatty acid content of the erythrocyte membrane have been described in various patients with neuroacanthocytosis and similar sydromes [4,10,20]. Apart from the fact that these changes mainly involve the poly-unsaturated fatty acids, they are difficult to put under a more specific common denominator; (2) manipulation of the phospholipid composition of the erythrocyte membrane and of their distribution over the outer and inner leaflets results in the formation of acanthocyte-like erythrocytes [5,6]; (3) dietary intake of essential fatty acids resulted in strong reduction of the pathological phenotype in a transgenic mouse model of Huntington's disease, which shares a number of symptoms with neuroacanthocytosis [14].

Since most cell shape phenomena are not easily explained by the coupled bilayer theory [19], changes in the lipid composition of the membrane may not be directly involved in the formation of acanthocytes. Indirectly, however, such changes may affect the structure of membrane proteins and result in altered function and/or increased proteolytic breakdown [2]. Breakdown may lead to the exposure of neoantigens and the generation of autoantibodies, as have been observed in some patients (Table 1). It is tempting to speculate that the postulated function of the XK protein as a transport protein [15] may include the transport of fatty acids and/or phospholipids across the membrane.

CONCLUSIONS

Changes in structure and function of erythrocyte band 3 seem to be a consistent and specific finding in acanthocytosis. More detailed analysis of these changes will be helpful in elucidating the molecular causes of neuroacanthocytosis. The observation that band 3 is also expressed in neurons of the brain, with a role in signalling of damaged and dying cells that is similar to that in erythrocytes [9,18], reinforces this conclusion. Concomitant fundamental research into the processes that cause acanthocytosis *in vitro* may be helpful in providing insight into the neurodegenerative mechanisms as well. However, the linkages between neuroacanthocytosis syndromes and the XK and CHAC proteins [15,23] are much more direct. This suggests that the elucidation of the physiological functions of these proteins is the shortest way towards a better understanding of the fundamental mechanisms underlying the pathophysiology of neuroacanthocytosis. Band 3 is likely to be involved in the regulation of these functions, in view of its central role in cell homeostasis.

REFERENCES

1. Alper SL (1991) The band 3-related anion exchanger (AE) gene family. *Ann Rev Physiol* **53:** 549-564.
2. Asano K, Osawa Y, Yanagisawa N et al (1985) Erythrocyte membrane abnormalities in patients with amyotrophic chorea with acanthocytosis. Part 2. Abnormal degradation of membrane proteins. *J Neurol Sci* **68:** 161-173.
3. Bennett V, Baines AJ (2001) Spectrin and ankyrin-based pathways: metazoan inventions for integrating cells into tissues. *Physiol Rev* **81:** 1354-1392.
4. Bird TD, Cederbaum S, Valpey RW, Stahl WL (1978) Familial degeneration of the basal ganglia with acanthocytosis: A clinical, neuropathological, and neurochemical study. *Ann Neurol* **3:** 253-258.

5. Boon JM, Smith BD (1999) Facilitated phospholipid translocation across vesicle membranes using low-molecular-weight synthetic flippases. *J Am Chem Soc* **121:** 11924-11925.
6. Boon JM, Smith BD (2001) Facilitated phosphatidylcholine flip-flop across erythrocyte membranes using low molecular weight synthetic translocases. *J Am Chem Soc* **123:** 6221-6226.
7. Bosman GJCGM, Kay MMB (1990) Alteration of band 3 transport protein by cellular aging and disease: erythrocyte band 3 and glucose transporter share a functional relationship. *Biochem Cell Biol* **68:** 1419-1427.
8. Bosman GJCGM, Bartholomeus IGP, De Man AJM, Van Kalmthout PJC De Grip WJ (1991) Erythrocyte membrane characteristics indicate abnormal cellular aging in patients with Alzheimer's disease. *Neurobiol Aging* **12:** 13-18.
9. Bosman GJCGM, Van Workum FPA, Renkawek K et al (1993) Proteins immunologically related to erythrocyte anion transporter band 3 are altered in brain areas affected by Alzheimer's disease. *Acta Neuropathol* **86:** 353-359.
10. Bosman GJCGM, Bartholomeus IGP, De Grip WJ, Horstink MWIM (1994) Erythrocyte anion transporter and antibrain immunoreactivity in chorea-acanthocytosis. A contribution to etiology, genetics, and diagnosis. *Brain Res Bull* **33:** 523-528.
11. Bosman GJCGM, Bartholomeus IGP, Horstink MWIM, De Grip WJ (2002) *unpublished*
12. Bossi D, Russo M (1996) Hemolytic anemias due to disorders of red cell membrane skeleton. *Mol Aspects Med* **17:** 171-188.
13. Bruce LJ, Kay MM, Lawrence C, Tanner MJ (1993) Band 3 HT, a human red-cell variant associated with acanthocytosis and increased anion transport, carries the mutation Pro-868 → Leu in the membrane domain of band 3. *Biochem J* **293:** 317-20.
14. Clifford JJ, Drago J, Natoli AL et al (2002) Essential fatty acids given from conception prevent topographies of motor deficit in a transgenic model of Huntington's disease. *Neuroscience* **109:** 81-88.
15. Ho M, Chelly J, Carter N, Danek A, Crocker P, Monaco AP (1994) Isolation of the gene for McLeod syndrome that encodes a novel membrane transport protein. *Cell* **77:** 869-880.
16. Kay MMB (1991) Drosophila to bacteriophage to erythrocyte: the erythrocyte as a model for molecular and membrane aging of terminally differentiated cells. *Gerontology* **37:** 5-32.
17. Kay MMB, Bosman GJCGM, Lawrence C (1988) Functional topography of band 3: a specific structural alteration linked to functional aberrations in human red cells. *Proc Natl Acad Sci USA* **85:** 492-496
18. Kay MMB, Goodman J, Lawrence C, Bosman GJCGM (1990) Membrane channel protein abnormalities and autoantibodies in neurological disease. *Brain Res Bull* **24:** 105-111.
19. Nakato M (2002) New insights into regulation of erythrocyte shape. *Curr Opin Hematol* **9:** 127-132.
20. Olivieri O, De Franceschi L, Bordin L et al (1997) Increased membrane protein phosphorylation and anion transport activity in chorea-acanthocytosis. *Haematologica* **82:** 648-653.
21. Oshima M, Osawa Y, Asano K, Saito T (1985) Erythrocyte membrane abnormalities in patients with amyotrophic chorea with acanthocytosis. Part 1. Spin labeling studies and lipid analyses. *J Neurol Sci* **68:** 147-160.
22. Pawloski JR, Hess DT, Stamler JS (2001) Export by red blood cells of nitricoxide bioavailability. *Nature* **409:** 622-626.
23. Rampoldi L, Dobson-Stone C, Rubio JP et al (2001) A conserved sorting-associated protein is mutant in chorea-acanthocytosis. *Nat Genet* **28:** 119-120.
24. Stevenson VL, Hardie RJ (2001) Acanthocytosis and neurological disorders. *J Neurol* **248:** 87-94.
25. Tanner MJ (2002) Band 3 anion exchanger and its involvement in erythrocyte and kidney disorders. *Curr Opin Hematol* **9:** 133-139.

CHAPTER 18

ERYTHROCYTE MEMBRANE ANION EXCHANGE ABNORMALITIES IN CHOREA-ACANTHOCYTOSIS: THE BAND 3 NETWORK

LUCIA DE FRANCESCHI AND ROBERTO CORROCHER
Department of Clinical and Experimental Medicine, Section of Internal Medicine, University of Verona, Italy

Abstract. The presence in peripheral circulation of red cells characterized by thorn or spur-like protrusion is known as acanthocytosis. This specific morphologic abnormality is found in a heterogeneous variety of disorders ranging from isolated acanthocytosis and spur cell anemia to neurological syndromes such as abetalipoproteinemia, McLeod syndrome and chorea-acanthocytosis. In the present paper, we have reviewed the principal studies on red cell membrane abnormalities in acanthocytic red cells associated with neurological disease, focusing on chorea-acanthocytosis. We finally present a functional model for acanthocytic red cells.

INTRODUCTION

The presence in the peripheral circulation of red cells characterized by thorn or spur-like protrusion is known as acanthocytosis [2]. This specific morphologic abnormality is found in a heterogeneous variety of disorders ranging from isolated acanthocytosis and spur cell anemia to neurological syndromes such as abetalipoproteinemia, McLeod syndrome and chorea-acanthocytosis [5,9]. Acanthocytic red cells display several abnormalities of their cell membrane.

THE NORMAL RED CELL MEMBRANE

In the normal red cell membrane (Figure 1) the submembrane spectrin-actin network is attached to the lipid bilayer throughout ankyrin and protein 4.1,

A. Danek (ed.), Neuroacanthocytosis Syndromes, 161–167.

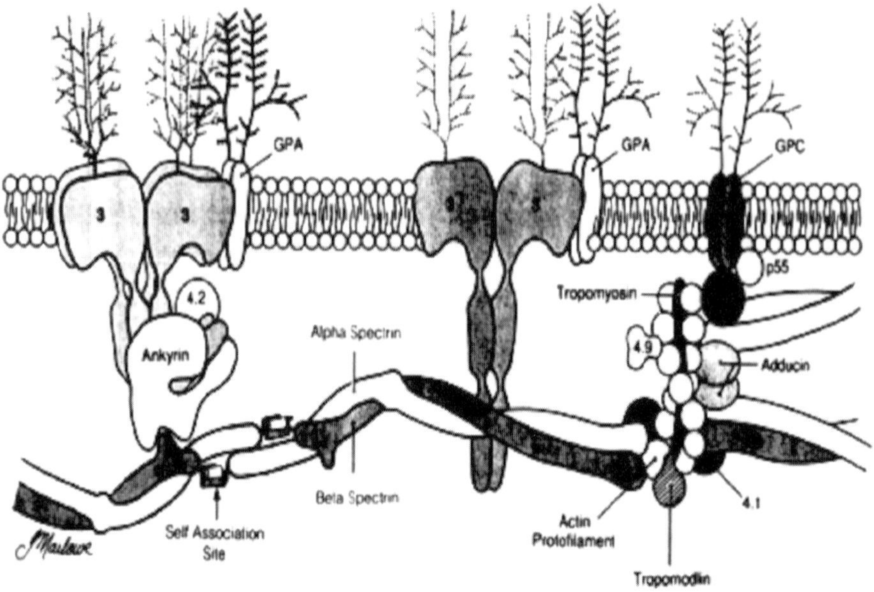

Figure 1. Schematic diagram of red cell membrane organization. The protein interaction crucial in regulating RBC membrane mechanical stability is shown: spectrin dimer-dimer interaction, spectrin-actin interaction, band 3-ankyrin-spectrin interaction, and glycophorin C-protein 4.1 interaction.

which in turn interacts with the transmembrane proteins: band 3, glycophorin A and glycophorin C [9,17].

Although the protein interactions involved in the attachment of the skeleton to the red cell membrane are relatively well characterized, the interaction between the skeletal proteins and the lipids of the inner bilayer is not direct and is not completely known [14,17]. Band 3 is the major transmembrane protein and is composed by two structurally and functionally distinct domains: the N-terminal region (43 kDa) which constitutes the cytoplasmatic portion binding ankyrin, glycolytic enzymes, hemoglobin, hemichrome, band 4.1 and band 4.2 (Figure 2). The band 3 N-terminal presents several phosphorylable sites, which may affect band 3 structure and function [14,17]. The C-terminal region (55 Kd) constitutes the trans-membrane domain, which crosses the lipid bilayer 12 to 14 times, forming the anion exchanger (Figure 2). Each band 3 binds only one molecule of ankyrin, about 40% of band 3 molecules are involved in anchoring the membrane skeleton [14,17].

Krupka and coworkers have proposed a working model for band 3 anion exchanger based on three steps. According to this functional model the crucial and limiting event is the translocation step in which the anion is trapped into the protein (Figure 3) [13]. Thus, any defects of band 3 protein sequence, of its

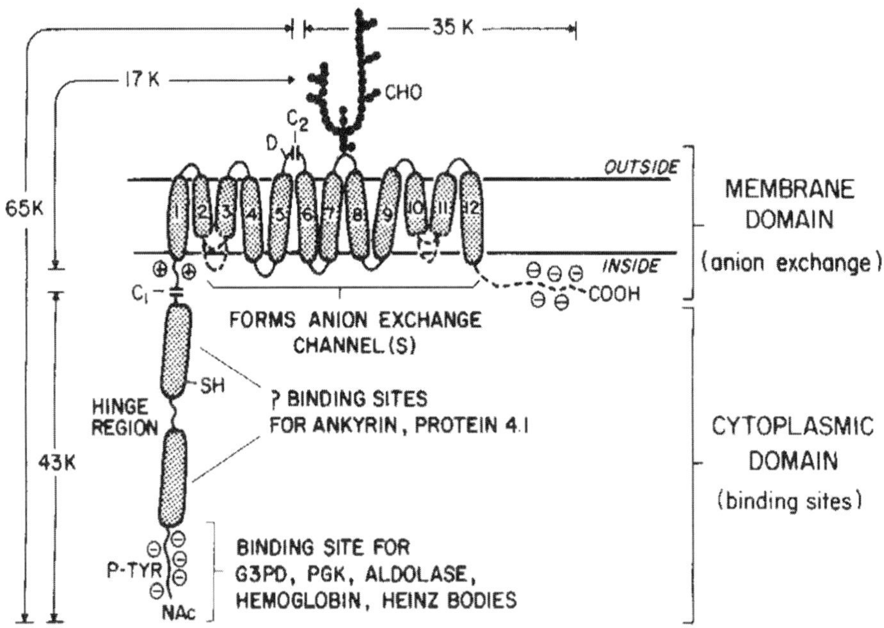

Figure 2. Schematic diagram of band 3 protein structure.

conformational state or of its phosphorylation/dephosphorylation state result in functional changes of band 3 anion exchanger activity. This functional model points out the importance of maintaining band 3 structure and also indicates how the band 3 conformational state is crucial for band 3 anion exchanger activity [13].

Studies in various hereditary hematological disorders, such as hereditary spherocytosis, ovalocytosis and congenital dyserythropoietic anemia type II, have clearly documented the important requirement for an optimal relation between membrane protein skeleton and lipid bilayer, in maintaining normal cell surface area [4,6,7,10].

THE ACANTHOCYTIC RED CELL MEMBRANE

Acanthocytic red cells show an abnormal membrane lipid composition, characterized in particular by an excess of membrane saturated fatty acid (sphingomyelin), which is also present when acanthocytic red cells are associated with neurological disorders such as chorea-acanthocytosis, abetalipoproteinemia and McLeod syndrome [5,11,19]. Studies of red cell morphology using electron microscopy have shown the presence in both abetalipoproteinemia and spur cell anemia of two distinct types of red cell morphology: the echinocytic red cells and the acanthocytic erythrocytes. Echinocytic transformation is reversed by

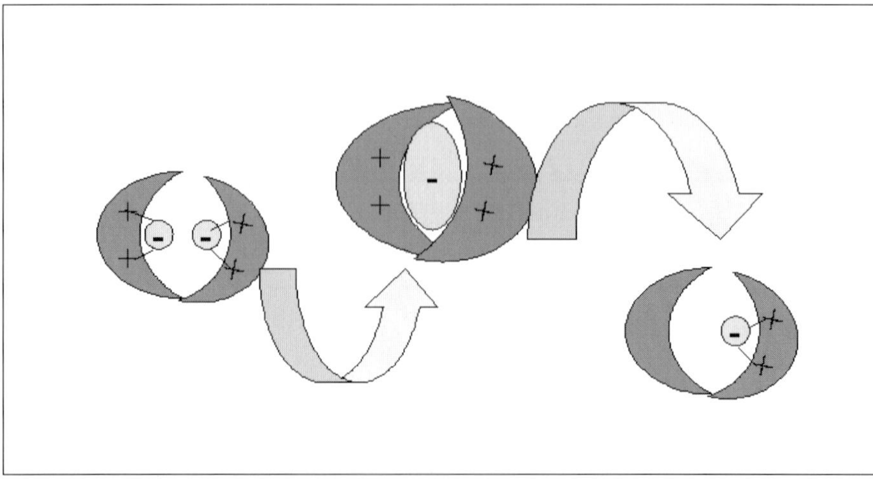

Figure 3. Functional model of band 3 anion exchanger according to the three step working model proposed by Krupka [13].

treatment with chlorpromazine, whose anti-echinocytic effect appears to result from differential expansion of the outer leaflet of the membrane lipid bilayer. This suggests a close and crucial relation between lipid bilayer and membrane protein cytoskeleton, in generating specific morphological erythrocytes abnormalities [5]. The authors proposed that echinocytes represent the first and reversible step which may be followed by irreversible generation of acanthocytes [5]. How the abnormalities in red cell membrane lipid composition generate acanthocytic red cells is, however, still under investigation.

Acanthocytic red cells have been studied both in isolated acanthocytosis and in chorea-acanthocytosis [3,11]. These studies focused mainly on band 3 structure and function. They showed that in isolated acanthocytosis RBC band 3 has an alteration in the restriction of rotational diffusion of band 3, increased anion transport activity and decreased ankyrin binding sites [3,11]. In one case of acanthocytosis, the red cells presented a mutated band 3 which was called band 3 HT due to the increase in the anion exchanger activity [3].

In chorea-acanthocytosis, the increase in band 3 anion exchanger activity was found to be associated with increased anti-band 3 antibodies. Physiologically, these are produced against older red cells in order to remove them from the peripheral circulation by macrophages [11]. Thus, an increase of anti-band 3 antibodies may reflect the abnormalities in conformational band 3 protein structure into the membrane, which permits the partial exposure of band 3, as usually happens in aged red cells [6,11,16,21]. However, the contribution of these different factors in generating acanthocytic red cells is still under investigation.

MODELLING THE MECHANISM OF ACANTHOCYTOSIS

Hereditary abnormalities of red cells may help to understand the pathophysiology of acanthocytosis. In ovalocytosis, a genetic disorder of the RBC membrane skeleton, mutation of band 3 induces a conformational change in its cytoplasmatic domain, leading to its entanglement in the skeletal network. This inhibits normal unwinding and stretching of the spectrin tetramers that is necessary for membrane extension, with an increase in membrane rigidity [14,17]. The functional model of ovalocytosis indicates that membrane rigidity is not a major determinant of erythrocyte survival and that abnormalities in the cytoplasmatic domain of an integral protein such as band 3 can profoundly affect membrane material behavior and protein function (i.e. band 3 anion exchanger activity) [14,17].

Studies in two other genetic disorders of red cells, β thalassemia and congenital dyserythropoietic anemia type II, have identified the presence of anti-band 3 antibody [7,16,21]. Both disorders are characterized by abnormal clusterization of band 3. In the first case this results from oxidative damage to the membrane, in the latter from absence of the glycosylated residue of band 3 [7,16,21]. In both, the presence of anti-band 3 IgG leads to complement activation via C5b and the premature removal of red cells from the peripheral circulation by macrophages [7,16,21]. These models suggest that conformational changes in band 3 domains might be responsible for abnormalities in anion exchanger activity and in production of anti-band 3 antibody.

In chorea-acanthocytosis the abnormality in band 3 anion exchanger activity has been shown to be associated with an increase of membrane serin-threonin phoshorylation (mainly band 3 and band 2), increased levels of band 3 tyrosin phosphorylation and increased casein-kinase membrane activity [18]. *In vitro* models have shown that changes in serin-threonin phosphorylation of the N-terminal cytoplasmatic domain of band 3 may affect the conformational structure of band 3, with modulation of anion exchange activity [1]. Furthermore, band 3 anion exchange activity is also modulated by altered tyrosin phosphorylation of band 3. In *in vitro* red cell models the increase in band 3 phosphorylation might produce conformational changes in the cytoplasmatic domain, modulating the binding of band 3 with other proteins of the cytoskeleton and contributing to *in vivo* regulation of red cell metabolism and cell shape [1,8,15,18]. The changes in band 3 phosphorylation as well as in membrane casein-kinase activity may be related to the abnormalities in band 3 protein conformation; this results from the abnormal relation between the lipid bilayer and the integral membrane proteins.

The ultrastructural changes of erythrocyte membrane skeletons in chorea-acanthocytosis and in McLeod syndrome demonstrated by electron microscopy show that the distribution of membrane skeleton is heterogeneous and it is mainly present in the more dense red cell fraction. In chorea-acanthocytosis, the red cells are more sensitive to energy dependent stress than normal red

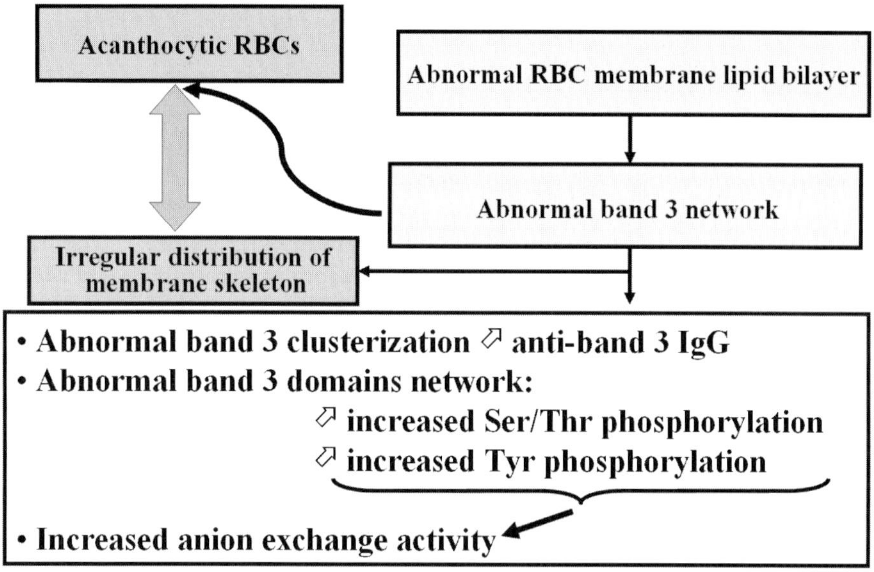

Figure 4. Functional model for acanthocytic red cells.

cells. Thus, Terada et al [20] have proposed that the differences in erythrocyte shape between chorea-acanthocytosis and hereditary spherocytosis may depend on the disruptive structure of the red cell skeleton in the first condition. In fact, electron microscopy indicates that the membrane protrusions depend upon the irregular distribution of the membrane skeleton network.

Based on the experimental evidence mentioned above we propose a functional model for acanthocytic red cells as follows (Figure 4). The abnormal membrane lipid bilayer affects the conformational structure of integral protein, such as band 3, resulting in abnormal distribution of membrane skeleton proteins. Band 3 may be abnormally clustered, causing the production of anti-band 3 IgG. The abnormal interaction of band 3 domains results in increased serine/threonine band 3 phosphorylation and increased band 3 tyrosin phosphorylation which, in turn, modulates the anion exchanger activity.

REFERENCES

1. Baggio B, Bordin L, Clari G, Gambaro G, Moret V (1993) Functional correlation between the Ser/Thr-phosphorylation of band 3 and band 3-mediated transmembrane anion transport in human erythrocytes. *Biochim Biophys Acta* **864:** 145-167.
2. Biemer JJ (1980) Acanthocytosis: Biochemical and physiological considerations. *Ann Clin Lab Sci* **10:** 238-239.
3. Bruce LJ, Kay MBK, Lawrence C, Tanner MJA (1993) Band 3 HT, a human red-cell variant associated with acanthocytosis and increased anion transport, carries the mutation Pro-868 → Leu in the membrane domain of band 3. *Biochem J* **293:** 317-320.

4. Bruce LJ, Ghosh S, King MJ, Layton DM, Mawby WJ, Stewart GW, Oldenborg PA, Delaunay J, Tanner MJA (2002) Absence of CD47 in protein 4.2-deficient hereditary spherocytosis in man: an interaction between the Rh complex and the band 3 complex. *Blood* **100:** 1878-1885.
5. Clark MR, Aminoff MJ, Tsun-Yee Chiu D, Kuypers FA, Friend DS (1989) Red cell deformability and lipid composition in two forms of acanthocytosis: enrichment of acanthocytic population by density gradient centrifugation. *J Lab Clin Med* **113:** 469-481.
6. De Franceschi L, Olivieri O, Miraglia del Giudice E, Perrotta S, Sabato V, Corrocher R, Iolascon A (1997) Membrane cation and anion transport in erythrocyte of hereditary spherocytosis. Effects of different membrane protein defects. *Am J Hematol* **55:** 121-128.
7. De Franceschi L, Turrini F, Miraglia del Giudice E, Perrotta F, Olivieri O, Corrocher R, Mannu F, Iolascon A (1998) Decreased band 3 anion transport activity and band 3 clusterization in congenital dyserythropoietic anemia type II. *Exp Hematol* **26:** 869-873.
8. Dekoswki SA Rybicki A, Drickamer K (1983) A tyrosine kinase associated with the red cell membrane phosphorylates band 3. *J Biol Chem* **258:** 275-2753.
9. Iolascon A, Miraglia del Giudice E, Camaschella C (1992) Molecular pathology of inherited erythrocyte membrane disorders: hereditary spherocytosis and elliptocytosis. *Haematologica* **77:** 60-72.
10. Jarolim P, Rubin HL, Lux SC, Cho MR, Brabec V, Derick Lh, Yi SL, Saad STO, Alper SL, Brugnara C, Golan DE, Palek J (1994) Duplication of 10 nucleotides in the erythroid band 3 (AE1) gene in kindred with hereditary spherocytosis and band 3 protein deficiency (band 3 PRAGA) *J Clin Invest* **93:** 121-127.
11. Kay MMB, Bosman GJ, Lawrence C (1988) Functional topography of band 3: specific structural alteration linked to functional aberrations in human erythrocytes. *Proc Natl Acad Sci USA* **85:** 492-496.
12. Kay MMB, Goodman J, Lawrence C, Bosman G (1990) Membrane channel protein abnormalities and autoantibodies in neurological disease. *Brain Res Bull* **24:** 105-111.
13. Krupka RM (1989) Role of substrate binding forces for mechanism of the anion exchanger of red cells. *J Membr Biol* **109:** 151.
14. Liu SC, Derick LH (1992) Molecular anatomy of red blood cell membrane skeleton: structure-function relationship. *Sem Hematol* **29:** 231-243.
15. Low PS, Allen DP, Zioncheck TF, Chari P, Willards BM, Geahlen RL, Harrison ML (1987) Tyrosine phosphorylation of band 3 inhibits peripheral protein binding. *J Biol Chem* **262:** 4592-4596.
16. Mannu F, Arese P, Cappellini MD, Fiorelli G, Cappadoro M, Giribaldi G, Turrini F (1995) Role of hemichrome binding to erythrocyte membrane in the generation of band 3 alteration in thalassemia intermedia erythrocytes. *Blood* **86:** 2014-2018.
17. Narla M, Chasis JA (1993) Red blood cell deformability, membrane material properties and shape: regulation by transmembrane skeletal and cytosolic proteins and lipids. *Sem Hematol* **30:** 171-192.
18. Olivieri O, De Franceschi L, Bordin L, Manfredi M, Miraglia del Giudice E, Perrotta S, De Vivo M, Guarini P, Corrocher R (1997) Increased membrane protein phosphorylation and anion transport activity in chorea-acanthocytosis. *Haematologica* **82:** 648-653.
19. Oshima M, Osawa Y, Asano K, Saito T (1985) Erythrocyte membrane abnormalities in patients with amyotrophic chorea with acanthocytosis. Part 1. Spin labeling studies and lipid analyses. *J Neurol Sci* **68:** 147-160.
20. Terada N, Fujii Y, Ueda H, Kato Y, Baba T, Hayashi R, Ohno S (1999) Ultrastructural changes of erythrocyte membrane skeletons in chorea-acanthocytosis and McLeod syndrome revealed by quick-freezing and deep-etching method. *Acta Haematologica* **101:** 25-31.
21. Turrini F, Mannu F, Cappadoro M, Ulliers D, Giribaldi G, Arese P (1994) Binding of natural occuring antibodies to oxidatively and non-oxidatively modified erythrocyte band 3. *Biochim Biophys Acta* **1190:** 297-303.

CHAPTER 19

THE SPECTRUM OF MUTATIONS AND POSSIBLE FUNCTION OF THE CHAC GENE

Carol Dobson-Stone, Luca Rampoldi, and Anthony P. Monaco

The Wellcome Trust Centre for Human Genetics, Oxford, UK

Abstract. The gene responsible for autosomal-recessive chorea-acanthocytosis, *CHAC*, spans 250 kilobases on chromosome 9q21. It is a large gene, comprising 73 exons, and it seems to be ubiquitously expressed. So far, 71 different mutations have been reported in *CHAC*, the majority of which are predicted to lead to a null allele. Analysis of the *CHAC* product, chorein, reveals that it contains several tetratricopeptide repeats, which are believed to be involved in protein-protein interactions. Chorein is homologous to Vps13p, implicated in protein trafficking in yeast. However, until functional studies are carried out, the possible function of chorein, and how its absence leads to the erythrocyte abnormalities and neurodegeneration that are characteristic of chorea-acanthocytosis, is open to speculation.

INTRODUCTION

In contrast to other neuroacanthocytosis syndromes, the genetics of chorea-acanthocytosis (ChAc, MIM 200150) have only recently been determined. Autosomal-dominant transmission of ChAc has been reported [9,10,20] but the majority of cases appear to show autosomal-recessive transmission, with several cases published from consanguineous families [2,1,13]. Rubio et al [14] performed linkage studies on 11 families segregating for ChAc. Using an autosomal-recessive model of inheritance, the disease was reported to be linked in all families to a 6 centiMorgan region of chromosome 9q21 flanked by the markers *GATA89a11* and *D9S1867*. This result was confirmed by homozygosity-by-descent analysis in offspring from inbred families. Four years later, we screened the probands from the same 11 families for mutations in a novel gene mapping to the ChAc critical region and found 16 different mutations

A. Danek (ed.), Neuroacanthocytosis Syndromes, 169–175.

[12]. In each of the probands we identified at least one mutation likely to disrupt or abolish gene function. Another group who independently identified the same gene, denoted *CHAC*, reported one additional deletion mutation in three Japanese families [17]. Recently, we screened 43 probands with ChAc and identified 57 *CHAC* disease mutations, 54 of which had not previously been reported [5]. This article reviews the spectrum of mutations in *CHAC*, and speculates on the function of its protein product, chorein.

CHAC GENE STRUCTURE AND EXPRESSION ANALYSIS

CHAC is organized in 73 exons in a genomic region of about 250 kilobases (Figure 1). Two splicing variants have thus far been identified: transcript A, which comprises exons 1-68 and 70-73, has an open reading frame of 9525 base pairs, encoding a 3174 amino-acid (aa) protein that was named chorein [17]. Transcript B includes exons 1-69 and encodes a 3095 aa protein which lacks the 111 aa encoded by exons 70 to 73 and has an additional 32 aa. Both transcripts have polyadenylation signals in their respective 3′ untranslated regions.

Expression analysis of the *CHAC* gene by Northern blot analysis and reverse transcriptase polymerase chain reaction (RT-PCR) experiments shows a signal in all the tissues analyzed, including the erythrocyte precursor cell line K562 and brain – i.e., the tissues mainly affected in ChAc, and suggests a ubiquitous pattern of expression. Since the signal detected in brain on Northern blot analysis was very weak, we investigated the possibility that *CHAC* might be differentially expressed in different brain regions. The analysis of a brain specific Northern blot (containing RNA from cerebellum, cerebral cortex, medulla, spinal cord, occipital pole, frontal lobe, temporal lobe and putamen) revealed a similarly weak expression in all the areas analyzed. By specific RT-PCR experiments we could detect similar levels of expression of both isoforms A and B in all tissues analyzed so far [12].

PREDICTED EFFECTS OF *CHAC* MUTATIONS

From 57 ChAc probands thus far screened, 71 different mutations have been identified [12,17,5]. As Figure 1 shows, they seem to be evenly distributed along the gene. These comprise 18 nonsense mutations, 29 small insertion/deletion mutations, three gross deletions, 16 splice-site mutations and five missense mutations.

The nonsense mutations all involve single base-pair changes, which alter the coding of a nucleotide triplet such that it now codes for a stop signal within the *CHAC* mRNA transcript. A transcript containing such a premature termination codon (PTC) is likely to be a target for a mechanism called nonsense-mediated mRNA decay (NMD). Any stop codon present upstream of the last intron is detected during mRNA processing and the transcript is flagged for early degradation [6]. This implies that *CHAC* mRNA containing such mutations will be degraded before being translated.

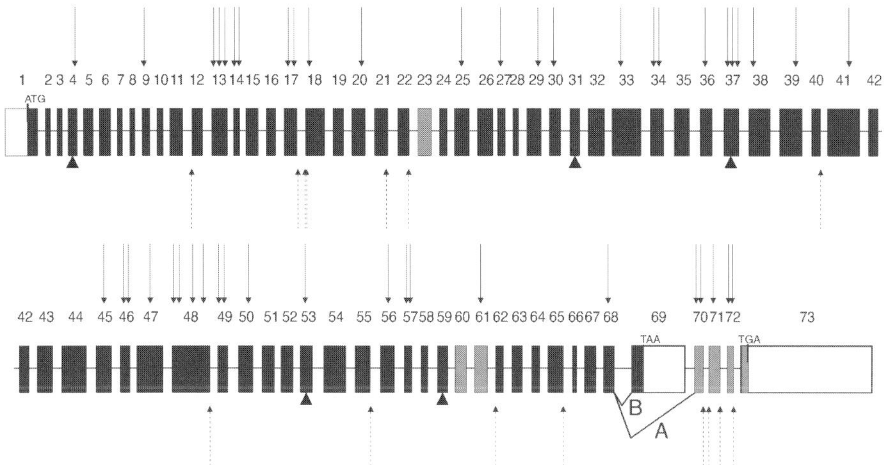

Figure 1. Distribution of *CHAC* mutations identified so far. The exonic structure of *CHAC* is illustrated. Arrows with solid lines indicate nonsense and insertion/deletion mutations leading to a premature termination codon. Arrows with dashed lines indicate splice-site mutations. Missense mutations are indicated by black triangles; exons found to be deleted in certain ChAc patients are colored grey.

Twenty-eight of the 29 small-scale *CHAC* nucleotide deletions or insertions involve addition or removal of one, two, four, five or 14 nucleotides from the transcript. Such changes shift the triplet reading frame so that it now codes for a number of erroneous amino acids before introducing a stop codon.

The remaining mutation, 2029_2031del3ins27 [5], results in the net addition of 24 nucleotides and thus 8 codons. This does not shift the reading frame: however, the third new codon that has been inserted is a stop codon. In both circumstances, the PTCs that are introduced will trigger NMD as explained above.

Analysis of the RNA from ChAc patients carrying a gross deletion of *CHAC* exons 60 and 61 revealed that exon 59 is spliced directly to exon 62, resulting in a net deletion of 260 bases and therefore a shift in the reading frame which introduces a PTC [17]. The other two gross deletions of exon 23 and of exons 70-73 have not been studied at the RNA level [5]. However, even if they do not cause an mRNA frameshift it is likely that such mutations, involving removal of hundreds of nucleotides, will result in production of non-functional protein.

Six of the 16 splice-site mutations identified so far are substitutions or deletions within the virtually invariant AG dinucleotide of the splice acceptor site at the 5 end of the intron. Another eight mutations affect the consensus GT dinucleotide of the splice donor site, at the 3′ end. The remaining two mutations are G>A substitutions at the 3′ terminal nucleotide of an exon. At this position, a G is normally present in 78% of splice-sites: only 9% of exons have an A at their 3′ end [16]. Although it has not been possible to analyse these lesions at the RNA level, it is likely that correct splicing will be markedly reduced or abolished in all three classes of mutation.

The five missense mutations identified so far are as follows: I90K, A1095P, S1452P, W2460R and Y2721C [12,5]. All are nonconservative amino-acid substitutions and all have been shown to be absent in at least 50 healthy control individuals. The substitutions all occur in positions that are reasonably well-conserved between putative chorein homologues: two of them (I90K and Y2721C) are located in the most highly-conserved amino- and carboxy-terminal regions, described below. For these reasons it is unlikely that they represent benign polymorphisms.

DISTRIBUTION OF *CHAC* MUTATIONS

Fifty out of 71 mutations (70%) are nonsense, insertion/deletion or gross deletion mutations, which are predicted to result either in absence of gene product or in production of a truncated, non-functional protein. This high proportion of null mutations is consistent with the recessive inheritance of ChAc and the theory that absence of functional chorein is the primary cause of the disease. The proportion of splice-site mutations (16/71, 23%) is also relatively high. This is probably a simple effect of the large number of splice sites, 143, present in the *CHAC* mRNA.

Perhaps the most functionally interesting of the mutational spectrum of any gene is that class of mutations which is not predicted to cause a gross effect, but involves substitution of a single amino-acid residue. The small proportion of these missense mutations (5/71, 7%) identified in *CHAC* perhaps reflects the relatively low degree of evolutionary conservation of the large central region of chorein (see below). It is possible that for much of the protein there is a degree of tolerance to amino-acid substitution. Unfortunately it cannot, however, be deduced that the sites of all the missense mutations *are* relevant to the function of chorein: some of the substitutions may simply cause misfolding of the protein, which would prevent chorein leaving the endoplasmic reticulum. Other missense mutations have been shown to trigger aberrant splicing of the mRNA transcript [4,19]. Functional analysis of these chorein mutants will be necessary to elucidate exactly how they differ from wild-type chorein.

CHOREIN: SEQUENCE ANALYSIS AND POSSIBLE FUNCTION

Computer analysis of chorein's sequence reveals the presence of ten putative tetratrico peptide repeats (TPRs) (see SwissProt accession number Q96RL7). The TPR is a 34 aa repeat that is widespread in evolution [7]. TPR-containing proteins are involved in a variety of cellular processes, such as cell-cycle control, transcription repression, stress response, protein kinase inhibition, mitochondrial and peroxisomal protein transport and neurogenesis. TPRs are predicted to form two amphipathic alpha-helices that can mediate protein/protein interaction. Interactions between TPR domains can occur intra- and intermolecularly. Furthermore, there are several examples of TPR/non-TPR interactions in the literature. The search for putative transmembrane domains with nine different programs gave inconclusive results, similarly to what had previously been reported for chorein's yeast homologue Vps13 (see below) [3].

Chorein is a novel, well conserved protein. It shows the same degree of sequence similarity (25-30% identities, 40-50% positives) with four proteins of comparable size: the protein VPS13 (3144 aa) of *S. cerevisiae* and its orthologue in *S. pombe* (fragments of 2503 and 600 aa) and the putative protein products T08G11.1 (3212 aa) and CG2093 (3242 aa), of *C. elegans* and *D. melanogaster* respectively. The sequence similarity significantly increases (up to 50% identities for the *Drosophila* and *C. elegans* proteins) in the amino and carboxy termini, where matches could also be found with a VPS13-like protein (3306 aa) of *Arabidopsis* and the protein TipC (3848 aa) of *Dictyostelium*. Taking into account the overall similarity and their unusually large size, it seems likely that these proteins are all descendants of a common ancestor. Even though the terminal regions do not contain any known functional domains, their strong conservation suggests that they might be important for the biological function of these proteins. By sequence similarity analysis we also identified a putative human chorein homologue on chromosome 15. The gene prediction sequence analysis of two overlapping BAC clones (1O10 and 16B9) identifies a gene with 76 exons encoding a protein of 3382 aa significantly similar to chorein (40% identities; 60% positives). Other hypothetical proteins, mainly matching in chorein's terminal regions, could be identified in *C. elegans*, *Drosophila*, *S. pombe* and *Arabidopsis*, suggesting that *CHAC* might be a member of a novel, evolutionarily conserved gene family. In addition to the homologies previously reported, we identified several ESTs from mouse, rat, *Bos Taurus*, *Gallus gallus* and *Canis familiaris*.

Among chorein's homologues only *vps13* and *tipC* are known genes. In *Dictyostelium tipC* mutants undergo abnormal tip formation and a defect in the signaling pathway responsible for directing morphogenesis beyond the aggregate state has been hypothesized [15]. In *S. cerevisiae vps* genes encode proteins required for proper protein trafficking from the Golgi to the vacuole. In particular, Vps13p (also reported as SOI1) has been shown to promote the cycling of Trans-Golgi Network (TGN) transmembrane proteins, such as Kex2p, Vps10p and Ste13p, between the TGN and the prevacuolar compartment (PVC), which corresponds to the multivesicular body/late endosome in animal cells. This is accomplished by antagonizing a retention signal at the TGN and, at the PVC, promoting the entry of proteins containing a retrieval signal in retrograde transport vesicles. Vps13p appears to exist in a high molecular weight heterooligomeric or homooligomeric complex peripherally associated with membranes. A computer search for transmembrane regions in Vps13p gave inconclusive results [3].

The existence of human, *D. melanogaster* and *C. elegans* homologues of Vps13p suggests a conservation of the mechanisms of TGN protein localisation and trafficking. It is possible that chorein is controlling one or more steps in the cycling of proteins (perhaps via TPR-mediated interactions?) through the TGN to early and late endosomes, lysosomes and the plasma membrane. Among the proteins whose trafficking could be modulated by chorein, furin-like proteases represent good candidates, since they are homologues of Kex2p protease of *S. cerevisiae*, and furin has been shown to cycle between the TGN, endosomes and the plasma membrane [11].

In McLeod syndrome, an X-linked disorder with acanthocytosis and very similar neurological findings to ChAc, the XK protein is defective. XK is a multiple transmembrane protein localized on the plasma membrane and covalently linked to the single pass membrane protein Kell, a member of the neprilysin subfamily of zinc metalloproteases. It has been demonstrated that Kell can cleave big endothelin-3 (big ET-3), yielding ET-3; and to a lesser extent big ET-1 and big ET-2 [8]. The fact that there is experimental evidence for a role of endothelins as basal ganglia neurotransmitters [18] and that Kell is expressed in brain might be important for the pathogenesis of the disorder. This also suggests a possible link with chorein's putative function, since endothelins are first synthesized as prepro-hormones that are intracellularly cleaved by furin-like proteases to give big ETs.

When XK is not functional, Kell's presence on the membrane is destabilized. Since Kell is associated with the underlying membrane skeleton this might lead to the structural changes thought to be related to acanthocyte formation. We hypothesize that the absence of functional chorein might lead to destabilization of erythrocyte plasma membrane structure due to failure in the localization or recycling of a plasma membrane protein(s).

Of course functional experiments are needed in order to assess chorein's intracellular localisation and to verify its possible role in protein trafficking.

However, the similarities at the clinical and neuropathological levels between chorea-acanthocytosis, McLeod syndrome and also Huntington's disease suggest that there may be a partial overlap in their pathogenetic processes.

REFERENCES

1. Alonso ME, Teixeira F, Jimenez G, Escobar A (1989) Chorea-acanthocytosis: report of a family and neuropathological study of two cases. *Can J Neurol Sci* **16:** 426-431.
2. Bird TD, Cederbaum S, Valey RW, Stahl WL (1978) Familial degeneration of the basal ganglia with acanthocytosis: a clinical, neuropathological, and neurochemical study. *Ann Neurol* **3:** 253-258.
3. Brickner JH, Fuller RS (1997) SOI1 encodes a novel, conserved protein that promotes TGN-endosomal cycling of Kex2p and other membrane proteins by modulating the function of two TGN localization signals. *J Cell Biol* **139:** 23-36.
4. D'Souza I, Poorkaj P, Hong M, Nochlin D, Lee VM, Bird TD, Schellenberg GD (1999) Missense and silent tau gene mutations cause frontotemporal dementia with parkinsonism-

chromosome 17 type, by affecting multiple alternative RNA splicing regulatory elements. *Proc Natl Acad Sci USA* **96:** 5598-5603.
5. Dobson-Stone C, Danek A, Rampoldi L et al (2002) Mutational spectrum of the *CHAC* gene in patients with chorea-acanthocytosis. *Eur J Hum Genet* **10:** 773-781.
6. Hentze MW, Kulozik AE (1999) A perfect message: RNA surveillance and nonsense-mediated decay. *Cell* **96:** 307-310.
7. Lamb JR, Tugendreich S, Hieter P (1995) Tetratrico peptide repeat interactions: to TPR or not to TPR? *Trends Biochem* **20:** 257-9.
8. Lee S, Lin M, Mele A, Cao Y et al (1999) Proteolytic processing of big endothelin-3 by the Kell blood group protein. *Blood* **94:** 1440-1450.
9. Levine IM, Estes JW, Looney JM (1968) Hereditary neurological disease with acanthocytosis. A new syndrome. *Arch Neurol* **19:** 403-409.
10. Metzer WS (1990) Neuroacanthocytosis with autosomal-dominant inheritance, normal serum CK and preserved myotatic reflexes. *Mov Dis* **5 Suppl 1:** 92 (Abstract).
11. Nakayama K (1997) Furin: a mammalian subtilisin/Kex2p-like endoprotease involved in processing of a wide variety of precursor proteins. *Biochem J* **327:** 625-635.
12. Rampoldi L, Dobson-Stone C, Rubio JP et al (2001) A conserved sorting-associated protein is mutant in chorea-acanthocytosis. *Nat Genet* **28:** 119-120.
13. Requena Caballero I, Arias Gomez M, Lema Devesa C, Sanchez Herrero J, Barros Angueira F, Coton Vilas JC (2000) Autosomal-recessive chorea-acanthocytosis linked to 9q21. *Neurologia* **15:** 132-135.
14. Rubio JP, Danek A, Stone C et al (1997) Chorea-acanthocytosis: genetic linkage to chromosome 9q21. *Am J Hum Genet* **61:** 899-908.
15. Stege JT, Laub MT, Loomis WF (1999) Tip genes act in parallel pathways of early *Dictyostelium* development. *Dev Genet* **25:** 64-77.
16. Stephens RM, Schneider TD (1992) Features of spliceosome evolution and function inferred from an analysis of the information at human splice sites. *J Mol Biol* **228:** 1124-1136.
17. Ueno S, Maruki Y, Nakamura M et al (2001) The gene encoding a newly discovered protein, chorein, is mutated in chorea-acanthocytosis. *Nat Genet* **28:** 121-122.
18. van den Buuse M, Webber KM (2000) Endothelin and dopamine release. *Prog Neurobiol* **60:** 385-405.
19. Vuillaumier-Barrot S, Barnier A, Cuer M, Durand G, Grandchamp B, Seta N (1999) Characterization of the 415G>A (E139K) PMM2 mutation in carbohydrate-deficient glycoprotein syndrome type Ia disrupting a splicing enhancer resulting in exon 5 skipping. *Hum Mutat* **14:** 543-544.
20. Walker RH, Morgello S, Davidoff-Feldman B, Melnick A, Walsh MJ, Shashidharan P, Brin MF (2002) Autosomal-dominant chorea-acanthocytosis with polyglutamine-containing neuronal inclusions. *Neurology* **58:** 1031-1037.

CHAPTER 20

IMMUNOHEMATOLOGY OF THE KELL AND KX BLOOD GROUP SYSTEMS

GEOFF DANIELS

Bristol Institute for Transfusion Sciences, Bristol, UK

Abstract. Kell is a complex blood group system comprising 24 antigens encoded by the *KEL* gene and located on a type II transmembrane glycoprotein with a large, highly folded C-terminal extracellular domain. The Kell system polymorphisms are associated with missense mutations representing single amino acid substitutions in the Kell glycoprotein. K_0 is a null phenotype in which no Kell system antigen is expressed. McLeod is a rare phenotype with an X-linked mode of inheritance in which all high frequency Kell system antigens are expressed weakly apart from Km (KEL20), which is absent. McLeod phenotype is a characteristic of McLeod neuroacanthocytosis, which results from absence of Kx from the red cells due to deletions or inactivating mutations of the X-linked gene, *XK*.

INTRODUCTION

One of the characteristics of McLeod syndrome, recognized before the association with neuroacanthocytosis, is the McLeod red cell phenotype, in which antigens of the Kell blood group system are expressed weakly and the Kx antigen of the Kx system is absent. This chapter is a review of the immunohematology of the Kell blood system and the serologically and biochemically related Kx system. For serological references, see Daniels, 2002 [4].

THE KELL BLOOD GROUP SYSTEM

Kell was the sixth blood group system to be discovered, and the first to be found as a result of the antiglobulin agglutination test. In 1946, Coombs et al identified K (KEL1) antigen on the red cells of about 9% of Caucasians. The high frequency allelic antigen, k (KEL2), was found 3 years later. The

A. Danek (ed.), Neuroacanthocytosis Syndromes, 177–186.

Table 1. Antigens of the Kell blood group system.

Number	Symbol	Allelic antigens	Frequency Caucasians	Frequency Africans	Molecular basis*	Exon
KEL1	K	k	9%	2%	M193 (T)	6
KEL2	k	K	> 99%	> 99%	T193 (M)	6
KEL3	Kp[a]	Kp[b], Kp[c]	2%	< 1%	W281 (R or Q)	8
KEL4	Kp[b]	Kp[a], Kp[c]	> 99%	> 99%	R281 (W or Q)	8
KEL5	Ku	–	> 99%	> 99%	Complex	
KEL6	Js[a]	Js[b]	< 1%	16%	P597 (L)	17
KEL7	Js[b]	Js[a]	> 99%	> 99%	L597 (P)	17
KEL10	Ul[a]	–	< 1%†	< 1%	V494 (E)	13
KEL11	K11	KEL17	> 99%	> 99%	V302 (A)	8
KEL12	K12	–	> 99%	> 99%	H548 (R)	15
KEL13	K13	–	> 99%	> 99%	Not known	
KEL14	K14	KEL24	> 99%	> 99%	R180 (P, H, or C)	6
KEL16	'k-like'	–	> 99%	> 99%	Not known	
KEL17	K17, Wk[a]	KEL11	< 1%	< 1%	A302 (V)	8
KEL18	K18	–	> 99%	> 99%	R130 (W or Q)	4
KEL19	K19	–	> 99%	> 99%	R492 (Q)	13
KEL20	Km	–	> 99%	> 99%	Complex	
KEL21	Kp[c]	Kp[a], Kp[b]	< 1%	< 1%	Q381 (R or W)	8
KEL22	K22	–	> 99%	> 99%	A322 (V)	9
KEL23	K23	–	< 1%	< 1%	R382 (Q)	10
KEL24	K24	KEL14	< 1%	< 1%	P180 (R)	6
KEL25	VLAN	–	< 1%	< 1%	R248 (Q)	8
KEL26	TOU	–	> 99%	> 99%	R406 (Q)	11
KEL27	RAZ	–	> 99%	> 99%	E299 (K)	8

*Shown in parentheses, amino acids associated with antigen-negative phenotype.
†Frequency of 3% in Finns.

Kell system now comprises 24 antigens, consisting of four pairs and one triplet of antithetical antigens, 10 antigens of high frequency and three of low frequency (Table 1). In the numerical terminology of the International Society of Blood Transfusion the antigens are numbered from KEL1 to KEL27, with three numbers obsolete (http://www.iccbba.com/page25.htm).

THE KELL GLYCOPROTEIN AND THE *KEL* GENE

The antigens of the Kell system are located on a type II red cell membrane glycoprotein consisting of an N-terminal cytoplasmic domain of 47 amino acids (or 28 amino acids if the codon for Met20 is used for translation initiation), a single membrane-spanning domain, and a large, 665-amino acid, C-terminal extracellular domain (Figure 1) [14]. The extracellular domain has six putative N-glycosylation sites (positions 94, 115, 191, 345, 627, and 724), though Asn724 is unlikely to be glycosylated as residue 725 is proline, which usually inhibits glycosylation. There are 15 extracellular cysteine residues, suggesting

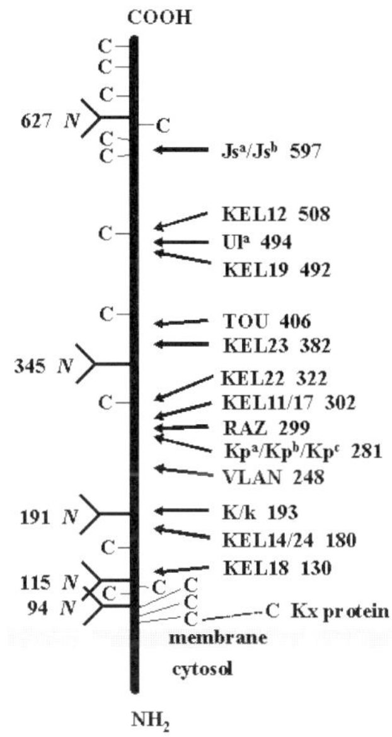

Figure 1. Linear representation of the Kell glycoprotein in the red cell membrane, showing the relative positions of the Kell system antigens, the five N-glycans (N), and 15 extracellular cysteine residues, including Cys72, which is linked to Cys347 of the Kx protein. The native Kell glycoprotein is highly folded.

the presence of seven intramolecular disulphide bonds, resulting in extensive folding of the molecule.

The *KEL* gene is located on chromosome 7q32-q36 and spans 21.5 kb organized into 19 exons of coding sequence [11]. Exon 1 encodes a possible translation-initiating methionine, exon 2 the cytoplasmic domain and a second possible translation initiation site at Met20, exon 3 the membrane spanning domain, and exons 4-19 the large extracellular domain.

THE KELL NULL PHENOTYPE, K_0

Like most blood group systems, Kell has a null phenotype (K_0), in which none of the antigens of the system is expressed. K_0 phenotype is very rare with a frequency of about 0.007 in Caucasians and Japanese. K_0 is usually revealed by the presence of anti-Ku (KEL5), an antibody to a non-polymorphic determinant on the Kell glycoprotein. Family studies have shown that K_0 results from apparent homozygosity for an amorph gene at the *KEL* locus. The molecular basis for K_0 has been determined for nine unrelated propositi, who are either homozygous for inactivating mutations in the *KEL* gene or heterozygous for two such mutations [7,16,26] (Table 2). When expressed in human embryonic

Table 2. *KEL* mutations responsible for K_0 phenotype [7,16,26].

Mutation		Origin
Cys83Stop	(exon 4)	Yugoslavia
Arg128Stop	(exon 4)	African Americans
Arg192Stop	(exon 6)	USA
Gln348Stop	(exon 9)	Portugal
Trp459Stop	(exon12)	Japan
Ser363Asn	(exon 10)	USA
Ser676Asn	(exon 18)	Israel, America
g → a, intron 3,5′ splice site		Réunion Island, USA
g → c, intron 3,5′ splice site		Taiwan
a → g, intron 5,3′ splice site		Japan

kidney cells, the Ser363Asn and Ser676Asn mutant proteins were retained in a pre-Golgi compartment and not transported to the cell surface [16].

THE KELL ANTIGENS

K (KEL1) and k (KEL2)

K has a frequency of about 9% in Caucasians, but is much less common in Africans and extremely rare in Eastern Asia and in native Americans. K achieves its highest frequency among people of the Arabian and Sinai peninsulas, where up to 25% may be K+. The k antigen has a high incidence in all populations; only about one in 500 Caucasians is K+ k−.

The K/k polymorphism results from a T698C transition within exon 6 of the *KEL* gene, which gives rise to Met193 in K and Thr193 in k [10]. This affects glycosylation of the molecule: in k, Asn-Arg-Thr193 is a consensus sequence for N-glycosylation of Asn191, whereas Asn-Arg-Met193 in K is not.

Four unrelated Japanese individuals with very weak K, no k, and weak expression of high frequency Kell-system antigens were homozygous for a *KEL* mutation encoding Thr193Arg [24]. Like Met193, which is usually associated with K, Arg193 would not support N-glycosylation of Asn191. Weakness of other Kell antigens was due to reduced quantity of Kell glycoprotein.

A Swiss-German K+ k+ blood donor with some degree of weakening of K, but normal expression of other Kell system antigens, was heterozygous for a mutation encoding Ser193 and for a normal *k* allele [18]. Like Thr193 of k, Ser193 would be expected to support glycosylation of Asn191, so K expression is surprising. In addition, flow cytometry suggested that the red cells expressed a homozygous dose of k. The significance of this mutation is still to be sorted out.

Kp^a (KEL3), Kp^b (KEL4), and Kp^c (KEL21)

Kp^a has a frequency of about 2% in Caucasians, but is rare in other populations. Kp^c is rare in all populations, but several examples of Kp(c+) phenotype [including Kp(a−b−c+) (Kp^c/Kp^c)] have been found in Japan. One English family with Kp(a−b+c+) (Kp^b/Kp^c) and a South American with Kp(a+b−c+) (Kp^a/Kp^c) phenotypes, have been identified. Kp^b is very common in all populations.

Kp^a and Kp^c alleles differ from the common allele, Kp^b, by single nucleotide changes at adjacent sites within codon 281 in exon 8 [12]. Kp^b has CGG encoding arginine, Kp^a has TGG encoding tryptophan, and Kp^c has CAG encoding glutamine.

Red cells of individuals homozygous for Kp^a, or heterozygous for Kp^a and K_0, have reduced levels of Kell glycoprotein and depressed Kell-system antigens. Expression of cDNA constructs in human embryonic kidney cells showed that the Kp^a mutation causes retention of most of the Kell glycoprotein in a pre-Golgi compartment [25].

Js^a (KEL6) and Js^b (KEL7)

Js^a is almost completely restricted to people of African origin, with an incidence of about 16% among African Americans, giving a frequency of 8% for the Js^a gene. Js^a is very rare in other ethnic groups. Js^b is of high frequency in all populations. The phenotype Js(a+b−) has not been reported in a person of non-African origin.

The Js^a/Js^b polymorphism is associated with two nucleotide changes in exon 17 of *KEL*, one encoding Pro597 in Js^a and Leu597 in Js^b, the other silent with both alleles encoding Leu633 [9]. The Leu597Pro substitution is between two cysteine residues and could affect disulphide bonding and, consequently, folding of the molecule.

Other Kell System Antigens

KEL17 (Wk^a) and KEL11 represent Arg302 and Val302, respectively, in the Kell glycoprotein [12]. KEL17 is present on the red cells of 0.3% of English blood donors, but only 0.1% of K+ donors. KEL11 is an antigen of very high frequency.

With a frequency of 2.6% in Finland, but almost unknown in other European populations, Ul^a (KEL10) is often considered a predominantly Finnish characteristic. However, 0.46% of Japanese and one of 12 Chinese were Ul(a+). An antibody antithetical to anti-Ul^a has not been found. Ul^a results from a Glu494Val substitution in the Kell glycoprotein [12].

KEL14 and KEL24 are antithetical antigens of very high and very low frequency, respectively. Loss of KEL14 and expression of KEL24 is associated with an Arg180Pro substitution [13]. Two unrelated Japanese negative for KEL14 revealed two other mutations: one encoding Arg180His; the other Arg180Cys [23].

KEL12, KEL19, KEL22, TOU (KEL26), and RAZ (KEL27) are all antigens of very high frequency, absence of which is associated with a single amino acid substitution in the Kell glycoprotein [8,15] (Table 1). Absence of the high frequency antigen KEL18 resulted from Arg130Trp and Arg130Gln substitutions in different KEL:-18 individuals [8]. Expression of the private antigens KEL23 and VLAN (KEL25) is also associated with single amino acid substitutions [8,15] (Table 1).

Spatial Arrangement of Kell Antigens on the Kell Glycoprotein

The Kell glycoprotein is highly folded and the shape of its large extracellular domain is unknown. Kell-system antigens are destroyed by disulphide-bond reducing agents and so must be dependent on the native conformation of the molecule. Some may also be discontinuous; that is, dependent on two linearly discrete regions of the polypeptide chain that come to proximity due to folding of the protein.

Competitive antibody-binding assays have been used in an attempt to identify clusters of antigen activity [3,17]. Very little correlation between spatial positioning as determined by competitive binding and linear positioning is obvious.

MCLEOD PHENOTYPE AND THE KX (XK1) ANTIGEN

In the McLeod phenotype, one of the characteristics of McLeod syndrome, high frequency Kell antigens are expressed weakly, the degree of depression of these antigens varying in different individuals. In some cases McLeod phenotype was initially mistaken for K_0. K is also weakly expressed when present. McLeod phenotype red cells lack the Kx and Km (KEL20) red cell antigens. When immunised by blood transfusion, individuals with McLeod phenotype may produce anti-Kx or anti-Km, but usually produce both. McLeod phenotype is very rare and no frequency estimate has been published. Two unrelated men with the McLeod phenotype were found by screening many thousands of donors from south-east England [22].

Anti-Kx recognizes the Kx protein, a polytopic polypeptide with 10 membrane-spanning domains. Kx is produced by *XK*, an X-linked gene at Xp21.1. In the red cell membrane, Kx protein is linked to the Kell glycoprotein through a single disulphide bond: Cys347 of the Kx protein to Cys72 of the Kell glycoprotein [19]. Because Kx is not produced by the *KEL* gene it does not belong to the Kell system and is the only antigen of the Kx system (XK1). Kx protein is present in reduced quantity on K_0 cells, yet Kx antigen is expressed strongly on K_0 cells, possibly because the absence of the Kell glycoprotein enhances binding of anti-Kx [1,16]. Like Kx, Km is absent from McLeod phenotype cells, but unlike Kx, is also absent from K_0 cells. Km is probably a discontinuous antigen, the product of interaction between Kell glycoprotein and Kx protein. Although Km has a Kell system number (KEL20), it could equally belong to the Kx system.

McLeod syndrome has an X-linked mode of inheritance and results from the absence of Kx protein, due to hemizygosity for deletions or inactivating mutations in *XK* [2,6,20,21]. These include the following:

- deletion of whole gene;
- deletion of exons 1, 2, or 3;
- Trp36Stop (exon 1);
- T deletion, codon 90 (exon 2);
- Arg133Stop (exon 2);
- Gln145Stop (exon 2);
- C insertion, codon 151 (exon 2);
- Arg222Gly (exon 3);
- TT deletion, codon 229 (exon 3);
- Trp236Stop (exon 3);
- G deletion, codon 257 (exon 3);
- CTCTA deletion, codon 285 (exon 3);
- Cys294Arg (exon 3);
- Gln299Stop (exon 3);
- 14 nucleotide deletion, codon 313 (exon 3);
- Trp314Stop (exon 3);
- T deletion, codon 338 (exon 3);
- g → c, intron 1, 5′ splice site;
- g → a, intron 2, 5′ splice site;
- g → a, at nucleotide 5, intron 2, 5′ splice site;
- g → a, intron 2, 3′ splice site.

Deletion of the whole gene often encompasses other genes, often resulting in chronic granulomatous disease, caused by deletion of *CYBB*.

Female carriers of *XK* mutations have two populations of red cells, one with normal Kell phenotype and one with McLeod phenotype. The proportion of McLeod phenotype red cells in female McLeod carriers usually varies from 5% to 85% and individual variation occurs within a family (Figure 2).

A mutation in the fifth nucleotide of the 5′ donor splice site of intron 2, a nucleotide that is 82% conserved in the splice consensus sequence, was found in a man with almost no Kell antigens on his red cells [5]. The mutation would be expected to cause some degree of abnormal splicing, but he did not have neuroacanthocytosis or muscle defects, possibly because there was some degree of normal *XK* RNA splicing. The extreme reduction in Kell antigen was attributed to the combined effects of Kx deficiency and homozygosity for a Kp^a allele of *KEL*, which inhibits trafficking of the Kell glycoprotein to the cell surface.

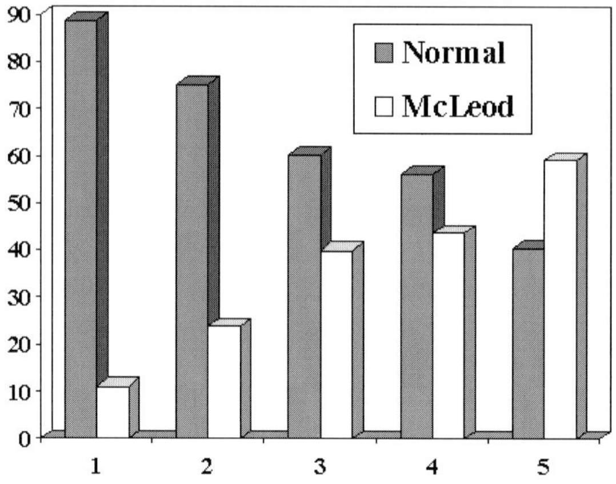

Figure 2. Histogram showing proportions of McLeod phenotype and normal Kell phenotype red cells in five heterozygous XK^0/XK women from the same family (Jung HH, Green C, Daniels G, unpublished observations, 2002).

OTHER DEPRESSED KELL PHENOTYPES

Gerbich-Negative Phenotypes

Another depressed Kell phenotype resulting from inheritance of a rare gene at a locus independent of *KEL* is that accompanying some Gerbich-negative phenotypes. Gerbich antigens are present on glycophorins C and D and encoded by a single gene on chromosome 2, *GYPC*. Red cells lacking Ge2 and Ge3 generally show at least some degree of weakening of Kell antigens, whereas those lacking only Ge2 do not. Red cells with the K_0, K_{mod}, and McLeod phenotypes have normal expression of Gerbich antigens. The biochemical nature of the phenotypic association between Gerbich and Kell is not understood, but probably involves the absence of exon 3 of *GYPC*, which codes for Ge3.

K_{mod}

K_{mod} is an umbrella term to describe phenotypes in which Kell antigens are expressed very weakly, often requiring adsorption/elution tests for detection, and in which Kx antigen expression is elevated. K_{mod} cells have reduced quantity of the Kell glycoprotein. Some K_{mod} individuals make an antibody that resembles anti-Ku, but differs from anti-Ku by being non-reactive with K_{mod} cells. K_{mod} is probably inherited as a recessive character, but the precise mode of inheritance has not been confirmed.

REFERENCES

1. Carbonnet F, Hattab C, Collec E, Le Van Kim C, Cartron J-P, Bertrand O (1997) Immunochemical analysis of the Kx protein from human red cells of different Kell phenotypes using antibodies raised against synthetic peptides. *Br J Haemat* **96:** 857-863.
2. Danek A, Rubio JP, Rampoldi L, Ho, M, Dobson-Stone C, Tison F, Symmans WA, Oechsner M, Kalckreuth W, Watt JM, Corbett AJ, Hamdalla HHM, Marshall AG, Sutton I, Dotti MT, Malandrini A, Walker RH, Daniels G, Monaco AP (2001) McLeod neuroacanthocytosis: genotype and phenotype. *Ann Neurol* **50:** 755-764.
3. Daniels G (2002) Section 4: antibodies to other blood group antigens. Coordinator's report. *Transfus Clin Biol* **9:** 75-80.
4. Daniels G (2002) *Human blood groups,* 2nd edn. Blackwell Science, Oxford.
5. Daniels GL, Weinauer F, Stone C, Ho M, Green CA, Jahn-Jochem H, Offner R, Monaco AP (1996) A combination of effects of rare genotypes at the *XK* and *KEL* blood group loci results in absence of Kell system antigens from the red blood cells. *Blood* **88:** 4045-4050.
6. Ho M, Chelly J, Carter N, Danek A, Crocker P, Monaco AP (1994) Isolation of the gene for McLeod syndrome that encodes a novel membrane transport protein. *Cell* **77:** 869-880.
7. Koda Y, Soejima M, Tsuneoka M, Yasumoto K, Higashitani T, Sagawa K, Kimura H (2002) Heterozygosity for two novel null alleles of the *KEL* gene causes the Kell-null phenotype in a Japanese woman. *Br J Haematol* **117:** 220-225.
8. Lee S (1997) Molecular basis of Kell blood group phenotypes. *Vox Sang* **73:** 1-11.
9. Lee S, Wu X, Reid M, Redman C (1995) Molecular basis of the K:6,-7 [Js(a+b−)] phenotype in the Kell blood groups system. *Transfusion* **35:** 822-825.
10. Lee S, Wu X, Reid M, Zelinski T, Redman C (1995) Molecular basis of the Kell (K1) phenotype. *Blood* **85:** 912-916.
11. Lee S, Zambas E, Green ED, Redman C (1995) Organization of the gene encoding the human Kell blood group protein. *Blood* **85:** 1364-1370.
12. Lee S, Wu X, Son S, Naime D, Reid M, Okubo Y, Sistonen P, Redman C (1996) Point mutations characterize *KEL10*, the *KEL3*, *KEL4*, and *KEL21* alleles, and the *KEL17* and *KEL11* alleles. *Transfusion* **36:** 490-494.
13. Lee S, Naime DS, Reid M, Redman C (1997) The *KEL24* and *KEL14* alleles of the Kell blood group system. *Transfusion* **37:** 1035-1038.
14. Lee S, Russo D, Redman C (2000) The Kell blood group system: Kell and XK membrane proteins. *Semin Hemat* **37:** 113-121.
15. Lee S, Reid ME, Redman CM (2001) Point mutations in KEL exon 8 determine a high-incidence (RAZ) and a low-incidence (KEL25, VLAN) antigen of the Kell blood group system. *Vox Sang* **81:** 259-263.
16. Lee S, Russo DCW, Reiner AP, Lee JH, Sy MY, Telen MJ, Judd WJ, Simon P, Rodrigues MJ, Chabert T, Poole J, Jovanovic-Srzentic S, Levene C, Yahalom V, Redman CM (2001) Molecular defects underlying the Kell null phenotype. *J Biol Chem* **276:** 27281-27289.
17. Petty AC, Green CA, Daniels GL (1997) The monoclonal antibody-specific immobilisation of erythrocyte antigens assay (MAIEA) in the investigation of human red cell antigens and their associated membrane proteins. *Transfus Med* **7:** 179-188.
18. Poole J, Daniels G, Hustinx H, Martin P, Green CA, Warke N, Bromilow I (2001) A novel *K* allele of the *KEL* gene. *Transfusion* **21:** 15S.
19. Russo D, Redman C, Lee S (1998) Association of XK and Kell blood group proteins. *J Biol Chem* **273:** 13950-13956.
20. Russo DCW, Lee S, Reid ME, Redman CM (2002) Point mutations causing the McLeod phenotype. *Transfusion* **42:** 287-293.
21. Singleton BK, Green CA, Renaud S, Fuhr P, Poole J, Daniels GL (2003) McLeod syndrome resulting from a novel *XK* mutation. *Br J Haematol* **122:** 682-685.

22. Swash M, Schwartz MS, Carter ND, Heath R, Leak M, Rogers KL (1983) Benign X-linked myopathy with acanthocytes (McLeod syndrome). Its relationship to X-linked muscular dystrophy. *Brain* **106:** 717-733.
23. Uchikawa M, Onodera T, Tsuneyama H, Nakajima K, Tadokoro K, Juji T, Ishijima A, Murata S (1999) Different point mutations in the same codon of KEL:-14 phenotype. *Transfusion* **39:** 50S.
24. Uchikawa M, Onodera T, Tsuneyama H, Enomoto T, Yuasa S, Nakajima K (2000) Molecular basis of unusual K_{mod} phenotype with $K+^{w}k-$. *Vox Sang* **78(suppl 1):** O011.
25. Yazdanbakhsh K, Lee S, Lu Q, Reid ME (1999) Identification in a defect in the intracellular trafficking of a Kell blood group variant. *Blood* **94:** 310-318.
26. Yu L-C, Twu Y-C, Chang C-Y, Lin M (2001) Molecular basis of the Kell-null phenotype. A mutation at the splice site of human *KEL* gene abolishes the expression of Kell blood group antigens. *J Biol Chem* **276:** 10247-10252.

CHAPTER 21

C. ELEGANS AS A DISEASE MODEL FOR NEUROACANTHOCYTOSIS

KELVIN WONG AND MICHAEL HENGARTNER

Institute of Molecular Biology, University of Zürich, Switzerland

Abstract. The core machinery responsible for the execution of the apoptotic program is conserved from invertebrates to humans [14,26,27]. The *Caenorhabditis elegans ced-8* gene was identified in a screen for mutants defective in *c*ell *d*eath (*ced*) [4]. Loss of CED-8 function results in delayed appearance and removal of cell corpses. Mutations in *ced-8* weakly protect from apoptotic cell death and enhance cell survival in sensitized genetic backgrounds [23]. The human homologue of CED-8, the XK protein, is thought to be a putative membrane associated transporter [12,16]. Loss of the XK protein is the cause of McLeod syndrome, a form of neuroacanthocytosis [12]. As such, *C. elegans* might be a useful tool in trying to understand the function of the human XK protein.

INTRODUCTION

Programmed cell death (apoptosis) is crucial for elimination of excess and/or damaged cells both during development and throughout adult life [1,15,20]. The universality of apoptosis is borne out by the fact that it has been observed in virtually all metazoan models [15]. Although many components of the signaling pathway and the core apoptotic machinery have been identified [8], the spatiotemporal regulation of programmed cell death needs elucidation.

The nematode *Caenorhabditis elegans* is an excellent model organism to use in the study of programmed cell death. *C. elegans* has a well defined and invariant cell lineage, a completely annotated genome and molecular and biochemical analyses are easily performed on this genetically tractable system. In the hermaphrodite, 131 of the 1090 somatic cells and approximately 300 (out of over 1000) germ cells are eliminated by programmed cell death [6,24,25,30].

A. Danek (ed.), Neuroacanthocytosis Syndromes, 187–195.

Genetic screens have identified *cell death* abnormal (*ced*) mutants that can be largely categorized into two groups. The core apoptotic machinery is encoded for by *ced-3*, *ced-4* and *ced-9* [2,9]. The *ced-3* gene encodes an interleukin converting enzyme-like caspase that acts cell autonomously and is responsible for either activating or inactivating proteins crucial for apoptosis [8,29]. The genes *ced-4* [28] and *ced-9* [9,11] encode proteins that are similar to mammalian Apaf-1 and the Bcl-2 family of proteins, respectively. CED-4 promotes the activation of CED-3 in cells that are destined to die. Conversely, CED-9 inhibits the activity of CED-4 in cells that need to survive.

Recognition and efficient removal of cell corpses requires the function of the genes *ced-1*, *ced-2*, *ced-5*, *ced-6*, *ced-7*, *ced-10* and *ced-12* [2,4,5,7]. Apoptotic cells are normally recognized and engulfed within an hour of their death. In contrast, in animals mutant for any one of these seven genes, many apoptotic cells are not engulfed and remain as persistent cell corpses for up to a few days.

The *ced-8* mutant was originally identified in a screen for engulfment defective mutants [3]. Indeed, *ced-8* mutants have a large number of cell corpses during late embryogenesis, a developmental stage at which all dead cells have already been removed in the wild type. However, subsequent analyses resolved the role of CED-8 protein to be in the regulation of the timing or kinetics of apoptosis rather than in the process of engulfment [23].

CELL CORPSE APPEARANCE IS DELAYED IN *CED-8* MUTANTS

Embryogenesis in *C. elegans* requires 14 hours at 25° C. Cell divisions occur largely in the first half of embryonic development whereas apoptosis peaks in mid-embryogenesis, corresponding to the 1.5 fold stage [25] (Figure 1).

However, in *ced-8* mutant embryos, the onset of cell corpse appearance is delayed and the peak of cell corpses is at the early three-fold stage [23] (Figure 2). The delayed appearance of cell corpses in a *ced-8* mutant was confirmed using genetic analyses with a *ced-7;ced-5* double mutant. In *ced-7; ced-5* double mutants, unengulfed cell corpses could be detected as early as the bean stage of embryogenesis, but the *ced-7;ced-5;ced-8* triple mutant only started to exhibit cell corpses at the early three-fold stage [23]. Additionally, the *ced-7;ced-5;ced-8* triple mutant did not display a higher total number of cell corpses when compared against the *ced-7;ced-5* double mutant [23]. This genetic analysis confirmed that the defect in *ced-8* mutants was specifically a retarded onset of cell corpses. Interestingly, not only do *ced-8* mutants show a delayed appearance of cell corpses, but also a much delayed clearance of these corpses [13] (Figure 2).

Thus, the kinetics of the whole cell death process appear to be slowed down in these mutants. A similar delay in all cell corpse appearance and removal can be observed in mutants harboring hypomorphic *ced-3* alleles (Figure 2), suggesting that CED-8 might act at a similar step in the apoptotic pathway as CED-3 and/or might be required for optimal CED-3 activity.

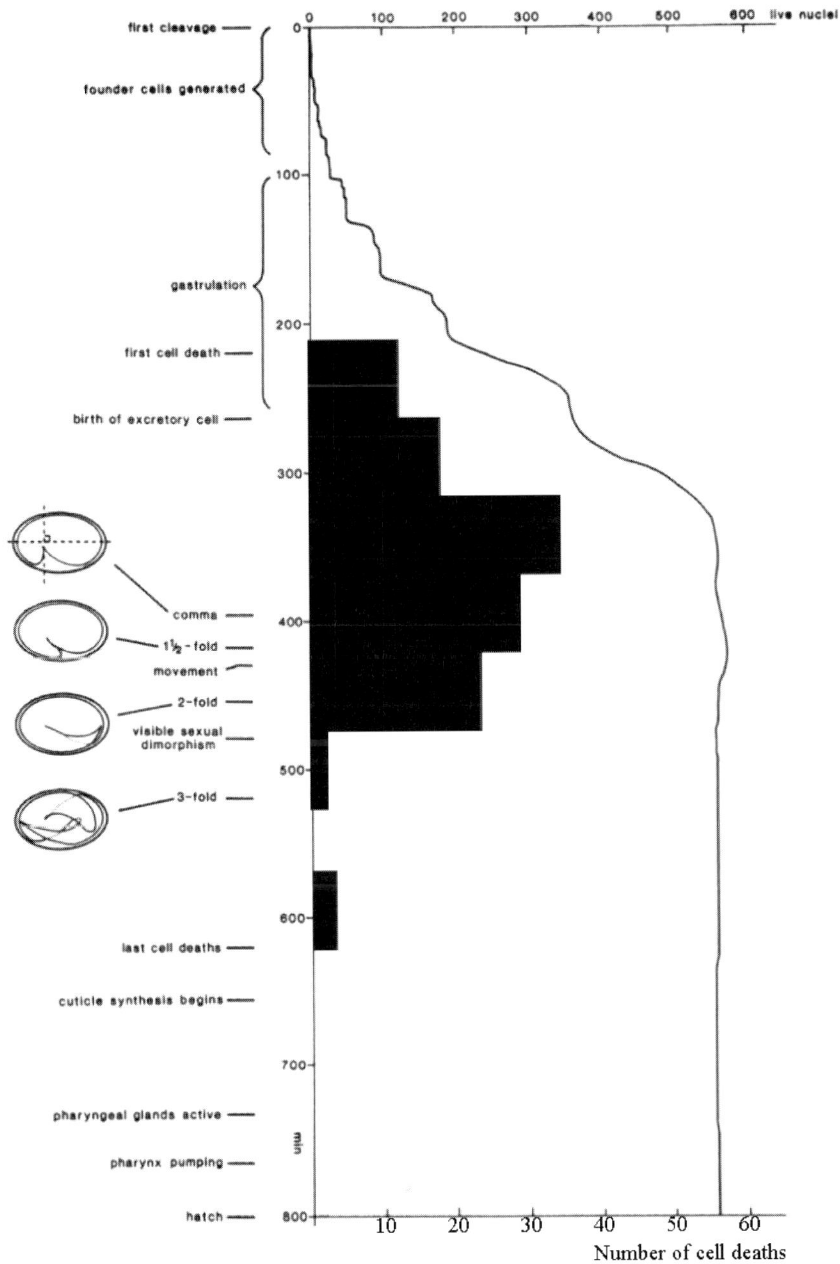

Figure 1. Cell proliferation and cell death during *C. elegans* embryonic development. Adapted from *Liu and Hengartner 1999* [19] and *Sulston et al 1983* [25].

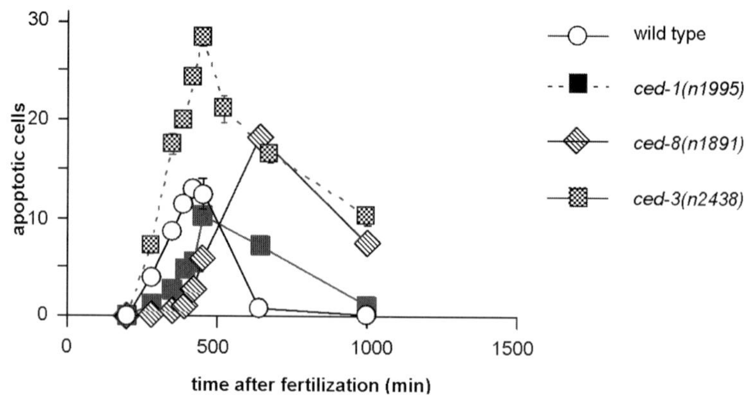

Figure 2. Delayed kinetics of apoptosis in *ced-8* and hypomorphic *ced-3* mutants. The number of cell corpses was determined at various embryonic development stages by Nomarski optics. Adapted from *Hoeppner et al 2001* [13].

DELAYED CELL DEATH KINETICS IN *CED-8* MUTANTS INCREASES CELL SURVIVABILITY

ced-8 mutants do not exhibit a strong increase in the number of ectopic cells, indicating that *ced-8* is not essential for cell death in *C. elegans* [23]. However, the similarity in the kinetics of cell death in *ced-8* and weak *ced-3* mutants raised the possibility that *ced-8* could have a weak pro-apoptotic activity that might be more readily observed in a sensitized genetic background. To test this possibility, a *ced-8* mutation was introduced into either a *ced-3* or *ced-4* background. While strong loss-of-function alleles of *ced-3* and *ced-4* allow the survival of all cells programmed to die, hypomorphic alleles allow the stochastic survival of a subset of these cells [10,22]. The *ced-8* mutation did not increase the total number of ectopically surviving cells in strong *ced-3* and *ced-4* loss-of-function alleles. However, the delayed kinetics of cell death did increase the total number of surviving cells in the background of weak *ced-3* and *ced-4* alleles [23]. Thus delaying the kinetics of cell death and/or engulfment in a cell that is slightly defective for apoptosis might give the cell the extra time it needs to enhance its chances of survival. Interestingly, a similar increase in survival has been observed in double mutants between hypomorphic *ced-3* alleles and mutations in any one of the seven genes that mediate the removal of apoptotic cells [13,21]. However, *ced-8* and the engulfment machinery appear to enhance survival in weak *ced-3* mutants through distinct mechanisms, as they can additively increase survival in a *ced-1*;*ced-3*;*ced-8* triple mutant.

The *ced-8* loss-of-function mutation could theoretically slow down the kinetics of the cell death process by either reducing the rate at which apoptosis occurs or by initiating the apoptotic program at a later stage of embryogenesis. As such, *in situ* TUNEL was performed as a more sensitive molecular means to judge the timing of onset of apoptosis. When a cell enters into the apoptotic program, it very rapidly initiates DNA degradation. This can be visualized by *T*dT-mediated d*U*TP *n*ick-*e*nd *l*abeling (TUNEL). Using TUNEL staining as a marker for cell death, Stanfield and Horvitz [23] determined that *ced-8* mutants initiated apoptosis only after a slight delay as compared to wild-type embryos. During embryogenesis, there were many more TUNEL positive cells in a *ced-8* mutant as compared to the number of observable cell corpses in a *ced-5;ced-7;ced-8* triple mutant. Thus it can be concluded that even though there is a slight delay in initiation of apoptosis in *ced-8* mutants, the major effect of the loss of CED-8 function is to slow down the progress of the apoptotic program. This conclusion is also supported by the slowed-down kinetics (Figure 2) of cell corpse removal in the *ced-8* mutants [23].

CED-8 FUNCTIONS DOWNSTREAM OF CED-9

Strong *ced-9* loss-of-function alleles result in maternal-effect lethality and sterility. Hypomorphic mutation in *ced-3* suppresses the *ced-9* loss-of-function defect, leading to the survival of some ectopic cells. When the *ced-8* mutation was introduced into the *ced-3;ced-9* double mutant, an even greater number of cells survived. Thus, CED-8 can be placed downstream of CED-9 function as it increases cell survival rates in the absence of CED-9.

CED-8 IS HOMOLOGOUS TO HUMAN XK PROTEIN

The *ced-8* gene encodes a 458 amino acid protein that bears significant similarity to the human XK and *Mus musculus* XK proteins as well as proteins from *Drosophila melanogaster*, *Ciona intestinalis* and *Anopheles gambiae* (Figure 3). These proteins contain 10 predicted transmembrane spanning regions. The human XK protein is predicted to be a membrane associated transporter [12]. In agreement with this, the CED-8::GFP fusion protein localizes to the plasma membrane of the cell.

Although the overall identity and similarity between CED-8 and the human XK protein is only about 19% and 37% respectively, there are two conserved stretches of amino acid residues (CED-8 amino acids 188-203 and 370-382) between CED-8 and all the proteins used for the multiple alignment (Human XK, I39284; *Mus musculus* XK, NP_075989; *Ciona intestinalis* COS41.5; *A. gambiae* EAA06621 and *Drosophila melanogaster* AAF57791). These regions might represent functionally important domains that are required for proper function of CED-8 or XK protein.

HUMAN XK PROTEIN AND MCLEOD SYNDROME

McLeod syndrome is a condition that is characterized by late onset neurodegeneration and erythrocyte acanthocytosis. Neuroimaging of McLeod patients

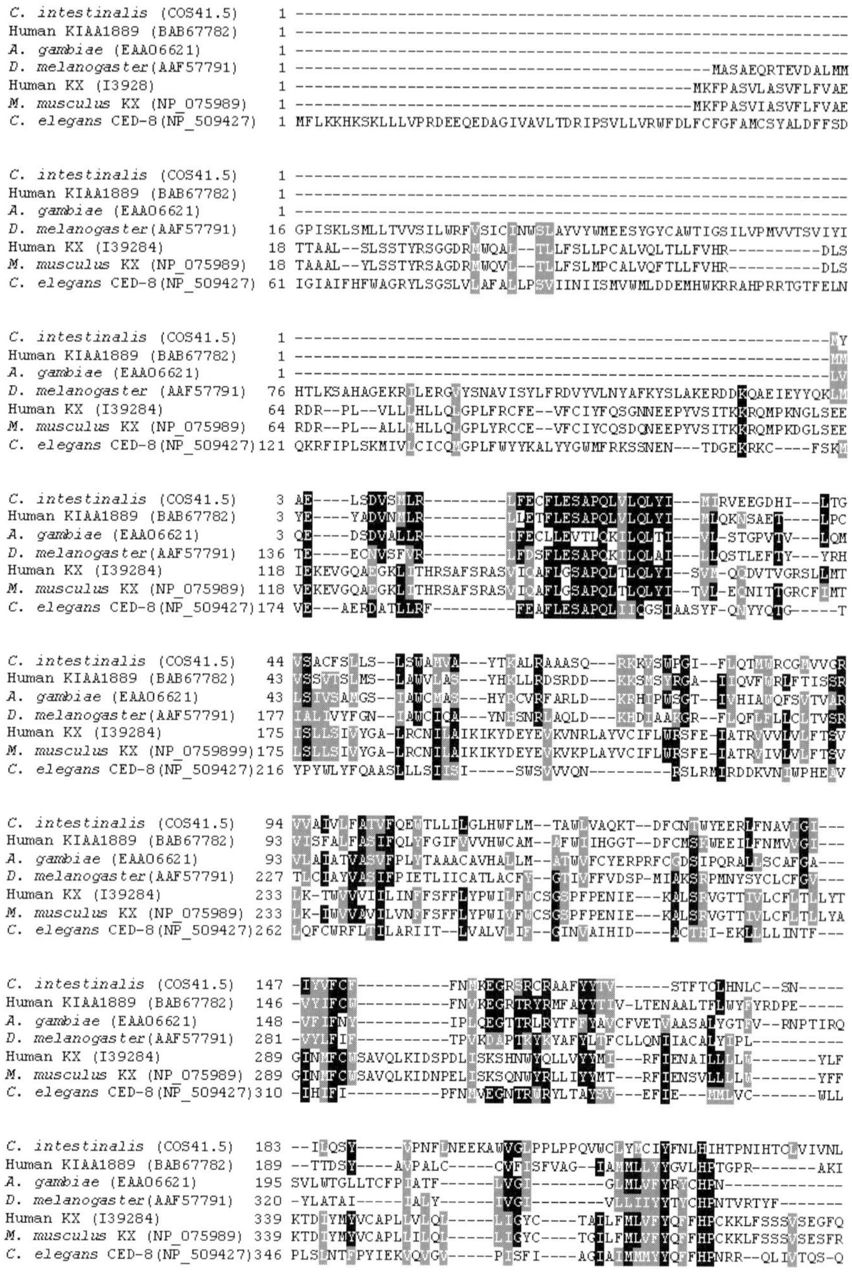

Figure 3. Multiple protein sequence alignment of *C. elegans* CED-8 (NP_509427), Human XK (I39284), *Mus musculus* XK (NP_075989), *Ciona intestinalis* COS41.5, *A. gambiae* EAA06621 and *Drosophila melanogaster* AAF57791 proteins. Identical amino acids shaded in black and conserved residues shaded in grey. Multiple protein sequence alignment was generated using the Pattern-Induced Multiple-sequence Alignment and Boxshade programs.

shows a similarity to early onset Huntington's disease with caudate nucleus atrophy. The human XK protein is known to be physically associated with the Kell glycoprotein by a single disulphide bond. Thus the absence of the XK protein results in the concomitant absence of the Kell antigen group. The Kell type II glycoprotein is predicted to be a member of the Neprilysin family of zinc endopeptidases that preferentially cleaves big endothelin-3 at Trp21-Ile22, creating bioactive endothelin-3. Endothelins can act as vasoconstrictors and mitogens and are involved in migration of neural crest derived cells. However it is interesting to note that while there is a pathology associated with the loss of the XK protein, there is no record of disease resulting from the loss of the Kell blood group antigen alone [17]. Since XK protein is thought to be a membrane transport protein, it might be plausible that it transports a factor(s) either into or out of the cell and the loss of this factor(s) is actually the cause of McLeod syndrome.

THE WORM AS A MODEL FOR NEUROACANTHOCYTOSIS

The loss of the XK protein in humans results in 15% of erythrocytes becoming acanthocytes, having finger like projections. The underlying cause of this might be due to dysfunctional membrane dynamics or deregulation of the erythrocyte cytoskeleton. The *ced-8* mutation in *C. elegans* caused a defect in the kinetics of apoptosis although there was no report of gross alteration of cell shape and/or cytoskeletal abnormalities. It would be interesting to know if the delay in the manifestation of cell corpse refractibilty in *ced-8* mutants was actually due to deregulation of membrane composition or cytoskeletal changes.

The deregulation of the rate of apoptosis in *ced-8* mutants might prove to be an interesting insight into the cause of neurodegeneration in McLeod syndrome patients. Although mutant, the XK protein is still expressed in certain McLeod patients. In fact there is high expression in the brain that correlates well with the neurodegeneration phenotype. It is still unknown if the XK protein is involved in the transport of factors out of the cell or the uptake of factors from the extracellular surroundings. However, the deregulation presumably affects the survivability of the neurons that express the XK protein. The pathology in McLeod patients might result from certain neurons that are required to undergo apoptosis doing so at a much reduced rate. Although there is data showing that the XK protein is involved in McLeod syndrome, there are also many outstanding questions to be answered.

It is quite possible that the cellular function of the CED-8 and XK proteins are evolutionarily conserved. This is reflected by the presence of homologues from worms to humans. In this aspect, the *C. elegans* might prove useful in the search for protein binding partners of CED-8, the cargo that it transports or other mutants that display genetic interaction with the *ced-8* mutant. Data garnered from such research might help in identifying useful leads to better understand human XK protein function.

Additionally, it would be interesting to analyze if the *ced-8* mutants display any neural defects. If the *ced-8* mutant had any defects in neural apoptosis and/or migration, it might serve as a bioassay in the search for the XK protein cargo or other possible therapeutic drugs that might lead to the cure of McLeod syndrome.

REFERENCES

1. Clarke PG, Clarke S (1996) Nineteenth century research on naturally occurring cell death and related phenomena. *Anat Embryol (Berl)* **193:** 81-99.
2. Ellis HM, Horvitz HR (1986) Genetic control of programmed cell death in the nematode *C. elegans*. *Cell* **44:** 817-829.
3. Ellis RE, Horvitz HR (1991) Two *C. elegans* genes control the programmed deaths of specific cells in the pharynx. *Development* **112:** 591-603.
4. Ellis RE, Jacobson DM, Horvitz HR (1991) Genes required for the engulfment of cell corpses during programmed cell death in *Caenorhabditis elegans*. *Genetics* **129:** 79-94.
5. Gumienny TL, Brugnera E, Tosello-Trampont AC, Kinchen JM, Haney LB, Nishiwaki K, Walk SF, Nemergut ME, Macara IG, Francis R, Schedl T, Qin Y, Van Aelst L, Hengartner MO, Ravichandran KS (2001) Ced-12/elmo, a novel member of the crkii/dock180/rac pathway, is required for phagocytosis and cell migration. *Cell* **107:** 27-41.
6. Gumienny TL, Lambie E, Hartwieg E, Horvitz HR, Hengartner MO (1999) Genetic control of programmed cell death in the *Caenorhabditis elegans* hermaphrodite germline. *Development* **126:** 1011-1022.
7. Hedgecock EM, Sulston JE, Thomson JN (1983) Mutations affecting programmed cell deaths in the nematode *Caenorhabditis elegans*. *Science* **220:** 1277-1279.
8. Hengartner MO (2000) The biochemistry of apoptosis. *Nature* **407:** 770-776.
9. Hengartner MO, Ellis RE, Horvitz HR (1992) *C. elegans* gene *ced-9* protects cells from programmed cell death. *Nature* **356:** 494-499.
10. Hengartner MO, Horvitz HR (1994) Activation of *C. elegans* cell death protein CED-9 by an amino-acid substitution in a domain conserved in *Bcl-2*. *Nature* **369:** 318-320.
11. Hengartner MO, Horvitz HR (1994) *C. elegans* cell death gene *ced-9* encodes a functional homolog of mammalian proto-oncogene *bcl-2*. *Cell* **76:** 665-676.
12. Ho M, Chelly J, Carter N, Danek A, Crocker P, Monaco, AP (1994) Isolation of the gene for McLeod syndrome that encodes a novel membrane transport protein. *Cell* **77:** 869-880.
13. Hoeppner DJ, Hengartner MO, Schnabel R (2001) Engulfment genes cooperate with *ced-3* to promote cell death in *Caenorhabditis elegans*. *Nature* **412:** 202-206.
14. Horvitz HR (1999) Genetic control of programmed cell death in the nematode *Caenorhabditis elegans*. *Cancer Res* **59:** 1701s-1706s.
15. Jacobson MD, Weil M, Raff MC (1997) Programmed cell death in animal development. *Cell* **88:** 347-354.
16. Khamlichi S, Bailly P, Blanchard D, Goossens D, Cartron JP, Bertrand O (1995) Purification and partial characterization of the erythrocyte Kx protein deficient in McLeod patients. *Eur J Biochem* **228:** 931-934.
17. Lee S, Russo DC, Reiner AP, Lee JH, Sy MY, Telen MJ, Judd WJ, Simon P, Rodrigues MJ, Chabert T, Poole J, Jovanovic-Srzentic S, Levene C, Yahalom V, Redman CM (2001) Molecular defects underlying the Kell null phenotype. *J Biol Chem* **276:** 27281-27289.
18. Lee S, Zambas ED, Marsh WL, Redman CM (1993) The human Kell blood group gene maps to chromosome 7q33 and its expression is restricted to erythroid cells. *Blood* **81:** 2804-2809.

19. Liu QA, Hengartner MO (1999) Human CED-6 encodes a functional homologue of the *Caenorhabditis elegans* engulfment protein CED-6. *Curr Biol* **9:** 1347-1350.
20. Liu QA, Hengartner MO (1999) The molecular mechanism of programmed cell death in *C. elegans. Ann N Y Acad Sci* **887:** 92-104.
21. Reddien PW, Cameron S, Horvitz HR (2001) Phagocytosis promotes programmed cell death in *C. elegans. Nature* **412:** 198-202.
22. Shaham S, Reddien PW, Davies B, Horvitz HR (1999) Mutational analysis of the *Caenorhabditis elegans* cell-death gene *ced-3. Genetics* **153:** 1655-1671.
23. Stanfield GM, Horvitz HR (2000) The *ced-8* gene controls the timing of programmed cell deaths in *C. elegans. Mol Cell* **5:** 423-433.
24. Sulston JE, Horvitz HR (1977) Post-embryonic cell lineages of the nematode, *Caenorhabditis elegans. Dev Biol* **56:** 110-156.
25. Sulston JE, Schierenberg E, White JG, Thomson JN (1983) The embryonic cell lineage of the nematode *Caenorhabditis elegans. Dev Biol* **100:** 64-119.
26. Vaux DL, Korsmeyer SJ (1999) Cell death in development. *Cell* **96:** 245-254.
27. White K, Grether ME, Abrams JA, Young L, Farrel K, Steller H (1994) Genetic control of programmed cell death in *Drosophila. Science* **264:** 677-683.
28. Yuan J, Horvitz HR (1992) The *Caenorhabditis elegans* cell death gene *ced-4* encodes a novel protein and is expressed during the period of extensive programmed cell death. *Development* **116:** 309-320.
29. Yuan J, Shaham S, Ledoux S, Ellis HM, Horvitz HR (1993) The *C. elegans* cell death gene *ced-3* encodes a protein similar to mammalian interleukin-1beta converting enzyme. *Cell* **75:** 641-652.
30. Zhou Z, Hartwieg E, Horvitz HR (2001) CED-1 is a transmembrane receptor that mediates cell corpse engulfment in *C. elegans. Cell* **104:** 43-56.

CHAPTER 22

THE KELL BLOOD GROUP PROTEIN, ITS RELATION TO XK AND ITS FUNCTION AS AN ENDOTHELIN-3-CONVERTING ENZYME

Colvin M. Redman, David C.W. Russo, Jeffrey J. Pu, and Soohee Lee

Lindsley F Kimball Research Institute of the New York Blood Center, New York, NY, USA

Abstract. Kell is a 93kDa type II membrane glycoprotein that exists in over 25 different polymorphic forms and, because various forms are immunogenic, it is considered to be an important blood group protein. On red cells, Kell protein is linked by a single disulfide bond to another protein, XK that is lacking in the McLeod phenotype. Kell and XK are preferentially expressed in erythroid tissue but are also present, in lesser amount, in a large number of other tissues including testis, brain, and skeletal muscle. Kell protein is an endothelin converting enzyme that preferentially activates endothelin-3 while XK has the structural characteristics of a membrane transporter.

INTRODUCTION

The relation between the Kell blood group protein and Kx, the antigen that is lacking on red cells of individuals exhibiting the McLeod syndrome, was first demonstrated by early serological studies. McLeod red cells are characterized by marked depression of all Kell antigens and complete absence of a single antigen termed Kx. Later biochemical studies showed that two distinct red cell membrane proteins, Kell and XK, are responsible for expressing Kx and the other Kell antigens and these two proteins are linked by a single disulfide bond. Kell is a 93kDa, highly polymorphic, membrane glycoprotein that expresses over twenty-three different antigens and XK is a 50.9kDa, hydrophobic protein that spans the membrane ten times and contributes the Kx blood group antigen [15,16].

A. Danek (ed.), Neuroacanthocytosis Syndromes, 197–203.

This review, while briefly describing the biochemistry and molecular biology of the Kell blood group system, will emphasize the role of Kell protein as an endothelin-3-converting enzyme and its relation to XK, whose absence is associated with the McLeod phenotype. More detailed discussions of the Kell blood group system are available in other recent reviews [9,10].

THE *KEL* GENE

A single gene expresses Kell membrane glycoprotein. In humans, *KEL* is located in chromosome 7q33 and in mice *Kel* is in proximal chromosome 6. Both in mice and humans the gene is organized into 19 exons with the transmembrane domain in exon 3 and the zinc-binding, enzymatic active site (H.E.L.L.H) in exon 16 [11,12]. The correct initiation methionine that signals the start of expression of human Kell protein has not been experimentally determined, since isolation of Kell protein from red cells yielded a blocked N-terminal amino acid and neither of the 2 methionine residues, encoded at the 5′ end of human Kell cDNA is preceded by a traditional Kozak sequence. However based on homology with mouse *Kel*, in which there is no possibility for initiating expression at the first methionine present in exon 1, human Kell protein may have a short 28-amino acid intracellular domain, rather than the larger, originally predicted, 47-amino acid segment. Further biochemical analysis of the N-terminal domain of human Kell protein is required to determine the correct N-terminal amino acid residue. However for convenience of numbering we continue to use the first possible initiation methionine in human Kell cDNA as the N-terminal amino acid.

KELL PROTEIN

Kell is a type II membrane glycoprotein with a single transmembrane region and a large, 665 amino acid extracellular domain that contains 15 cysteine residues [13]. As will be discussed later Cys72 forms a disulfide bond with XK protein and the other cysteine residues presumably assist in maintaining intramolecular disulfide bonds that stabilize a folded structure. The extracellular domain contains a zinc-binding consensus sequence that is common to all zinc endopeptidases at amino acid residues 581 to 585 and 5 probable N-linked sugar moieties at asparagines residues 94, 115, 191, 345 and 627. Kell has several putative phosphorylation sites and is phosphorylated *in vitro* by casein kinase I and II [2]. However, it is not known if *in vivo* phosphorylation occurs or plays a significant physiological role. The C-terminus of Kell has a consensus sequence, CXXX, that is a possible isoprenylation site but based on analogy with studies on endothelin converting enzyme, this cysteine residue is not likely to be derivatized [14].

A prominent feature of Kell protein is that it is a highly polymorphic protein and alloantibodies are commonly produced when unmatched blood is transfused. Over 25 different Kell phenotypes have been described and the molecular basis of most of them has been determined [7]. In most cases single base mutations in the coding region lead to amino acid substitutions which presumably present different epitopes. As will be discussed later, Kell is homologous with the M13 family of zinc endopeptidases and the majority of amino acid substitutions occur in a domain which is least conserved in the M13 family of proteins. One exception is the KEL6 or Js^a phenotype that is more prevalent in persons of African heritage than in Caucasians. The KEL 6 phenotype involves a Leu597Pro substitution close to the enzyme active site located at 581-585. However KEL 6 red cells and recombinant Kell protein with the KEL6 mutation have similar endothelin converting enzyme activity as wild-type Kell protein (our unpublished studies).

The crystal structure of Kell protein has not been determined but based on homology with neprilysin, which has been crystallized, Kell is likely to be similar with two multiple linked extracellular globular domains [1,17].

HOMOLOGY WITH THE M13 FAMILY OF ZINC ENDOPEPTIDASES

The neprilysin or M13 family of zinc endopeptidases are type II integral membrane glycoproteins whose physiological functions are to activate, inactivate and degrade a number of bioactive peptides. In addition to Kell, other related mammalian zinc endopeptidases in the M13 family are neprilysin, the endothelin converting enzymes (ECE-1 and ECE-2), PEX and XCE [23]. Kell has 32 to 36% amino acid identity with neprilysin and ECE-1 in a C-terminal domain (residues 550-732 of Kell) that contains the zinc binding enzyme active site. The M13 proteins also share 10 conserved cysteine residues indicating structural similarities.

The physiological roles of the M13 family of enzymes are varied. Neprilysin is widely distributed occurring largely in lymphoid progenitor cells, brain, kidney and intestine and is involved in the activation and inactivation of a number of regulatory peptides in the nervous, cardiovascular, immune and inflammatory systems. Its wide substrate specificity includes the chemotactic and natriuretic peptides, tachykinins, and enkephalins [22]. The two endothelin converting enzymes, ECE-1 and ECE-2, on the other hand, although they also have wide tissue distribution, have limited substrate specificity, preferentially cleaving inactive big-endothelin-1, which is a peptide of 41 amino acids, at Trp21-Val22, yielding a 21 amino acid bioactive peptide, termed endothelin-1. ECE-1 and ECE-2 also activate, to a less extent, endothelin-2 and endothelin-3. ECE-2, which is a misnomer since like ECE-1 it also preferentially activates endothelin-1, has an acidic pH optimum and is thought to have an intracellular rather than a plasma membrane function. ECE-1, like neprilysin, can also cleave bradykinin [5]. PEX, another M13 endopeptidase, is mostly expressed in bone tissue and the recombinant protein degrades parathyroid-derived peptides. Mutations in PEX affect renal phosphate transport and cause X-linked hypophosphatemic rickets [18]. XCE is preferentially expressed in the central nervous system but its physiological role is not yet known [24].

KELL IS AN ENDOTHELIN-3 CONVERTING ENZYME

Our studies with both a recombinant truncated form of Kell protein, lacking the intracellular and transmembrane domains, and with red cells, showed that Kell is an endothelin-3-converting enzyme. Kell preferentially and specifically cleaves big endothelin-3 at Trp21-Ile22 producing bioactive endothelin-3. To a less extent Kell also cleaves big endothelin-1 and big endothelin-2 at Trp21-Val22 yielding endothelin-1 and endothelin-2 [8]. Thus there is marked difference between Kell and ECE-1 in that ECE-1 mostly activates endothelin-1 while Kell preferentially activates endothelin-3. There is however overlap. The Km value of recombinant Kell with big endothelin-3 as substrate is $0.33\,\mu$M while the Km for big endothelin-1 and big endothelin-3 are 60 to 130 times higher. An enzyme that preferentially activates endothelin-2 has not been described.

In contrast to ECE-1 that has a neutral pH optimum, both recombinant Kell and wild-type red cells have an acidic pH optimum, pH $5.5 - 6.0$, that resembles ECE-2. This is surprising since an acidic pH optimum is common to enzymes involved in the intracellular processing of secreted proteins or in endocytic mechanisms rather than in the functions of ecto-enzymes. This raises the question as to the role of Kell on red cells. Since K_0 red cells that lack Kell protein have no endothelin converting enzyme activity it appears that Kell is the only red cell protein capable of activating the endothelins (for the Kell null phenotype see also Chapter 20). Therefore, if the endothelin converting enzyme activity of red cells is physiologically relevant it probably acts in segregated environments that are acidic, rather than in the general blood circulation.

ASSOCIATION OF KELL AND XK PROTEINS

Biochemical evidence for the covalent association of Kell and XK proteins was first obtained in red cells by immunoprecipitation of Kell and XK with specific antibodies in non-reduced conditions and the demonstration that XK was co-isolated with Kell. Studies with the rare K_0 and McLeod red cells showed that XK, but not Kell/XK complex, could be isolated from K_0 cells and that the Kell/XK complex is absent in McLeod red cells. These studies indicated that Kell and XK are linked by disulfide bonds. It was also shown that while K_0 cells have increased Kx antigen they parenthetically contain less XK protein than red cells with common Kell phenotype, suggesting that Kell protein may occlude domains on XK that express Kx antigen [3,6]. Taken together with the original observation that McLeod red cells that lack XK have reduced amount of Kell, these data suggest that the Kell/XK complex is stable and that absence of either Kell or XK leads to reduced amount of the companion protein, probably due to degradation.

Combination of gene mutations may also affect the expression of both Kell and XK. Red cells of the KEL3 (Kp^a) phenotype exhibit weak expression of all Kell antigens but express normal levels of Kx antigen. In one reported case, a KEL6 homozygote expressed an unusual diminished amount of Kell and Kx antigens [4]. This rare phenotype was caused by a single base mutation at a donor splice consensus sequence in *XK* intron 2, again indicating that diminished expression of XK leads to a reduced amount of Kell membrane protein.

The association of Kell and XK has been studied in transfected cells programmed to co-express both proteins. These studies support the view that co-expression of Kell and XK is not necessary for placement of the proteins on the plasma membrane. In addition, expression of Kell by itself allows the surface expression of well-defined Kell antigens as detected by specific antibodies suggesting that Kell can travel to the plasma membrane in the absence of XK and can assume proper configuration. However, co-expression of Kell and XK in transfected cells does lead to the formation of a Kell/XK complex and the association commences soon after the proteins are expressed, while the nascent proteins are present in the endoplasmic reticulum [19].

Site directed mutagenesis of selected cysteine residues and co-expression of recombinant Kell and XK by COS cells showed that XK Cys 342, present on the small fifth extracellular loop is linked to Kell Cys 72 [20]. XK Cys 342 is the only extracellular cysteine residue and Kell Cys 72 is located on the extracellular domain close to the transmembrane region. Kell Cys 72 occurs in a cluster of 5 cysteine residues and the other 4 cysteine residues are conserved in the M13 family of zinc endopeptidases. It is of interest that Kell is the only member of the M13 family that is linked to another protein. ECE-1 is linked to itself as a dimer and the other family members are monomers.

EXPRESSION IN NON-ERYTHROID TISSUES

Kell is primarily expressed in erythroid tissues but it is also expressed in near equal amounts in testis. Northern blot analysis and PCR-screening of cDNA from different tissues also detected Kell expression in many tissues including different parts of the brain, skeletal muscle, heart, stomach, lymph node and in a large number of other tissues. Kell/ XK complex is not limited to red cells and this complex was detected in skeletal muscle although the study could not determine the percent of these proteins that are linked or remain unattached [21].

UNRESOLVED ISSUES

It is clear that on red cells Kell and XK exist predominantly as a two-protein complex linked by a disulfide bond and that there is very little free Kell and XK on the membrane. However, this has not been established for non-erythroid tissues. In skeletal muscle, for example, there is a Kell/XK complex but since there is only a small amount of Kell and XK in this tissue a direct Western immunoblot assay could not be performed and the ratio of free to complexed proteins could not be determined. There are indications however that free Kell and XK may occur in this tissue. XK appears to be expressed in greater amount than Kell and immunocytochemistry of skeletal tissue indicates that Kell and XK may be located in separate cellular locations. Similar unequal distributions of Kell and XK may occur in brain and other non-erythroid tissues.

It is not known if Kell is the only enzyme that preferentially activates endothelin-3. Kell and ECE-1 have overlapping enzyme functions in that they both activate endothelin-1, -2 and -3 although Kell prefers endothelin-3 and ECE-1 prefers endothelin-1. This may explain why K_0 persons that lack Kell protein do not have adverse clinical symptoms since ECE-1 may compensate for lack of Kell. On the other hand it is also possible that Kell is not the only endothelin-3 converting enzyme. Another protein with endothelin-3 converting activity has been reported to be present in the iris of the eye. We also do not know if the big-endothelins are the only substrate for Kell.

A critical unresolved issue is whether Kell and XK have complementary functions. Kell's enzyme function is well established but the function of XK is unknown and its role as a membrane transporter is based solely on its structural similarity with other transport proteins. The fact that Kell and XK are covalently linked, at least on red cells, suggests a complementary role but this has not been established.

Acknowledgements

This study was supported by the National Institutes of Health Specialized Center of Research (SCOR) grant in Transfusion Medicine and Biology HL54459.

REFERENCES

1. Bur D, Dale GE, Oefner C (2001) A three-dimensional model of endothelin-converting enzyme (ECE) based on the X-ray structure of neutral endopeptidase 24.11 (NEP). *Protein Eng* **14:** 337-341.
2. Carbonnet F, Hattab C, Cartron JP et al (1998) Kell and Kx, two disulfide-linked proteins of the human erythrocyte membrane are phosphorylated *in vivo*. *Biochem Biophys Res Comm* **247:** 569-575.
3. Carbonnet F, Hattab C, Collec E et al (1997) Immunochemical analysis of the Kx protein from human red cells of different Kell phenotypes using antibodies raised against synthetic peptides. *Brit J Haematol* **96:** 857-863.

4. Daniels GL, Weinauer F, Stone C et al (1996) A combination of the effects of rare genotypes at the *XK* and *KEL* blood group loci results in absence of Kell system antigens from the red blood cells. *Blood* **88:** 4045-4050.
5. Kedzierski RM, Yanagisawa M (2001) Endothelin system: the double-edged sword in health and disease. *Annu Rev Pharmacol Toxicol* **41:** 851-876.
6. Khamlichi S, Bailly P, Blanchard D et al (1995) Purification and partial characterization of the erythrocyte Kx protein deficient in McLeod patients. *Eur J Biochem* **228:** 931-934.
7. Lee S (1997) Molecular basis of Kell blood group phenotypes. *Vox Sang* **73:** 1-11.
8. Lee S, Lin M, Mele A et al (1999) Proteolytic processing of big endothelin-3 by the Kell blood group protein. *Blood* **94:** 1440-1450.
9. Lee S, Russo D, Redman C (2000) Functional and structural aspects of the Kell blood group system. *Transfus Med Rev* **14:** 93-103.
10. Lee S, Russo D, Redman CM (2000) The Kell blood group system: Kell and XK membrane proteins. *Sem Hematol* **37:** 113-121.
11. Lee S, Russo DC, Pu J et al (2000) The mouse Kell blood group gene (*Kel*): cDNA sequence, genomic organization, expression, and enzymatic function. *Immunogenetics* **52:** 53-62.
12. Lee S, Zambas E, Green ED et al (1995) Organization of the gene encoding the human Kell blood group protein. *Blood* **85:** 1364-1370.
13. Lee S, Zambas ED, Marsh WL et al (1991) Molecular cloning and primary structure of Kell blood group protein. *Proc Nat Acad Sci USA* **88:** 6353-6357.
14. MacLeod KJ, Fuller RS, Scholten JD et al (2001) Conserved cysteine and tryptophan residues of the endothelin-converting enzyme-1 CXAW motif are critical for protein maturation and enzyme activity. *J Biol Chem* **276:** 30608-30614.
15. Marsh WL, Redman CM (1987) Recent developments in the Kell blood group system. *Transfusion Med Rev* **1:** 4-20.
16. Marsh WL, Redman CM (1990) The Kell blood group system: a review. *Transfusion* **30:** 158-167.
17. Oefner C, D'Arcy A, Hennig M et al (2000) Structure of human neutral endopeptidase (Neprilysin) complexed with phosphoramidon. *J Mol Biol* **296:** 341-349.
18. Rowe PS, Oudet CL, Francis F et al (1997) Distribution of mutations in the PEX gene in families with X-linked hypophosphataemic rickets (HYP). *Hum Mol Genet* **6:** 539-549.
19. D, Lee S, Redman C (1999) Intracellular assembly of Kell and XK blood group proteins. *Biochim Biophys Acta* **1461:** 10-18.
20. Russo D, Redman C, Lee S (1998) Association of XK and Kell blood group proteins. *J Biol Chem* **273:** 13950-13956.
21. Russo D, Wu X, Redman CM et al (2000) Expression of Kell blood group protein in nonerythroid tissues. *Blood* **96:** 340-346.
22. Turner AJ, Brown CD, Carson JA et al (2000) The neprilysin family in health and disease. *Adv Exp Med Biol* **477:** 229-240.
23. Turner AJ, Tanzawa K (1997) Mammalian membrane metallopeptidases: NEP, ECE, KELL, and PEX. *FASEB J* **11:** 355-364.
24. Valdenaire O, Richards JG, Faull RLM et al (1999) XCE, a new member of the endothelin-converting enzyme and neutral endopeptidase family, is preferentially expressed in the CNS. *Mol Brain Res* **64:** 211-221.

CHAPTER 23

ENDOTHELINS AS BASAL GANGLIA TRANSMITTERS

Maarten van den Buuse
Behavioural Neuroscience Laboratory,
Mental Health Research Institute, Parkville, Melbourne, Australia

Abstract. The role of endothelins in the central nervous system has been relatively little studied and most work has focused on possible ischemic/neurotoxic effects. We provide evidence that endothelins play a physiological role in the striatum, stimulating dopamine release. In addition, other studies suggest that endothelins may have growth factor properties in striatum and other parts of the brain. Loss of endothelin production could therefore have adverse effects on motor control and could induce neurodegeneration.

INTRODUCTION

Endothelins are 21-amino acid peptides that form a distinct family, the most important members of which are endothelin-1, endothelin-2, and endothelin-3. Ever since the first publication on endothelin [29], it has been clear that endothelins are present in the central nervous system. All components of the endothelin system, including gene expression, the endothelin precursors ("big-endothelin"), endothelin processing enzymes (endothelin converting enzyme), and endothelin-immunoreactivity have been demonstrated in different areas of the brain. In addition, endothelins are released by neurons or glial cells and exert physiological effects on target cells in the brain mediated through the endothelin receptors ET_A and ET_B. Recent reviews are available that provide a comprehensive overview of the different elements of endothelin neurotransmission in the brain [22,23]. In this paper, we will review the most recent evidence on effects of endothelin in the basal ganglia and summarize our own findings that suggest that endothelins may be neurotransmitters in this region, particularly acting to stimulate dopamine release.

A. Danek (ed.), Neuroacanthocytosis Syndromes, 205–212.

ENDOTHELINS IN NEUROPATHOLOGY

There has been considerable interest in the possible role of endothelins in stroke and ischemic damage in the basal ganglia [24] and endothelin receptor antagonists may be useful for acute treatment of stroke. In addition, endothelins may be involved in neurodegenerative diseases. For example, endothelin-1 levels were increased in frontal and occipital lobes of post-mortem brains of patients with Alzheimer's disease [17], whereas ET_B receptor expression was reduced [14]. Endothelin-1 converting enzyme (ECE-1) has furthermore been shown to be involved in degradation of beta-amyloid [5]. Surprisingly, another study found that cerebrospinal fluid (CSF) levels of endothelin-1 were reduced in patients with Alzheimer's disease [30].

Little is known about the possible involvement of endothelins in other neurological or psychiatric disorders. Kraus and co-workers compared patients with subarachnoid hemorrhage, head injury and intractable epilepsy and found increased CSF levels of endothelin-3, but not endothelin-1, only in the subarachnoid hemorrhage group [15]. One early study reported significantly reduced CSF levels of endothelin-1 in patients with depressive illness [11]. We obtained preliminary evidence showing reduced levels of ET_B receptors in caudate nucleus of patients with Parkinson's disease [26] (see below).

The observation, that the Kell protein shows ECE-3 properties (for references, see [2] and Chapter 22), has sparked interest in the possible role of endothelins in diseases such as McLeod syndrome in which the XK protein, which binds the Kell protein, is absent. However, the involvement of endothelins in McLeod syndrome is poorly understood, in part because the physiological role of endothelins in the basal ganglia has been relatively little studied.

ENDOTHELINS IN THE BASAL GANGLIA

Several papers have shown endothelin gene expression in human and rat basal ganglia. For example, in human brain, endothelin gene expression was found in striatum and some dopaminergic cells of the substantia nigra [9,20]. We [28] and others [8,20] have shown significant levels of predominantly endothelin-3 in rat striatum. Endothelin-3 was present particularly in acetylcholine- and somatostatin-producing cell bodies [8], rather than in dopaminergic projections [28]. This is in contrast to endothelin receptors in striatum, a significant proportion of which are present on dopaminergic terminals. As shown by membrane binding studies using selective endothelin receptor antagonists, rat striatum contains a homogeneous population of ET_B receptors [21,25]. Prior lesions of the nigrostriatal system by micro-injection of 6-hydroxydopamine caused a marked 53% reduction of the B_{max} of [^{125}I]-endothelin-1 binding with little change in the K_d [25]. Because these lesions are known to cause in excess of 90% depletion of dopamine and ET_B binding was not reduced to the same extent, it is likely that there are two populations of ET_B receptors in striatum, one present on dopaminergic terminals [25] and one present on other neuronal or glial elements [12]. Preliminary autoradiography studies in human brain would support this. In human caudate nucleus, globus pallidus and putamen, binding of [^{125}I]-endothelin-1 could be displaced almost completely by the selective ET_B receptor antagonist BQ-788, suggesting that, as in rat brain, basal ganglia endothelin binding is almost exclusively to ET_B receptors [26]. In patients with Parkinson's disease, specific binding in the caudate nucleus was reduced by 31% [26]. While further studies are needed to confirm and extend this observation, these results would suggest that, as in rat striatum, a significant proportion of ET_B receptors in basal ganglia are present on dopaminergic terminals.

ENDOTHELINS AND BRAIN DOPAMINE ACTIVITY

The presence of endothelin expression and endothelin receptors in striatum raises the question as to the physiological role of these peptides in this part of the brain. Unfortunately, most studies use relatively high doses of endothelins. As we have postulated before [23], the effects of endothelins in striatum are likely a continuum from physiologically relevant neuromodulation at low local concentrations, reversible hyperexcitation and hypermetabolism at higher concentrations, to irreversible neurotoxic damage at high doses. For example, it has been shown that micro-injection of 430 pmol of endothelin-1 into the striatum of rats caused marked vasoconstriction resulting in ischemic infarction of striatal tissue [7,16]. Also micro-injection of 40 pmol of endothelin resulted in reduced local cerebral blood flow [19]. It is therefore possible that ischemia is involved in the effect of higher doses of endothelins on dopaminergic activity, as it has been shown that in hypoxic ischemia, dopamine release is markedly enhanced. However, the presence of ET_B receptors on dopaminergic projections (see above) argues against a non-specific, ischemic mechanism of action of endothelins on dopamine release. Furthermore, we found that much lower doses of endothelins, even below 1 pmol, stimulated dopaminergic activity. This effect was shown both on behavioral consequences of dopamine release, as well as directly on extracellular dopamine concentrations in striatum.

In order to quantify the behavioral effect of low doses of endothelin in the striatum [27], we used rats that received unilateral 6-hydroxydopamine lesions of the nigrostriatal bundle. These rats showed marked contralateral turning behavior when treated with the dopaminergic receptor agonist apomorphine, indicating an imbalance in dopaminergic regulation of locomotor activity. Micro-injection of endothelin-1 into the intact ventral striatum of these rats caused a dose-dependent increase of ipsilateral turning (Figure 1). Such an effect would also be expected with systemic treatment of a dopamine releasing drug, such as amphetamine, and this prompted us to suggest that endothelin-1 was causing dopamine release in the striatum [27]. Indeed, systemic pretreatment with the dopamine D_2 receptor agonist raclopride com-

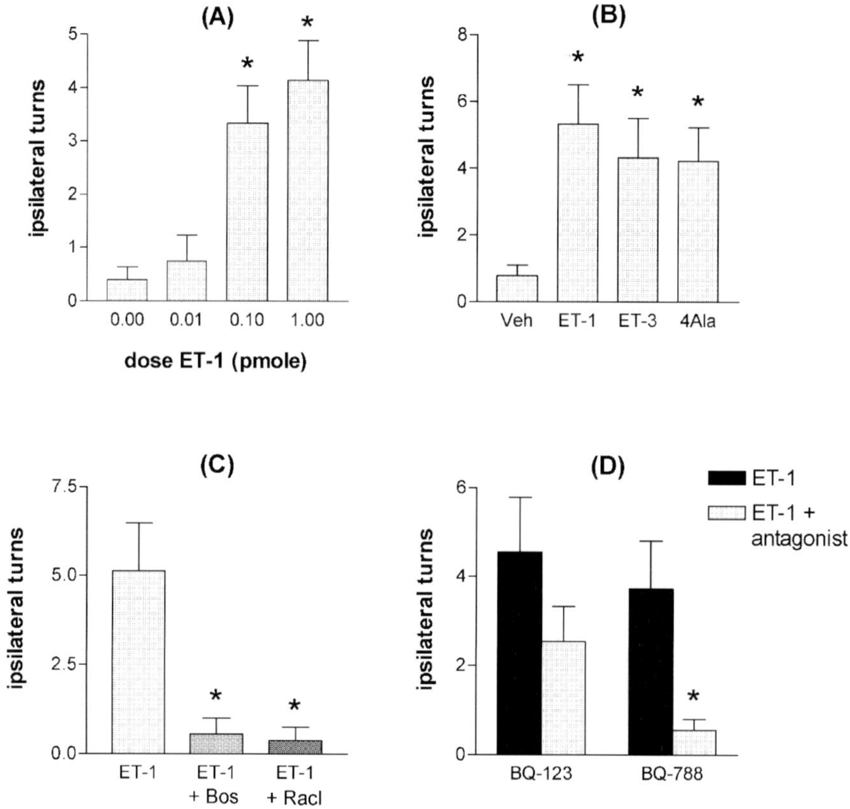

Figure 1. The effect of intrastriatal injection of endothelins on locomotor activity of rats. Behavior was quantified as the number of ipsilateral turns during a 5 min observation period. Injection of endothelin-1 dose-dependently increased the number of turns (panel A). The effect of 1 pmol of endothelin-1 could be mimicked with a similar dose of endothelin-3 and the selective ET_B receptor agonist $[Ala^{1,3,11,15}]$endothelin-1 (4Ala, panel B). Systemic pretreatment with the dopamine D_2 receptor antagonist raclopride (0.1 mg/kg) or the non-selective endothelin receptor antagonist bosentan (1 nmol intrastriatally) (panel C). Intrastriatal injection of the ET_B receptor antagonist BQ-788, but not the ET_A receptor antagonist BQ-123, significantly reduced the effect of endothelin-1 (panel D). For further details see [27].

pletely blocked the behavioral effect of intrastriatal endothelin-1 (Figure 1). The effect of endothelin-1 in the striatum was mediated by ET_B receptors, as it could be mimicked by intrastriatal injection of endothelin-3 and the selective ET_B receptor agonist $[Ala^{1,3,11,15}]$endothelin-1 (Figure 1). Moreover, both the non-selective endothelin receptor antagonist bosentan and the selective ET_B receptor antagonist BQ-788, but not the ET_A receptor antagonist BQ-123, significantly inhibited the effect of endothelin-1 (Figure 1). Taken together, these *in vivo* data suggested that endothelin injection into the striatum caused ac-

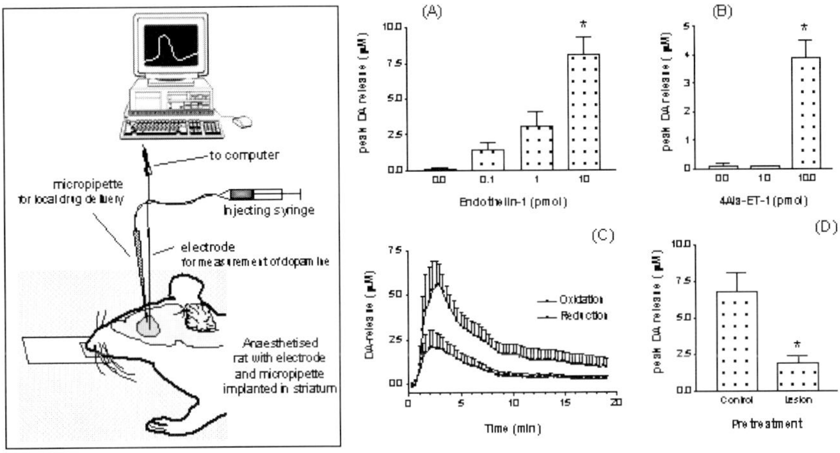

Figure 2. The effect of endothelin on extracellular levels of dopamine in ventral striatum as measured with chronoamperometry. A micro-electrode/micro-pipette assembly was stereotaxically implanted in the ventral striatum of anesthetized rats (left panel). Micro-injection of endothelin-1 caused a dose-dependent increase in extracellular dopamine concentrations (panel A). This effect was mimicked by the ET_B receptor agonist $[Ala^{1,3,11,15}]$endothelin-1 (panel B). The effect of endothelin-1 was maximal at 4 min after injection into the striatum (panel C) and was markedly reduced in striatum that was depleted of dopamine by prior 6-hydroxydopamine lesions of the nigrostriatal system (panel D).

tivation of ET_B receptors on dopaminergic terminals (see above), leading to dopamine release which, through activation of dopamine D_2 receptors, caused behavioral activation.

We confirmed this hypothesis by direct measurements of extracellular dopamine concentration using chronoamperometry [25]. This voltammetry technique utilizes a carbon fiber micro-electrode implanted into the ventral striatum. By applying a set positive potential, dopamine and other oxidizable compounds are made to liberate electrons, resulting in a measurable current directly proportional to the concentration of the compound. Micro-injection of endothelin-1 directly into the striatum, close to the tip of the electrode, caused a marked and dose-dependent increase in current, with a peak at 4 min (Figure 2A, B). This response was subject to desensitization, as a second injection of endothelin-1, one hour later, caused a markedly smaller response (not shown). The specificity of the voltammetry response was illustrated by the finding that similar micro-injection of endothelin-1 into ventral striatum, that had been depleted of dopamine by a prior 6-hydroxydopamine lesion, had little effect (Figure 2). As with turning behavior, the effect of endothelin-1 could be mimicked by local injection of the ET_B receptor agonist $[Ala^{1,3,11,15}]$endothelin-1 (Figure 2B).

CONCLUSIONS AND FUTURE STUDIES

The results obtained with voltammetry and behavioral methods, together with the binding and radio-immunoassay data, make it highly likely that endothelins are produced in the striatum and directly act on ET_B receptors located on dopaminergic terminals to increase dopamine release. Thus, endothelins could play a role in modulating on-going dopaminergic activity and, as such, be involved in movement control. Loss of endothelin production through loss of gene expression or changes in enzyme processing capacity, such as in McLeod syndrome, could then be responsible for disturbances in movement control.

The identity and physiological role of the other population of ET_B receptors, present in the striatum but not located on dopaminergic neurons, remains to be shown. It is likely that these receptors are located on glial cells [4,18]. It is interesting to note that growth factor properties have been attributed to endothelins, particularly in situations of nervous system injury. For example, endothelins promote the activation of glial cells in striatum, an effect mediated by ET_B receptors [13]. ET_B receptor expression is increased in glial cells in response to nerve injury [18]. Conversely, rats with mutations in the ET_B receptor gene, the 'spotting lethal' rats (see [3]), show increased rates of apoptosis in the dentate gyrus of the hippocampus [6]. A similar result was obtained from tissue obtained from rabbit and human brain infected with pneumococcal meningitis, which causes marked down-regulation of ET_B receptor expression [6]. Astrocytes from nervous tissue of endothelin-1 knockout mice showed increased vulnerability to hypoxic/ischemic injury [10], suggesting a neuroprotective action of endothelins. Thus, loss of endothelin production in the brain could lead to reduced neuroprotection and compensatory responses in neurodegenerative diseases.

Currently, there are several other knock-out models available to study the importance of different components of the endothelin system. For example, endothelin-2, endothelin-3, ET_A receptor, ET_B receptor, and ECE-1 knockout mice have been developed [31]. However, while most studies on these models have focused on neurodevelopmental and cardiovascular aspects [1,31], little is known on effects of these mutations in the CNS. Moreover, marked systemic and developmental problems make these mouse models difficult to study. Future studies will hopefully provide knockout models where the mutation is only expressed in adulthood or it is localized only to certain brain areas.

Acknowledgements

The work presented in this paper was largely performed under my supervision by Dr. Kim Webber at the Baker Medical Research Institute, Prahran, Australia, as part of the requirements for her degree of Doctor of Philosophy (Ph.D.). These studies were supported by a block institute grant of the National Health and Medical Research Council of Australia to the Baker Medical Research Institute.

REFERENCES

1. Berthiaume N, Yanagisawa M, Labonte J, D'Orleans-Juste P (2000) Heterozygous knock-out of ET_B receptors induces BQ-123-sensitive hypertension in the mouse. *Hypertension* **36:** 1002-1007.
2. Danek A, Tison F, Rubio J, Oechsner M, Kalckreuth W, Monaco AP (2001) The chorea of McLeod syndrome. *Mov Dis* **16:** 882-889.
3. Dembowski C, Hofmann P, Koch T, Kamrowski-Kruck H, Riedesel H, Krammer HJ, Kaup FJ, Ehrenreich H (2000) Phenotype, intestinal morphology, and survival of homozygous and heterozygous endothelin B receptor–deficient (spotting lethal) rats. *J Pediatr Surg* **35:** 480-488.
4. Desagher S, Cordier J, Glowinski J, Tence M (1997) Endothelin stimulates phospholipase D in striatal astrocytes. *J Neurochem* **68:** 78-87.
5. Eckman EA, Reed DK, Eckman CB (2001) Degradation of the Alzheimer's amyloid beta peptide by endothelin-converting enzyme. *J Biol Chem* **276:** 24540-24548.
6. Ehrenreich H, Nau R, Dembowski C, Hasselblatt M, Barth M, Hahn A, Schilling L, Siren A L, Bruck W (2000) Endothelin B receptor deficiency is associated with an increased rate of neuronal apoptosis in the dentate gyrus. *Neuroscience* **95:** 993-1001.
7. Fuxe K, Kurosawa N, Cintra A, Hallström Å, Goiny M, Rosén L, Agnati LF, Ungerstedt U (1992) Involvement of local ischemia in endothelin-1 induced lesions of the neostriatum of the anesthetized rat. *Exp Brain Res* **88:** 131-139.
8. Fuxe K, Tinner B, Staines W, Hemsén A, Hersh L, Lundberg JM (1991) Demonstration and nature of endothelin-3-like immunoreactivity in somatostatin and choline acetyl-transferase-immunoreactive nerve cells of the neostriatum of the rat. *Neurosci Lett* **123:** 107-111.
9. Giaid A, Gibson SJ, Herrero MT, Gentleman S, Legon S, Yanagisawa M, Masaki T, Ibrahim N, Roberts GW, Rossi ML, Polak JM (1991) Topographical localisation of endothelin mRNA and peptide immunoreactivity in neurones of the human brain. *Histochemistry* **95:** 303-314.
10. Ho MC, Lo AC, Kurihara H, Yu AC, Chung SS, Chung SK (2001) Endothelin-1 protects astrocytes from hypoxic/ischemic injury. *FASEB J* **15:** 618-626.
11. Hoffman A, Keiser HR, Grossman E, Goldstein DS, Gold PW, Kling M (1989) Endothelin concentration in cerebrospinal fluid in depressive patients. *Lancet* **2:** 1519.
12. Hori S, Komatsu Y, Shigemoto R, Mizuno N, Nakanishi S (1992) Distinct tissue distribution and cellular localization of two messenger ribonucleic acids encoding different subtypes of rat endothelin receptors. *Endocrinology* **130:** 1885-1895.
13. Ishikawa N, Takemura M, Koyama Y, Shigenaga Y, Okada T, Baba A (1997) Endothelins promote the activation of astrocytes in rat neostriatum through ET_B receptors. *Eur J Neurosci* **9:** 895-901.
14. Kohzuki M, Onodera H, Yasujima M, Itoyama Y, Kanazawa M, Sato T, Abe K (1995) Endothelin receptors in ischemic rat brain and Alzheimer brain. *J Cardiovasc Pharmacol* **26:** S329-S331.
15. Kraus GE, Bucholz RD, Yoon KW, Knuepfer MM, Smith KR (1991) Cerebrospinal fluid endothelin-1 and endothelin-3 levels in normal and neurosurgical patients: a clinical study and literature review. *Surg Neurol* **35:** 20-29.
16. Kurosawa M, Fuxe K, Hallström Å, Goiny M, Cintra A, Ungerstedt U (1991) Responses of blood flow, extracellular lactate, and dopamine in the striatum to intrastriatal injection of endothelin-1 in anesthetized rats. *J Cardiovasc Pharmacol* **17:** S340-S342.
17. Minami M, Kimura M, Iwamoto N, Arai H (1995) Endothelin-1-like immunoreactivity in cerebral cortex of Alzheimer-type dementia. *Prog Neuropsychopharmacol Biol Psychiatry* **19:** 509-513.

18. Nakagomi S, Kiryu-Seo S, Kiyama H (2000) Endothelin-converting enzymes and endothelin receptor B messenger RNAs are expressed in different neural cell species and these messenger RNAs are coordinately induced in neurons and astrocytes respectively following nerve injury. *Neuroscience* **101:** 441-449.
19. Park L, Thornhill J (2000) Hypoxic modulation of striatal lesions induced by administration of endothelin-1. *Brain Res* **883:** 51-59.
20. Takahashi K, Ghatei MA, Jones PM, Murphy JK, Lam HC, O'Halloran DJ, Bloom SR (1991) Endothelin in human brain and pituitary gland: comparison with rat. *J Cardiovasc Pharmacol* **17:** S101-S103.
21. Tayag EC, Jeng AY, Savage P, Lehmann JC (1996) Rat striatum contains pure population of ET_B receptors. *Eur J Pharmacol* **300:** 261-265.
22. Van den Buuse M (1997) Effects of endothelins on the nervous system. In: *Huggins JP, Pelton JT (eds) Endothelins in biology and medicine.* CRC Press, Boca Raton, pp 223-255.
23. Van den Buuse M, Webber KM (2000) Endothelin and dopamine release. *Progress Neurobiol* **60:** 383-403.
24. Volpe M, Cosentino F (2000) Abnormalities of endothelial function in the pathogenesis of stroke: the importance of endothelin. *J Cardiovasc Pharmacol* **35:** S45-S48.
25. Webber KM, Pennefather JN, Head GA, Van den Buuse M (1998) Endothelin induces dopamine release from rat striatum via endothelin-B receptors. *Neuroscience* **86:** 1173-1180.
26. Webber KM, Pennefather JN, Howells DW, Porritt MJ, Dean R, Van den Buuse M (1998) Endothelin ET-B receptors in rat and human striatum: association with dopaminergic terminals. *Naunyn-Schmied Arch Pharmacol* **358:** R46 (abstract P35.73).
27. Webber KM, Van den Buuse M (1996) Intrastriatal injection of endothelin evokes dopaminergic turning behaviour in rats through activation of the ET_B receptor. *Brain Res* **724:** 180-185.
28. Webber KM, Wallace CA, Smith AI, Van den Buuse M (1998) Endothelin interactions with brain dopamine systems. *J Cardiovasc Pharmacol* **31:** S373-S375.
29. Yanagisawa M, Kurihara H, Kimura S, Tomobe Y, Kobayashi M, Mitsui Y, Yazaki Y, Goto K, Masaki T (1988) A novel potent vasoconstrictor peptide produced by vascular endothelial cells. *Nature* **332:** 411-415.
30. Yoshizawa T, Iwamoto H, Mizusawa H, Suzuki N, Matsumoto H, Kanazawa I (1992) Cerebrospinal fluid endothelin-1 in Alzheimer's disease and senile dementia of Alzheimer's type. *Neuropeptides* **22:** 85-88.
31. Zollmann FS, Paul M (2000) Transgenic models for the study of endothelin function in the cardiovascular system. *J Cardiovasc Pharmacol* **35:** S13-S16.

CHAPTER 24

SUBSTRATES FOR TRANSGLUTAMINASE-CATALYZED CROSS-LINKING: RELEVANCE TO PATHOGENESIS OF HUNTINGTON'S DISEASE AND CHOREA-ACANTHOCYTOSIS

MARIAROSA A.B. MELONE[1] AND GIANFRANCO PELUSO[2]
[1]Department of Neurological Sciences, Second University of Naples, School of Medicine and [2]National Institute of Cancer "Fondazione G. Pascale"; Naples, Italy

Abstract. Protein aggregates are a hallmark of polyglutamine diseases, including Huntington's disease (HD). The transglutaminases, a class of Ca^{2+}-dependent cross-linking enzymes, could be directly involved in the disease mechanisms [14,15]. In fact, we have shown that when tTGase is activated by a calcium ionophore, the expanded huntingtin fragment is more abundant and insoluble high molecular weight (Mw) aggregates in HD fibroblasts appear together with evidence of apoptosis. Interestingly, in chorea-acanthocytosis (ChAc) a rare autosomal-recessive disorder without CAG repeats, we found increased amounts of tTGase products in muscle and erythrocytes. Furthermore, immunohistochemistry demonstrated abnormal accumulation of tTGase products, as well as of proteinaceous bodies in ChAc muscles.

INTRODUCTION

A growing number of neurodegenerative disorders, including Huntington's disease (HD) and several forms of spinocerebellar ataxia are currently known to be caused by expansion of CAG-triplet repeats coding for polyglutamine (poly-Q) stretches [7]. These stretches of repeats are inherently unstable, and this instability promotes expansion. Although the genes causing these disorders do not share homology except for the CAG repeats, there are several striking similarities and probably common pathogenic mechanisms among these disorders.

A. Danek (ed.), Neuroacanthocytosis Syndromes, 213–221.

All nine polyglutamine disorders known so far are characterized by progressive, bilateral and symmetric "systems" neuronal loss that typically begins in mid-life. Despite the ubiquitous expression of all nine genes, only a certain subset of neurons is vulnerable to degeneration. Expansions of CAG repeats correlate with severity and early onset. The polyglutamine stretches lead to a pathological "gain-of-function" which is the core of the pathogenesis [34].

Investigations have focused on three different characteristics of the mutant protein: subcellular localization, proteolytic processing, and the formation of intranuclear inclusions. The intranuclear neuronal inclusions are now strong candidates for a central role in the pathogenesis of all of the glutamine-repeat diseases [6,8,10,34]. In considering the polyglutamine segment in huntingtin (htt), Perutz [33] predicted that individual molecules could interact to form a tight "polar zipper" structure based on a β-sheet conformation for the glutamine tracts. Green [15] proposed that the poly-Q domain in the expressed protein of CAG repeat diseases could result in increased substrate activity for tissue transglutaminase (tTGase). This is a Ca^{2+}-dependent cross-linking enzyme which has been implicated in a number of cellular processes including cell attachment, bone development, axonal growth and regeneration, as well as cell growth regulation and apoptosis [12,13,16,17,28,32]. Furthermore, crosslinking may change the solubilities and interactions of proteins by reaction with tTGase and could adversely affect the functions and viability of the neuron [13,16]. In an in vitro model of HD, we have demonstrated that when tTGase is activated by calcium ionophore, the expanded htt fragment is more abundant and insoluble high molecular weight aggregates in HD fibroblasts appear [31]. Interestingly, we also demonstrated tTGase involvement in chorea-acanthocytosis (ChAc) [27], which is not a CAG triplet repeat disease. Enhanced creatine kinase concentration is a constant feature of the condition [11]. In fact, we found increased amounts of tTGase-derived N^{ϵ}-(-γ-glutamyl)-lysine isopeptide cross-links in erythrocytes and muscle of ChAc patients. Furthermore, immunohistochemistry demonstrated abnormal accumulation of tTGase products, as well as of proteinaceous bodies in ChAc muscle [27].

A NOVEL PROPERTY CONFERRED BY POLYGLUTAMINE: A TRIGGER FOR A PATHOGENIC CASCADE?

The polyglutamine segment seems to cause the mutant protein to adopt an unusual conformation. But how could poly-Q cause brain pathology? Once synthesized, the abnormal protein may remain partially or completely unfolded and may form microaggregates with itself and with other cell constituents. In particular, the glutamine tract, when present in a short protein fragment, promotes self-aggregation, with a conversion to an insoluble β-sheet amyloid structure [6,9,10,19,36].

Such a property was predicted by Perutz [33] and experimentally demonstrated using amino-terminal fragments representing 35% of huntingtin (htt). In this circumstance, the glutamine tract is freed from constraints imposed by the remaining $\sim 3,000$ amino acids of the protein. Indeed, it is likely that in the native protein, the propensity for insoluble conversion may give way to abnormal interactions with other cellular constituents. Similarly, the abnormal conformation of the glutamine tract could affect the inherent activity of the host protein.

Consequently, whereas pathogenesis is clearly triggered by the presence of the abnormal glutamine tract in each disorder, the disease pathway in each may begin differently: (i) through an effect on the function of the host protein, (ii) through an effect on existing or new interactions of the full-length host protein, or (iii) through the process or consequences of forming insoluble poly-Q amyloid [34]. Once neuronal homeostasis has been disrupted, it is likely that many other downstream changes occur. These changes may be shared across diseases, they may be poly-Q independent and some may lead to cell death.

On the basis of the postulated mechanisms, two hypotheses could be proposed to explain how expanded poly-Q domains form insoluble aggregates. The expanded poly-Q repeats may interact with each other through a polar zipper and thus contribute to aggregate formation [33]. Green stated that "organisms cannot tolerate proteins with very long stretches of glutamines" and he proposed a mechanism of aggregation based on the role of TGase in the growth of the mammalian epidermis [15]. Most residues of the epidermal protein involucrin are glutamines and some of these are present in tandem repeats. Such repeats form good substrates for TGase-catalysed crosslinking to glycine ethylester in proteins associated with three different ataxias, and the rate of crosslinking increases with the length of the glutamine repeats [22].

DOES TG-ASE ACTIVITY PLAY A COMMON ROLE IN HD AND IN CHOREA-ACANTHOCYTOSIS?

The TGases function as Ca^{2+}-dependent acyl transferases that catalyze the formation of an amide bond between γ-carboxamide groups of peptide-bound glutamine residues and either the ϵ-amino groups of lysine residues in proteins or the amino groups from polyamines [16]. The reaction results in the post-translational modification of proteins by establishing ϵ-(γ-glutamyl)-lysine and N,N-bis(γ-glutamyl)-polyamine isodipeptide linkages [16]. The cross-linking activity of tissue transglutaminase (tTGase) is highly regulated at the post-transcriptional level by GTP, polyamines, and nitric oxide [16,28,32]. tTGase is undetectable in most mammalian cells and its mRNA is specifically transcribed as a consequence of the induction of apoptosis [13].

Retinoic acid (RA) and its various synthetic analogs affect mammalian cell growth, differentiation, and apoptosis. The ability of RA to protect against apoptosis is linked to the expression of active TGase [1]. tTGase-dependent formation of covalent cross-links in apoptotic cells leads to the polymerization of substrate proteins that can be dismantled only by the proteolytic degradation of the protein chains [13].

tTGase activity has been measured in numerous different neural tissues including brain, spinal cord, peripheral nerve and superior cervical ganglia [for a review see 17]. The role of tTGase in neural functioning is likely to be quite complex. tTGase-mediated protein crosslinks have been reported to form in response to long term potentiation in rat hippocampal slices stimulated by high frequency. Therefore, tTGase may be involved in synaptic function through stabilization of newly formed macromolecular complexes [12].

tTGase activity increases during development and there is significant evidence that tTGase may be involved in mediating neurite outgrowth and/or neuronal differentiation [17,32]. Conversely, it has also been hypothesized that deregulation of neuronal tTGase may play a part in the pathogenesis of neurodegenerative disorders, such as HD and other CAG diseases, all characterized by the presence of extremely insoluble protein complexes in the brain [4,6,14,22,26,31].

Huntingtin (htt) is a large protein of more than 3,100 amino acids that bears no close similarity to any other protein [9]. The mutant version displays a disproportionate decrease in electrophoretic mobility in SDS polyacrylamide gels due to the expanded glutamine tract, suggesting an unusual conformation in this region [2]. The first experimental clues that the novel property of mutant htt can indeed promote self-aggregation came from transgenic mice overexpressing only exon 1 of the HD gene with an extreme CAG repeat [8]. Subsequently, anti-htt-reactive nuclear inclusions were detected in HD postmortem brain [23].

Abnormal protein aggregation has been postulated to explain the molecular basis for the poly-Q diseases [13,25,34]. In HD, the inclusions are enriched in medium-sized neurons of the striatum and neurons of layers III, V, and VI of the cerebral cortex, which are all selectively vulnerable in HD [23]. The inclusions appear to be relatively circumscribed and to exclude DNA [34], suggesting that they do not physically disrupt the chromatin in the nucleus. They may, however, have more subtle effects on gene transcription. For instance, it is conceivable that they could alter transcription of genes required for neuronal viability, perhaps including glutamate receptors, mitochondrial proteins, or other molecules previously implicated in neuronal vulnerability.

Dystrophic neurites that label for the N-terminal htt antibody have been observed in HD [10]. This raises the possibility that deposition of poly-Q outside of the nucleus may also be important, at least in HD. Presumably, the abnormal molecules accumulate and form inclusions due to the cell's failure to rapidly degrade these polypeptides.

Accordingly, treatment of the cells with proteasome inhibitors results in greater accumulation of poly-Q polypeptides, increase in number of inclusion bodies, and enhanced apoptosis [5]. So, it appears that multiple factors, including cell type and level of expression, may regulate a cell's susceptibility to form aggregates, as well as the form and subcellular localization of these aggregates. Therefore, it has been proposed that the HD mutation, by virtue of the expanded glutamine tract, confers a new and deleterious physical property on mutant htt. In particular, pathological-length polyglutamine expansions, such as those that occur in the htt protein, are excellent substrates for tTGase in vitro, and therefore contribute to aggregate formation in HD brain [2,6,9].

In favor of this hypothesis, we have shown that the formation of covalent cross-links between polyglutamine peptides and polyamines yields high-Mw aggregates – a process that is favored with longer polyglutamines. In the presence of tTGase, purified glyceraldehyde-3-phosphate dehydrogenase (a key glycolytic enzyme that binds tightly to the polyglutamine domains of both htt and dentato-rubro-pallido-Luysian atrophy proteins) is covalently attached to polyglutamine peptides in vitro, resulting in the formation of high-Mw aggregates.

In addition, endogenous glyceraldehyde-3-phosphate dehydrogenase of a Balb-c 3T3 fibroblast cell line overexpressing human tTGase, forms cross-links with a Q60 polypeptide added to the cell homogenate [14]. Mutant htt is expressed throughout the body, in the cytoplasm of neurons and non-neuronal cells alike, both in cells vulnerable to degeneration and those resistant to the effects of the defect. Interestingly, a previous study reported an increase in tTGase activity in lymphocytes from HD patients as compared to healthy individuals [4].

Recently, we have set up a cellular model of aggregate formation by using dermal fibroblasts of heterozygous HD patients cultured in different conditions. In particular, we examined the effects of all-trans retinoic acid (RA) and calcium ionophore (CI) both on cell tTGase expression and activation and on aggregate and apoptosis induction. RA treatment of cultured cells increases tTGase expression, both in normal and pathological cells. Whereas, when tTGase is activated the expanded htt fragment is more abundant and insoluble high-Mw aggregates in HD fibroblasts appear. In this regard, the use of RA and CI significantly contributed to aggregate formation. Besides, we showed an increase of apoptotic nuclei, of bax, and of bcl-2 expression. Caspase transcript was not detected in control fibroblasts while low levels of caspase 6 and 3 were detected in HD fibroblasts. In particular, levels of caspase 3 protein significantly correlated with apoptotic nuclei ($p < 0.05$) and with bax expression ($p < 0.01$). These data indicate that HD fibroblasts, simultaneously with aggregate formation, modulate the expression of caspases and that caspase 3 is involved in cell death [31].

If tTGase contributes to the modifications of the htt protein that occur in HD, it is reasonable to hypothesize that tTGase activity and tTGase expression are altered in HD-affected brain as compared to healthy individuals. Indeed, tTGase levels and tTGase activity were increased significantly in the striatum and the superior frontal cortex, two brain areas preferentially affected in HD. Moreover in the cerebellum, a brain area spared in HD [23], tTGase activity and tTGase levels were not significantly increased as compared to controls [25].

Other direct evidence for the participation of tTGase (or other transglutaminases) in HD patients in vivo was reported by Jeitner et al [21] who demonstrated that levels of N^{ϵ}-(-γ-glutamyl)-lysine isopeptide – a "marker" of the transglutaminase reaction – were elevated in the cerebrospinal fluid of HD patients.

Given the similarities of the leading symptoms, chorea and basal ganglia degeneration, the pathogenetic mechanism behind HD and ChAc could quite possibly be very similar. The data we present here is the first direct report of abnormal accumulation of tTGase products in erythrocytes and muscles obtained from ChAc patients, as well as of proteinaceous bodies in ChAc muscles [27]. The *CHAC* gene encodes a conserved protein that is probably involved in protein sorting, or more generally, in cell trafficking and several mutations have been identified in ChAc individuals [35,38].

In previous experiments, we have shown, by Western blot analysis of erythrocyte membranes, an upward shift of band 3 molecular weight [26]. Since the Gln-30 residue of human band 3 anion transporter is a primary site for glutamine-lysine cross-linking by tTGases [29], we suggested that tTGases could be responsible for polyamination of band 3. Indeed, structural and functional band 3 alterations had previously been described in the erythrocytes of ChAc patients [3,30].

Subsequently, we found that N^{ϵ}-(-γ-glutamyl)-lysine isopeptide was 25 times more abundant in ChAc erythrocytes than in controls, and suggested that the excessive production of the isopeptide causes the membrane dysmorphism typical of acanthocytosis. We further suggested that the presence of measurable amounts of isopeptide in control erythrocytes may be a marker of tTGase activity in the physiological turnover of erythrocytes.

Creatine kinase serum levels are constantly elevated to a value that could not be attributed to the chronic denervation alone or to the choreic and jerky movements. This finding suggests a sarcolemmal alteration, possibly mediated by tTGases in the myofibres. In fact, we observed amorphous elements in degenerating myofibers and at the paranuclear level in ChAc muscle. The ultrastructural characteristics of these bodies strongly suggest that they are constituted by protein aggregates. Axonal polymorphic inclusions have been reported previously in peripheral nerve of ChAc subjects [18], yet so far no inclusion bodies in ChAc muscle.

We think that tTGase is involved in the production of these inclusion bodies. In fact, our immunohistochemical and biochemical studies showed inclusion bodies in muscles that contained increased levels of cross-links of N^{ϵ}-(-γ-glutamyl)-lysine. Western blot of the muscles probed with tTGase antibody and immunohistochemistry showed higher tTGase levels in ChAc muscle with respect to controls. Therefore, it is conceivable that in ChAc the excess of cross-linking by the isopeptide bond could cause conformational changes, which, at a certain stage of the disease, induce membrane distortions leading to the release of creatine kinase and to the acanthocytic features of red blood cells.

CONCLUSIONS

The presence of intranuclear inclusions in the affected neuronal cells of poly-Q diseases has suggested a mechanism for pathogenesis based on protein misfolding and aggregation. tTGase may be critical in the pathogenesis, via cross-linking huntingtin. Indeed, neuronal intranuclear and cytosolic inclusions composed of mutant htt are present in HD brains, and tTGase is seen in nuclei and in cytoplasm isolated from the brains of HD patients [23].

In our opinion, increased htt (or htt fragment containing the poly-Q domain) in the nucleus, increased ability of the poly-Q domains to act as substrate, increased Ca^{2+} levels and inherent tTGase activity in their combination contribute to increased cross-linking of proteins in HD brain.

At first, the proteasome machinery can recognize and degrade the cross-linked proteins, but over time the machinery may be overwhelmed and protein aggregates will accumulate. From these data, tTGase may be a potential therapeutic target in the treatment of poly-Q diseases [20]. Indeed, the intraperitoneal administration of the competitive tTGase inhibitor, cystamine, to transgenic mice expressing exon 1 of huntingtin containing an expanded poly-Q repeat, alters the course of their HD-like disease [24].

Neurological findings of ChAc closely resemble those observed in HD. The remarkable increase of N^{ϵ}-(-γ-glutamyl)-lysine cross links in ChAc muscle and erythrocytes implicates tTGase in the pathogenetic process. We do not know the mechanism responsible for the excess production of tTGase in ChAc, but may speculate that it results from dysregulation of the enzyme consequent to over-stimulation by its numerous substrates, including the conserved sorting-associated protein that is mutated in ChAc [35,38].

Alternatively, it is not inconceivable that interactions between, for example, chorein and band 3, may negatively affect cell survival. Certainly, when hyperactivity of tTGase has been initiated, it proceeds rapidly and intensely with formation and deposition of its insoluble products [13]. So, the possibility that tTGase is also involved in ChAc pathogenesis, makes tTGase an attractive target for possible intervention and might perhaps even be combined with caspase inhibitors.

REFERENCES

1. Antonyak MA, Singh US, Lee DA, Boehm JE, Combs C, Zgola MM, Page RL, Cerione RA (2001) Effects of tissue transglutaminase on retinoic acid-induced cellular differentiation and protection against apoptosis *J Biol Chem* **276:** 33582-33587.
2. Aronin N, Chase K, Young C, Sapp E, Schwarz C, Matta N, Kornreich R, Landwehrmeyer B, Bird E, Beal MF et al (1995). CAG expansion affects the expression of mutant Huntingtin in the Huntington's disease brain. *Neuron* **15:** 1193-1201.
3. Asano K, Osawa Y, Yanagisawa N, Oshima M (1985) Erythrocyte membrane abnormalities in patients with amyotrophic chorea with acanthocytosis. *J Neurol Sci* **68:** 161-173.
4. Cariello L, de Cristofaro T, Zanetti L, Cuomo T, Di Maio L, Campanella G, Rinaldi S, Zanetti P, Di Lauro R, Varrone S (1996) Transglutaminase activity is related to CAG repeat length in patients with Huntington's disease. *Hum Genet* **98:** 633-635.

5. Chai Y, Koppenhafer SL, Shoesmith SJ, Perez MK, Paulson HL (1999) Evidence for proteasome involvement in polyglutamine disease: Localization to nuclear inclusions in SCA3/MJD and suppression of polyglutamine aggregation in vitro. *Hum Mol Genet* **8:** 673-682.
6. Cooper AJ, Sheu KF, Burke JR, Strittmatter WJ, Gentile V, Peluso G, Blass JP (1999) Pathogenesis of inclusion bodies in (CAG)n/Qn-expansion diseases with special reference to the role of tissue transglutaminase and to selective vulnerability. *J Neurochem* **72:** 889-899.
7. Cummings CJ, Zoghbi HY (2000) Trinucleotide repeats: Mechanisms and pathophysiology. *Annu Rev Genomics Hum Genet* **1:** 281-328.
8. Davies SW, Turmaine M, Cozens BA, DiFiglia M, Sharp AH, Ross CA, Scherzinger E, Wanker EE, Mangiarini L, Bates GP (1997) Formation of neuronal intranuclear inclusions underlies the neurological dysfunction in mice transgenic for the Hd mutation. *Cell* **90:** 537-548.
9. Davies S, Ramsden DB (2001) Huntington's disease. *Mol Pathol* **54:** 409-413.
10. DiFiglia M, Sapp E, Chase KO, Davies SW, Bates GP, Vonsattel JP, Aronin N (1997) Aggregation of huntingtin in neuronal intranuclear inclusions and dystrophic neurites in brain. *Science* **277:** 1990-1993.
11. Feinberg TE, Cianci CD, Morrow JS, Pehta JC, Redman CM, Huima T, Koroshetz WJ (1991) Diagnostic tests for choreoacanthocytosis. *Neurology* **41:** 1000-1006.
12. Festoff BW, Suo Z, Citron BA (2001) Plasticity and stabilization of neuromuscular and CNS synapses: Interactions between thrombin protease signaling pathways and tissue transglutaminase. *Int Rev Cytol* **211:** 153-177.
13. Fesus L, Davies PJ, Piacentini M (1991) Apoptosis: Molecular mechanisms in programmed cell death. *Eur J Cell Biol* **56:** 170-177.
14. Gentile V, Sepe C, Calvani M, Melone MA, Cotrufo R, Cooper AJ, Blass JP, Peluso G (1998) Tissue transglutaminase-catalyzed formation of high-molecular-weight aggregates in vitro is favored with long polyglutamine domains: A possible mechanism contributing to CAG-triplet diseases. *Arch Biochem Biophys* **352:** 314-321.
15. Green H (1993) Human genetic diseases due to codon reiteration: relationship to an evolutionary mechanism *Cell* **74:** 955-956.
16. Greenberg CS, Birckbichler PJ, Rice RH (1991) Transglutaminases: Multifunctional cross-linking enzymes that stabilize tissues. *FASEB J* **5:** 3071-3077.
17. Hand D, Perry MJ, Haynes LW (1993) Cellular transglutaminases in neural development. *Int J Dev Neurosci.* **11:** 709-720.
18. Hardie RJ, Pullon HW, Harding AE, Owen JS, Pires M, Daniels GL, Imai Y, Misra VP, King RH, Jacobs JM, et al (1991) Neuroacanthocytosis. A clinical, haematological and pathological study of 19 cases. *Brain* **114:** 13-49.
19. Huang CC, Faber PW, Persichetti F, Mittal V, Vonsattel JP, MacDonald ME, Gusella JF (1998) Amyloid formation by mutant huntingtin: Threshold, progressivity and recruitment of normal polyglutamine proteins. *Somat Cell Mol Genet* **24:** 217-233.
20. Igarashi S, Koide R, Shimohata T, Yamada M, Hayashi Y, Takano H, Date H, Oyake M, Sato T, Sato A, Egawa S, Ikeuchi T, Tanaka H, Nakano R, Tanaka K, Hozumi I, Inuzuka T, Takahashi H, Tsuji S (1998) Suppression of aggregate formation and apoptosis by transglutaminase inhibitors in cells expressing truncated DRPLA protein with an expanded polyglutamine stretch. *Nat Genet* **18:** 111-117.
21. Jeitner TM, Bogdanov MB, Matson WR, Daikhin Y, Yudkoff M, Folk JE, Steinman L, Browne SE, Beal MF, Blass JP, Cooper AJ (2001) N(epsilon)-(gamma-L-glutamyl)-L-lysine (GGEL) is increased in cerebrospinal fluid of patients with Huntington's disease. *J Neurochem* **79:** 1109-1112.

22. Kahlem P, Terre C, Green H, Djian P (1996) Peptides containing glutamine repeats as substrates for transglutaminase-catalyzed cross-linking: relevance to diseases of the nervous system. *Proc Natl Acad Sci USA* **93:** 14580-14585.
23. Karpuj MV, Garren H, Slunt H, Price DL, Gusella J, Becher MW, Steinman L (1999) Transglutaminase aggregates huntingtin into nonamyloidogenic polymers, and its enzymatic activity increases in Huntington's disease brain nuclei *Proc Natl Acad Sci USA* **96:** 7388-7393.
24. Karpuj MV, Becher MW, Springer JE, Chabas D, Youssef S, Pedotti R, Mitchell D, Steinman L (2002) Prolonged survival and decreased abnormal movements in transgenic model of Huntington disease, with administration of the transglutaminase inhibitor cystamine. *Nat Med* **8:** 143-149.
25. Lesort M, Tucholski J, Miller ML, Johnson GV (2000) Tissue transglutaminase: a possible role in neurodegenerative diseases. *Prog Neurobiol* **61:** 439-463.
26. Melone MAB, Peluso G, Gentile V, Di Fede G, Petillo O, D'Amico B, Cotrufo R (1998) Tissue transglutaminase (tTGase) a key enzyme in the pathogenesis of Huntington chorea and neuroacanthocytosis. *Mov Dis* **13(Suppl. 2):** 283.
27. Melone MAB, Di Fede G, Peluso G, Lus G, Di Iorio G, Sampaolo S, Capasso A, Gentile V, Cotrufo R (2002) Abnormal accumulation of tTGase products in muscle and erythrocytes of chorea-acanthocytosis patients *J Neuropathol Exp Neurol* **61:** 841-848.
28. Monsonego A, Shani Y, Friedmann I, Paas Y, Eizenberg O, Schwartz M (1997) Expression of GTP-dependent and GTP-independent tissue-type transglutaminase in cytokine-treated rat brain astrocytes. *J Biol Chem* **272:** 3724-3732.
29. Murthy SN, Wilson J, Zhang Y, Lorand L (1994) Residue Gln-30 of human erythrocyte anion transporter is a prime site for reaction with intrinsic transglutaminase. *J Biol Chem* **269:** 22907-22911.
30. Oshima M, Osawa Y, Asano K, Saito T (1985) Erythrocyte membrane abnormalities in patients with amyotrophic chorea with acanthocytosis. Part 1. Spin labeling studies and lipid analyses. *J Neurol Sci* **68:** 147-160.
31. Peluso G, Giordano A, Jori FP, Petillo O, Margarucci S, Grippo P, Cotrufo R, Melone MAB (2002) Tissue Transglutaminase: A possible role in Huntington disease. *Clin Neuropathol* **21:** 128.
32. Perry MJ, Haynes LW (1993) Localization and activity of transglutaminase, a retinoid-inducible protein, in developing rat spinal cord. *Int J Dev Neurosci* **11:** 325-337.
33. Perutz M (1994) Polar zippers: their role in human disease. *Protein Sci* **3:** 1629-1637.
34. Perutz MF, Windle AH (2001) Cause of neural death in neurodegenerative diseases attributable to expansion of glutamine repeats. *Nature* **412:** 143-144.
35. Rampoldi L, Dobson-Stone C, Rubio JP, Danek A, Chalmers RM, Wood NW, Verellen C, Ferrer X, Malandrini A, Fabrizi GM, Brown R, Vance J, Pericak-Vance M, Rudolf G, Carre S, Alonso E, Manfredi M, Németh AH, Monaco AP (2001) A conserved sorting-associated protein is mutant in chorea-acanthocytosis. *Nat Genet* **28:** 119-120.
36. Scherzinger E, Lurz R, Turmaine M, Mangiarini L, Hollenbach B, Hasenbank R, Bates GP, Davies SW, Lehrach H, Wanker EE (1997) Huntingtin-encoded polyglutamine expansions form amyloid-like protein aggregates in vitro and in vivo. *Cell* **90:** 549-558.
37. Singaraja R, Kazemi-Esfarjani P, Devon R, Kim SU, Bredesen DE, Tufaro F, Hayden MR (1998) Length of huntingtin and its polyglutamine tract influences localization and frequency of intracellular aggregates. *Nat Genet* **18:** 150-154.
38. Ueno S, Maruki Y, Nakamura M, Tomemori Y, Kamae K, Tanabe H, Yamashita Y, Matsuda S, Kaneko S, Sano A (2001) The gene encoding a newly discovered protein, chorein, is mutated in chorea-acanthocytosis. *Nat Genet* **28:** 121-122.

CHAPTER 25

HUNTINGTON'S DISEASE ANIMAL MODELS: WHAT LESSONS CAN BE LEARNED FOR RESEARCH ON NEUROACANTHOCYTOSIS SYNDROMES?

CHRISTOPH M. KOSINSKI
Neurologische Klinik, Universitätsklinikum Aachen, Germany

Abstract. The basis for unraveling the pathophysiological mechanisms in Huntington's disease (HD) was the identification of the HD gene defect and the subsequent development of genetically modified animal models of the disease. In particular, studies in transgenic mice have led to identification of some hallmarks of the disease pathophysiology. In addition, researchers working with HD transgenic mice have learned about a number of limitations in mimicking human hereditary neurodegenerative disease genotype and phenotype in mice. Knowledge about these findings in HD will allow researchers on neuroacanthocytosis syndromes to avoid some pitfalls in developing mouse models and also to identify possible common pathways in striatal neurodegeneration.

HUNTINGTON'S DISEASE: RELATION TO NEUROACANTHOCYTOSIS

Huntington's disease (HD) is an autosomal-dominantly inherited neurodegenerative disease. From the perspective of a neurologist HD in many cases is the most obvious differential diagnosis of neuroacanthocytosis syndromes (see also Chapter 2). Similar to these, HD starts mostly in midlife and shows a varying picture of a hyperkinetic movement disorder together with psychiatric symptoms (personality changes, mood disorder, psychosis) and a cognitive decline that finally leads to dementia. On neuropathological examination there is progressive neuronal loss and gliosis most prominently in the striatum which is in many respects identical to the changes in neuroacanthocytosis (see also Chap-

A. Danek (ed.), Neuroacanthocytosis Syndromes, 223–231.

ter 11). In HD, however, other brain regions, especially the cerebral cortex, show massive neurodegeneration in late stages of the disease.

Consequences of HD Gene Defect Identification for HD Research

The gene defect for HD has been identified in 1993 as a CAG triplet expansion in the coding region (exon 1) of a so far unknown protein called "huntingtin". Expansion over 38 triplet repeats invariably leads to disease with an earlier disease onset found in patients with longer CAG repeat length [6]. Since the gene defect has become known, it was possible to produce a number of useful genetic disease models which allow us to study the disease mechanisms and which have led to a tremendous increase in our knowledge about HD pathophysiology in recent years. A number of promising treatment strategies could be deduced from this knowledge which can now be tested preclinically in the HD transgenic mouse models. With this strategy, a number of drugs were already identified for which large clinical trials for HD are currently established.

Why Caring about HD Research When Working on Neuroacanthocytosis Syndromes?

For the emerging field of molecular research on neuroacanthocytosis syndromes the knowledge about HD pathophysiology can be of importance in two ways.
- There is the question of a shared pathophysiological pathway in which all these diseases might end. The reason to believe in the possibility of such a common pathway is that all these diseases lead to a preferential degeneration of the same very specific type of neurons, i.e. the striatal projection neurons.
- There are various pitfalls that researchers working on HD genetic models ran into which might also easily occur while developing genetic models for neuroacanthocytosis syndromes and which could perhaps be avoided.

The gene defects for two of the neuroacanthocytosis syndromes have been identified, the gene for the X-linked McLeod syndrome and the *CHAC* gene for autosomal-recessive chorea-acanthocytosis linked to chromosome 9 [5,12,16]. For both syndromes, the development of genetically modified mouse models is on the way (authors' personal communication). Thus, this chapter will focus on experiences with mouse models in HD research.

THE DIVERSITY OF GENETIC MODELS FOR HD

With the knowledge of the gene defect in HD, various genetic models to study HD pathophysiology have been developed. It is very important to understand that each model proved highly valuable for the study of certain disease aspects and that none of the models was able to answer all questions. Only the parallel evaluation of disease aspects in these various models led to the emergence of a conclusive picture of the HD mechanism.

These models include *in vitro models* of either transient or stable transfected cell lines in which the HD gene defect induces increased sensitivity to various forms of cell stress and early apoptotic cell death. These models are very useful for the study of short term consequences of the gene defect on cell metabolism.

- *Yeast* models turned out to be particularly helpful for identifying proteins which interact with the HD gene product ("huntingtin") and for studying the effect of the HD gene defect on these protein interactions.
- Transfer of the HD gene defect was also achieved in *C. elegans* worms and *Drosophila* fruit flies.

These organisms have very simple nervous systems, their entire genome has been identified and transfer of the HD gene defect leads to neurodegeneration. These models have proven extremely helpful in identifying genetic modifiers of the disease [15].

HD MOUSE MODELS

Of great importance for HD research in the last decade was the development of genetic mouse models. The obvious advantage of mouse models is to study the consequences of the gene defect in a mammalian brain, also allowing the complex intercellular interactions which might contribute to the disease. The aim was to achieve animals that imitate all characteristics of the human disease. Ideally, such models should carry a gene defect identical to the one in HD patients. They should develop a progressive phenotype which mimics the various aspects of neurologic, psychiatric, and cognitive dysfunctions and which leads to a limited life span. Finally, the brain morphology should demonstrate the typical characteristics of neurodegeneration with gliosis in the striatum. None of the current available disease models, however, fulfills all of these criteria and, as we have learned from the current mouse models, it is very unlikely that any mouse model will ever be able to do so. The reasons can be understood from a genotype – phenotype comparison of the various models.

Genotype - Phenotype Correlation in HD Mouse Models

HD mouse models vary in a number of aspects which include length of the CAG repeat, length of the huntingtin protein expressed together with the repeat, expression rate of the mutant protein depending on the promotor used, and last but not least, the differences in the genetic background due to various strain lines used. Thus it is not very surprising that HD mice also differ profoundly in their phenotypic presentation.

Nevertheless some rules can be drawn from the comparison of the various HD mouse models; they can be categorized into three groups:

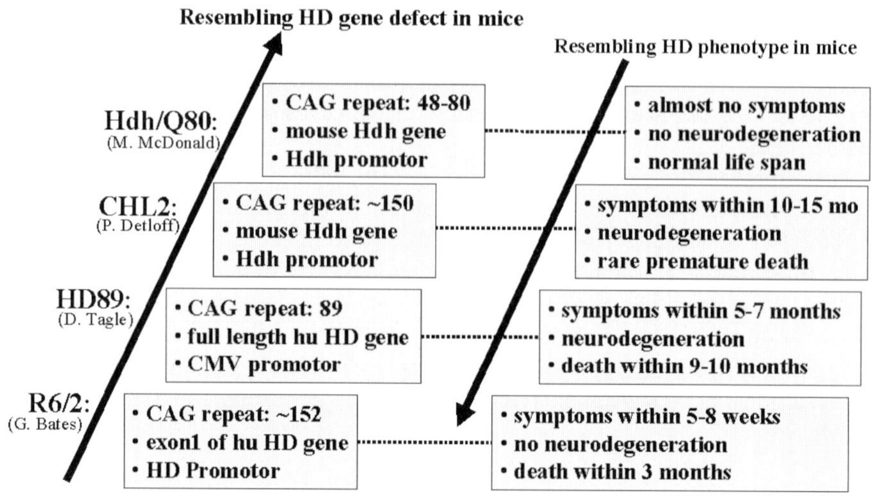

Figure 1. Genotype – Phenotype correlation in HD mouse models.

- There are in fact some models which mimic the gene defect of the human disease perfectly in the mouse genome. These are so called knock in mice in which an expanded CAG repeat with a repeat length in the typical range of human HD has been introduced into the mouse HD gene homologue (hdh). These mice, however, develop no neurological phenotype, have a normal life span (around 2 years) and show almost no signs of striatal neuronal degeneration [18].
- The other extreme are transgenic mice with a massively increased CAG repeat length (which in humans can only be found in a small minority of patients showing juvenile disease onset) that is expressed together with only a small part (exon 1) of the human HD gene. These mice develop a progressive neurological phenotype and die after a short period of time (3-4 months). Although some important aspects of the morphological characteristics can be seen in these mice (intranuclear aggregates) there are no signs of neurodegeneration nor is there gliosis in striatum [10].
- In between these two groups, there are some transgenic and knock in mice with intermediate characteristics. These mice might express the full huntingtin protein but with a still considerably expanded CAG repeat or with a very high expression level of the disease gene. These mice develop a mild disease phenotype and show only a somewhat shortened life span. Neuronal degeneration with striatal atrophy and signs of striatal gliosis can, however, be detected [13].

Figure 1 lists some of the HD mouse models with phenotype – genotype correlation. In summary, it can be said that the imitation of a human hereditary disease which usually does not start before the age of 35 years has some obvious limitations in mice with a life expectancy of not more than two years. Some exaggeration of human disease conditions is necessary in order to induce an onset of disease within this short life time. Extreme exaggeration with very short disease duration (as described for the second group above), however, can lead to death of the transgenic mice before the process of neurodegeneration has taken place.

Behavioral Characteristics of HD Transgenic Mice

The behavioral characteristics of HD transgenic mice have been tested most extensively in the R6/2 model since these mice develop the most obvious phenotype. In the first report it was described that they initially appear normal until some gait abnormalities start to occur around 8 to 10 weeks of age [10]. Lifting these mice at the tail induces a dystonic behavior of the hindlimbs which was called "*clasping*". The mice also develop progressive weight loss. Diabetes mellitus is present in about 25%. With further disease progression some mice show epileptic seizures. These mice were reported to rarely live longer than 16 weeks. From this report the development of a progressive neurological disease in these mice was obvious but the similarities to the typical features of human HD seemed very limited.

Only further behavioral testing allowed an advanced characterization of this mouse model. One of the major symptoms of human HD, the *loss of motor coordination* can be tested very effectively with the rotarod test which allows an easy monitoring of the progressive decline of these mice. With this test the onset of motor problems can be observed as early as 6 weeks of age [1].

Open field tests are well suited to evaluate *spontaneous explorative behavior*. R6/2 mice very early (around 4 weeks of age) show hyperactivity most closely related to the hyperkinetic state in human disease. Later in the disease, they become clearly hypoactive, similar to late stages of human HD in which parkinsonian and dystonic features often occur [8].

Using tests of *spatial orientation and memory* such as the Morris water-maze it was demonstrated that R6/2 mice indeed have an impairment both in orientation and memory as early as 3 weeks of age [7].

Behavioral testing in these mice allowed not only to detect specific characteristics of human HD such as hyper-/hypokinesia, motor incoordination or cognitive decline but also of subtle deficits that were detected much earlier than by mere observation. Applying the same tests in other transgenic animals, subtle changes were demonstrated in lines of mice which had previously been regarded as asymptomatic.

It can be concluded that certain tests have now been clearly characterized in HD transgenic mice to detect striatal dysfunction and that they are well suited for the characterization of a neurological phenotype in neuroacanthocytosis models. Given the variability of clinical presentation in human neuroacanthocytosis, it will be necessary to explicitly look for subtle changes in corresponding mouse models.

Morphological Characteristics of HD Transgenic Mice

In the first report on the R6/2 mice the authors described a generalized brain atrophy. Yet they had not detected signs of neuronal loss or gliosis in striatum or other brain regions [10]. One year later, however, the same group discovered ubiquitinated *neuronal intranuclear inclusions* (*NII*) for the first time [4]. The major component of these amyloid-like aggregates are the polyglutamine containing fragments of mutant huntingtin which – as we know now – tend to aggregate with each other like polar zippers. Based on this finding, NII have now been described not only in human HD and a number of HD transgenic mouse models but also in various other human CAG triplet diseases and are thought to be a hallmark in the pathophysiology of these diseases.

It is noteworthy that even very thorough examination with the use of stereological techniques was unable to demonstrate cell loss in R6/2 striatum even in very late disease stages shortly before death. Thus, the phenotype of these mice cannot be explained by neuronal degeneration; instead, neuronal dysfunction can precede neurodegeneration for a long period.

Classical neuropathological techniques fail to detect neuronal abnormalities in HD transgenic mice. If only these techniques are applied, the brains of many HD transgenic mice look completely normal: This might also be encountered in mouse models of neuroacanthocytosis.

Other techniques turned out to be extremely helpful for demonstration of neuronal dysfunction in HD mice. Receptor autoradiography can demonstrate *specific changes in neurotransmitter receptor expression* (affecting dopamine D1 and D2 receptors, adenosine A2a receptors, metabotropic glutamate receptors and others) already at early disease stages. These alterations not only replicate many findings known from human HD but also occur in parallel with the progression of the neurological phenotype of these mice [2].

CURRENT CONCEPTS OF HD PATHOPHYSIOLOGY

The gene defect for HD was identified almost 10 years ago but we still do not have a complete understanding of the mechanism that leads to neurodegeneration. There are, however, a number of key steps within the pathophysiological cascade which are now replicated in a number of different disease models. These results have been accepted by a broad group of researchers. Some of these theories will briefly be mentioned.

Toxic Fragment Hypothesis

Huntingtin mainly is a cytoplasmic protein. For disease pathology it is necessary that the mutant polyglutamine containing fragment of the protein gets cleaved from the remainder of the protein [17]. This cleavage most likely is carried out by *caspases*, special proteases which are also activated in the apoptotic cell death cascade. The mutant huntingtin fragment then gets translocated into the nucleus by unknown mechanisms (Figure 2).

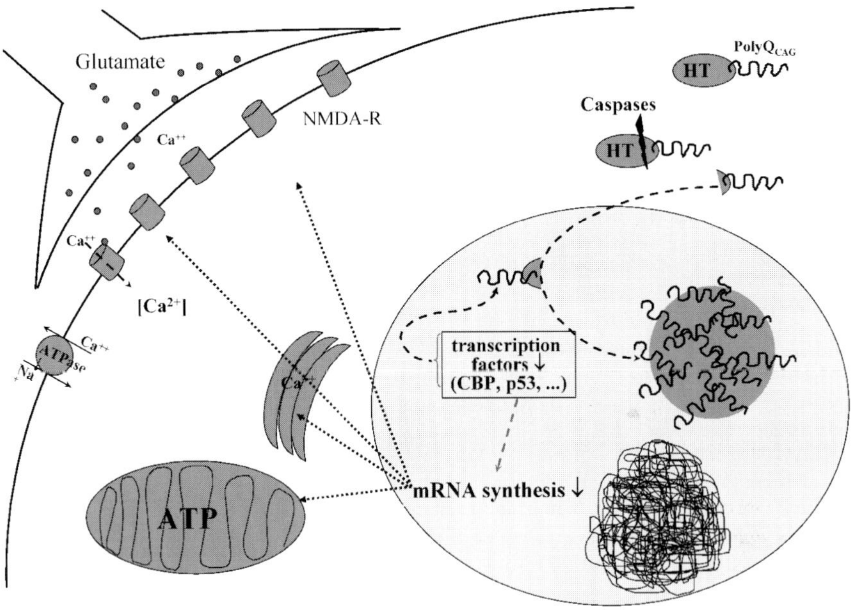

Figure 2. Key steps in HD pathophysiology.

Polyglutamine Aggregation Theory

As has first been suggested by Max Perutz, polyglutamine chains may form hairpins once they contain more than 40 repeats – which is the threshold for most CAG triplet diseases – and tend to aggregate with themselves and other polyglutamines [11]. Scherzinger et al [14] have shown that this aggregation takes place also in a cell free medium and the aggregates show amyloid-like features identical to the NII found in the brains of HD transgenic mice and human patients (Figure 2).

Transcriptional Dysregulation and Neuronal Dysfunction in HD

As has been mentioned earlier, HD mice develop early changes in the expression of neurotransmitter receptors [2]. It could be demonstrated that this is not only true for the receptor protein: there is also a decrease in mRNA expression for these receptors. *Microchip array studies* have shown very elegantly in various HD mouse lines that these changes are only part of a very complex but specific alteration in the neuronal expression pattern which occurs very early in the disease [9]. As a consequence, neuronal dysfunction develops at multiple cell sites affecting calcium homeostasis, mitochondrial energy metabolism, neurotransmitter dysfunction and various further processes (Figure 2).

The specificity of these changes in transcriptional regulation has focused current research on the role of transcription factors. In fact, it was demonstrated that a number of *transcription factors* including CREB binding protein (CBP) or p53 form aggregates with mutant huntingtin and can be found in NII of HD mice. Therefore it is highly likely that transcriptional dysregulation in HD is a consequence of inhibition of transcription factors through incorporation into nuclear aggregates [3] (Figure 2).

REFERENCES

1. Carter RJ, Lione LA, Humby T, Mangiarini L, Mahal A, Bates GP, Dunnett SB, Morton AJ (1999) Characterization of progressive motor deficits in mice transgenic for the human Huntington's disease mutation. *J Neurosci* **19:** 3248-3257.
2. Cha JH, Kosinski CM, Kerner JA, Alsdorf SA, Mangiarini L, Davies SW, Penney JB, Bates GP, Young AB (1998) Altered brain neurotransmitter receptors in transgenic mice expressing a portion of an abnormal human huntington disease gene. *Proc Natl Acad Sci USA* **95:** 6480-6485.
3. Cha JH (2000) Transcriptional dysregulation in Huntington's disease. *Trends Neurosci* **23:** 387-392.
4. Davies S, Turmaine M, Cozens B, DiFiglia M, Sharp A, Ross C, Scherzinger E, Wanker E, Mangiarini L, Bates G (1997) Formation of neuronal intranuclear inclusions (NII) underlies the neurological dysfunction in mice transgenic for the HD mutation. *Cell* **90:** 537-548.
5. Ho M, Chelly J, Carter N, Danek A, Crocker P, Monaco AP (1994) Isolation of the gene for McLeod syndrome that encodes a novel membrane transport protein. *Cell* **77:** 869-880.
6. Huntington's Disease Collaborative Research Group (1993) A novel gene containing a trinucleotide repeat that is expanded and unstable on Huntington's disease chromosomes. *Cell* **72:** 971-983.
7. Lione LA, Carter RJ, Hunt MJ, Bates GP, Morton AJ, Dunnett SB (1999) Selective discrimination learning impairments in mice expressing the human Huntington's disease mutation. *J Neurosci* **19:** 10428-10437.
8. Lüesse H, Schiefer J, Sprünken A, Puls C, Block F, Kosinski CM (2001) Evaluation of R6/2 transgenic mice for therapeutic studies in Huntington's disease: Behavioral testing and impact of diabetes mellitus. *Behav Brain Res* **126:** 185-195.
9. Luthi-Carter R, Strand A, Peters NL, Solano SM, Hollingsworth ZR, Menon AS, Frey AS, Spektor BS, Penney EB, Schilling G, Ross CA, Borchelt DR, Tapscott SJ, Young AB, Cha JH, Olson JM (2000) Decreased expression of striatal signaling genes in a mouse model of Huntington's disease. *Hum Mol Genet* **9:** 1259-1271.
10. Mangiarini L, Sathasivam K, Seller M, Cozens B, Harper A, Hetherington C, Lawton M, Trottier Y, Lehrach H, Davies S, Bates G (1996) Exon 1 of the HD gene with an expanded CAG repeat is sufficient to cause a progressive neurological phenotype in transgenic mice. *Cell* **87:** 493-506.
11. Perutz M (1996) Glutamine repeats and inherited neurodegenerative diseases: molecular aspects. *Curr Opinion Struct Biol* **6:** 848-858.
12. Rampoldi L, Dobson-Stone C, Rubio JP, Danek A, Chalmers RM, Wood NW, Verellen C, Ferrer X, Malandrini A, Fabrizi GM, Brown R, Vance J, Pericak-Vance M, Rudolf G, Carre S, Alonso E, Manfredi M, Nemeth AH, Monaco AP (2001) A conserved sorting-associated protein is mutant in chorea-acanthocytosis. *Nat Genet* **28:** 119-120.

13. Reddy PH, Williams M, Charles V, Garrett L, Pike-Buchanan L, Whetsell WO, Miller G, Tagle DA (1998) Behavioural abnormalities and selective neuronal loss in HD transgenic mice expressing mutated full-length HD cDNA. *Nat Genet* **20:** 198-202.
14. Scherzinger E, Lurz R, Turmaine M, Mangiarini L, Hollenbach B, Hasenbank R, Bates G, Davies S, Lehrach H, Wanker E (1997) Huntingtin-encoded polyglutamine expansions form amyloid-like protein aggregates in vitro and in vivo. *Cell* **90:** 549-558.
15. Sipione S, Cattaneo E (2001) Modeling Huntington's disease in cells, flies, and mice. *Mol Neurobiol* **23:** 21-51.
16. Ueno S, Maruki Y, Nakamura M, Tomemori Y, Kamae K, Tanabe H, Yamashita Y, Matsuda S, Kaneko S, Sano A (2001) The gene encoding a newly discovered protein, chorein, is mutated in chorea-acanthocytosis. *Nat Genet* **28:** 121-122.
17. Wellington C, Hayden M (1997) Of molecular interactions, mice and mechanisms: New insights into Huntington's disease. *Curr Opin Neurol* **10:** 291-298.
18. Wheeler VC, Auerbach W, White JK, Srinidhi J, Auerbach A, Ryan A, Duyao MP, Vrbanac V, Weaver M, Gusella JF, Joyner AL, MacDonald ME (1999) Length-dependent gametic CAG repeat instability in the Huntington's disease knock-in mouse. *Hum Mol Genet* **8:** 115-122.

CHAPTER 26

MOTOR DEFICITS AS BIOMARKERS IN HUNTINGTON'S DISEASE: PERSPECTIVES FOR NEUROACANTHOCYTOSIS SYNDROMES

RALF REILMANN
Department of Neurology, Westfälische Wilhelms-Universität, Münster, Germany

Abstract. Objective measures to assess the progression of Huntington's Disease (HD) are desirable. The neurophysiological analysis of isometric grip forces during grasping with a precision grip was used to identify motor deficits correlating with the severity of HD. The amount of grip force variability while holding an object was correlated to the total motor score of the Unified Huntington's Disease Rating Scale (UHDRS). Grip force variability increased in all HD patients during a three year follow-up. It assessed disease progression more sensitively than the UHDRS. In addition, a method for the assessment of involuntary choreatic movements is introduced. Possible applications of the techniques as biomarkers in HD and in studies on patients with neuroacanthocytosis syndromes are discussed.

INTRODUCTION

Cognitive decline, psychosocial impairments and the development of motor deficits are typical findings in patients with Huntington's Disease (HD) [25]. Besides the characteristic involuntary movements, depicted as "*chorea*", various other motor impairments, such as bradykinesia, dystonia, sequencing deficits, motor impersistence, and, particularly in later stages, akinesia, are common [7,22,33]. Recent studies presented evidence for the presence of subtle motor deficits in otherwise clinically presymptomatic carriers of the HD gene [30,31,49]. The impairments in motor function are attributed to the widespread and progressive neuronal degeneration in the brains of patients with HD which is observed even before clinical symptoms become apparent [1-5,46,51].

A. Danek (ed.), Neuroacanthocytosis Syndromes, 233–242.

At present, no causal treatment for HD is available. Advances in genetics and molecular biology, however, have generated a number of different approaches to possible therapeutic interventions [13]. Some of these have already been transferred into clinical trials, others are currently tested in animal models [6,36]. The quality, sensitivity, and validity of the methods used to assess possible treatment effects are crucial for the success of future clinical trials which in HD is particularly important, since the number of patients available for clinical trials is limited due to the fairly low incidence. This also applies to patients with neuroacanthocytosis syndromes.

In the setting of clinical trials, patients with HD so far have commonly been assessed by the use of clinical scales, such as the *Unified Huntington's Disease Rating Scale* (UHDRS) [27]. Unfortunately, these scales present serious limitations: (1) the categorical structure of the tests will not detect small changes in performance, thus limiting their sensitivity; (2) categorical data do not allow for the application of powerful statistical analysis which is only possible with continuous, numerical, objective data; (3) errors of the individual rater influence the reliability which is particularly critical in multi-center trials. Thus, more objective and quantitative measures that can serve as biomarkers of disease stage and of disease progression are desirable.

With respect to the different clinical symptoms of HD and neuroacanthocytosis syndromes, the assessment of motor deficits may provide an approach to identify objective and quantitative measures for preclinical evaluation and objective follow-up. The progressive degeneration of brain areas involved in motor control has been described in several studies in HD [2-5,32,51]. It is hypothesized that the severity of motor dysfunction is dependent upon the degree of the neurodegeneration as observed morphologically. Neurophysiological paradigms capable of assessing motor dysfunction quantitatively should be suitable to serve as objective biomarkers in HD and possibly also in neuroacanthocytosis syndromes. Ideally, these paradigms should: (1) assess a functionally relevant motor task; (2) require a complex integration of sensorimotor circuits in order to detect deficits caused by the widespread neurodegeneration as early as possible; (3) be non-invasive; (4) carry no risk for the patients; and (5) be usable in an outpatient setting.

The coordination and development of grip forces in precision grasping between the thumb and index finger is a highly complex motor task [20]. It requires the sensorimotor integration of several sensory modalities and the cooperation of several motor areas in the brain to generate adequate motor output (i.e. scaling of grip forces) [28]. In addition, it constitutes a functionally relevant task needed for successful manipulation of objects in everyday life. The physiology of grip force control in precision grasping has well been studied in humans [29]. Adaptation to different object conditions such as weight, surface structure, and perturbations have led to robust findings [20,28]. This paradigm has successfully been applied for the evaluation of motor dysfunction in various movement disorders, including Parkinson's Disease [15,17-19,21,23],

Multiple System Atrophy [38], dystonia [39,47], and Tourette's Syndrome [16]. Prior evidence for impaired coordination of grip forces in HD was reported after sudden perturbations [14]. So far the coordination of isometric grip forces has not been investigated in patients with neuroacanthocytosis syndromes.

In our group, the coordination of grip forces during precision grasping in patients with HD has been investigated with the following objectives: (1) to identify impairments in the force coordination during different tasks involving precision grasping in cross-sectional studies; (2) to identify force measures that correlate to the severity of HD as assessed clinically in the UHDRS; (3) to test, in a follow-up study, whether these measures progress in the individual disease course, based on the assumption that they might be capable to assess motor deficit progression and the disease process itself. In addition, development of a method to assess the amount of involuntary choreatic movements was desirable to determine the role of chorea in HD and its impact on other motor measures.

The following sections will summarize the most important results of these studies. In the final section I will discuss the potential of the new methods with respect to applications in patients suffering from neuroacanthocytosis syndromes.

IMPAIRED COORDINATION OF GRIP FORCES IN HD AND CORRELATION TO CLINICAL DEFICITS

The coordination of grip force generation was compared between patients with HD and control subjects [22]. Subjects were asked to grasp and lift an object equipped with force transducers and to hold it stable in the air next to a marker for several seconds (Figure 1A). Grip (normal) and load (tangential) forces of both digits and the object's position in space were recorded (Figure 1B). The object's weight (200 g, 400 g, 800 g) and surface texture (sandpaper or rayon) were modified to test the adaptation of grip forces to different sensory stimuli [22]. In a second experiment, the same subjects were asked to grasp and transport the object, again the object's weight and surface texture were altered [40].

The results indicated that subjects with symptomatic HD show impaired initiation and delayed transitions between movement sequences as well as in different phases of the force development during initiation of the grasp [22,40]. In addition, they produce higher and very variable isometric grip forces in the static holding phase compared to controls [22]. The increase in grip force variability assessed as the coefficient of variation of grip forces was one of the most robust findings across all object conditions (see Figure 1C). Increased variability of grip forces was also found when the patients transported the object [40]. In contrast, subjects with HD demonstrated preserved anticipatory scaling of force development based on the object's expected physical properties (planning) and adjustment of the force to the object's actual physical properties such as weight and surface texture (sensorimotor integration) [22,40].

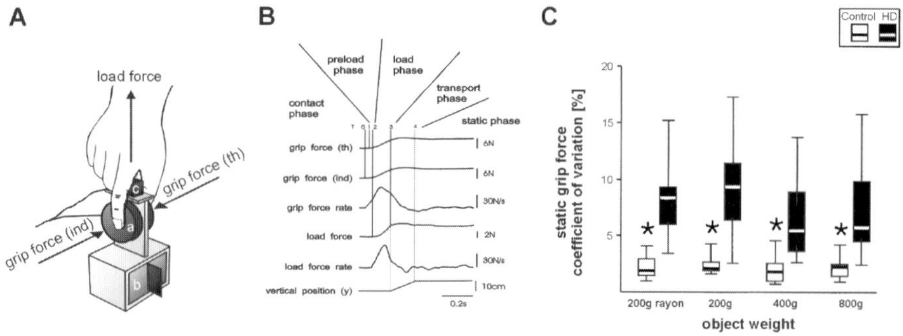

Figure 1. (A): Grip instrument with force transducers, a; weight of the object adjustable to 200 g or 800 g, b; electromagnetic position sensor, c. (B): Sample recording, showing the grip force of each finger, the mean grip force rate, mean load force, mean load force rate, vertical position and the phases evaluated. (C): Box-whisker plots indicating the medians and quartiles (boxes) as well as the minimum and maximum values (whiskers) of the coefficient of variation (mean/standard deviation ×100) in grip force during the static phase for each object condition (sandpaper contact surfaces used except where noted). Asterixes above the box-whisker plots indicate a significant difference ($*p < 0.05$) from the subjects with Huntington's Disease for that condition. (*Figures reproduced and modified from Gordon AM, Quinn L, Reilmann R, Marder KM, Exp Neurol, 2000; 163:136-148 with kind permission of Elsevier Science Publisher, Orlando, FL, USA.*)

The observed deficits generally were unrelated to the overall disease severity as assessed clinically in the UHDRS [27]. However, the variability of grip and load forces in the static holding phase was correlated with the deficits seen in the total motor score and the functional capacity score of the UHDRS [22]. This observation was independent of the level of chorea observed in the UHDRS. These results implied that patients more severely affected by HD, i.e. patients in more advanced stages of the disease, might exhibit more grip force variability than patients in earlier stages.

It was hypothesized that grip force variability might increase during the course of HD within individual subjects, suggesting that it could serve as a biomarker for the stage and progression of motor deficits and the disease process. In order to test this hypothesis, grip force variability and other measures of grip force coordination were analysed in a follow-up study in patients with HD.

ASSESSMENT OF PROGRESSION IN HD USING GRIP FORCE ANALYSIS

The coordination of isometric grip forces in precision grasping and the clinical performance in the UHDRS were assessed in ten patients with HD with a mean follow up of three years [44]. Patients performed the same paradigm as described in section 2. Grip and load forces and the object's 3D position (x-,

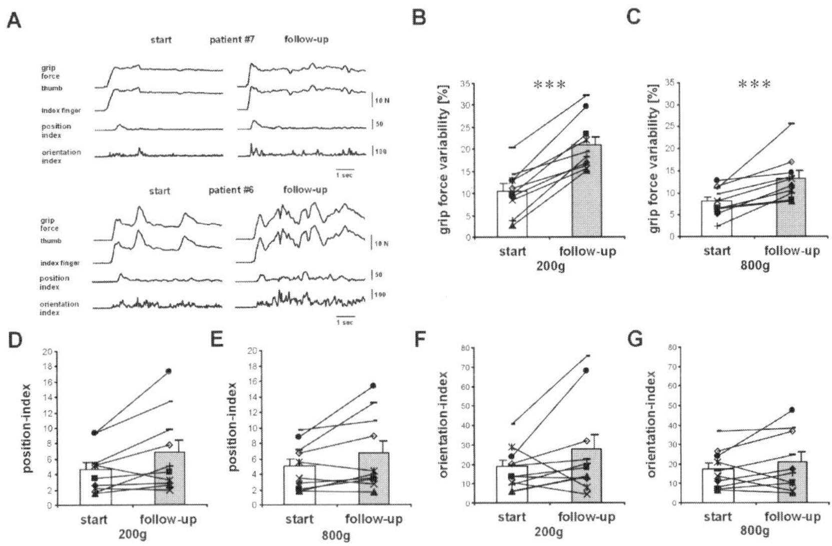

Figure 2. (A): Grip forces, position index, and orientation index during a typical trial with the 800 g object weight at the start compared to the follow-up assessment for two representative patients with Huntington's Disease. (B, C): Grip force variability for the two object weights 200 g and 800 g at the start and follow-up assessment. (D, E): position index and (F, G): orientation index of the object at start and follow-up (shown for the 200 g and the 800 g weight). Bars represent means ± SEM. Symbols represent individual patients ($p \leq 0.001$). (*Figures reproduced and modified from Reilmann et al, Neurology, 2001; 57:920-924 with kind permission of Lippincott Williams & Wilkins Publishers, Baltimore, MD, USA.*)

y-, z-position and roll-, pitch-, yaw-orientation) were recorded [44]. The performance of patients was compared between the start and follow-up assessment.

Isometric grip force variability in the static holding phase increased significantly across patients with HD during the follow-up period (Figure 2A-C) [44]. This increase was seen in each of the patients and in all object conditions. The amount of chorea, as assessed either by the UHDRS or by using the objective recordings of the object's 3D-position and orientation [45] (see section 4), did not increase significantly during the follow-up (Figure 2D-G). Accordingly, correlation analysis revealed that the increase in grip force variability could not be explained by changes in the amount of chorea [44]. Instead the motor variability may represent a distinct motor deficit in HD possibly caused by impaired feedback control [49] or impaired hand motor control [8,22].

In conclusion, grip force variability provided a measure for the deterioration of motor function in all HD patients in this study. Since the increase in force variability should be caused by the progressive neurodegeneration [4,5], grip force variability could be a candidate for a biomarker of disease progression in HD. In contrast, a simultaneous progression of motor deficits could not reliably be detected using the UHDRS motor scores in this study [44]. In order to document significant changes in the UHDRS motor scores in longitudinal studies, large sample sizes and long-term follow-up periods are required [35,48]. Thus, grip force variability might constitute a more sensitive marker to assess disease progression and motor deficits than the UHDRS.

NEUROPHYSIOLOGICAL ASSESSMENT OF CHOREA IN HD

Involuntary movements are a characteristic clinical sign in HD. The severity of chorea usually increases in early to mid stages and decreases towards final stages of HD [25]. Clinical observation suggests that the amount of chorea is very variable within subjects depending on psychological factors like stress and mood [25]. Therefore chorea is not a suitable marker to assess the stage and progression of HD. However, an objective investigation of the role of chorea and its impact on other motor symptoms was desirable.

An objective assessment of chorea should be performed in a controlled and functionally meaningful setting. Usually chorea at rest does not disturb the patients [50]. In contrast, chorea is impairing if it interferes with the performance of voluntary motor activity, e.g. the manipulation of objects. Manipulating objects, e.g. holding a glass of water, is an important everyday task. Chorea interfering with this task may result in spilling of the water or even in dropping the glass. The previously applied paradigm of grasping, lifting and holding an object was used to assess the amount and quality of involuntary movements occurring during this task [45].

While eighteen patients with HD were holding an object, the x-, y-, z-position and the roll-, pitch-, yaw-orientation of the object were recorded [45]. Derivatives of the position and orientation channels were calculated and their absolute values were added to create a "position-index" and an "orientation-index" (see Figure 2A) [44]. Both indices were indicative of the amount of movement that occurred while patients were instructed to hold the object stable. The mean values of both the position- and orientation-index were correlated to the UHDRS chorea scores [45]. This suggests that both indices provide a quantitative and objective measure of the amount of chorea in HD with higher indices indicating more severe chorea.

NEUROPHYSIOLOGICAL ASSESSMENT OF MOTOR DEFICITS IN NEUROACANTHOCYTOSIS SYNDROMES

Patients with neuroacanthocytosis syndromes clinically exhibit motor symptoms similar to those seen in HD [41]. Since the methods described above were successfully applied to investigate motor impairments in patients with HD and other movement disorders, it seemed reasonable to expect that they would also elucidate deficits of motor function in patients with neuroacanthocytosis syndromes.

Chorea-Acanthocytosis

As indicated by its name, chorea-acanthocytosis (MIM 200150) is usually associated with hyperkinetic movements. Choreatic movements are seen in about 85% of the patients and usually develop in the third or fourth decade of life [42]. The hyperkinesia is similar to the chorea observed in HD, characterized by abrupt, irregular, involuntary movements of variable severity mainly affecting the limbs [24]. In later stages, many of the patients eventually show signs of ataxia, they develop dystonia, motor tics, and Parkinsonian features [24,42].

The degeneration observed in the brains of these patients is focused on the basal ganglia, particularly the caudate nucleus and putamen [41]. Thus, impairments in grip force coordination are likely to be similar to those seen in HD [22,44]. However, since HD is associated with a significant neurodegeneration in the cortex [32,46], which is not seen in chorea-acanthocytosis [41], differences in the pathology of grip force coordination may be observed. These differences could offer valuable insights into the normal physiology of grip force coordination. Since the neurodegeneration in chorea-acanthocytosis is progressive, motor deficits assessable by the above test battery might be possible candidates for biomarkers.

McLeod Syndrome

McLeod syndrome (MIM 110900) is associated with hyperkinetic choreiform movements [41] in more than 94% of patients [9]. They commonly develop in the fifth decade and are progressive, particularly affecting the limbs, the trunk and the face [10]. Signs of dystonia eventually develop in about half of the patients. Also Parkinsonian features occur, but are less common [9,41].

Neurophysiological assessment of chorea in McLeod syndrome also should be possible as in HD [45]. The neurodegeneration observed in the brains of these patients is most pronounced in the caudate nucleus and the putamen [41] suggesting deficits in grip force coordination similar to those observed in HD [22,44]. In addition, PET analysis indicated hypometabolism in the frontal and parietal cortex suggesting possible impairment of motor control beyond that observed in basal ganglia degeneration [12]. Since the neurodegeneration progresses in the course of McLeod syndrome, motor deficits associated with the progression of the disease may be identified and serve as valuable biomarkers.

Abetalipoproteinemia

The main motor symptom of abetalipoproteinemia (MIM 200100) is a progressive ataxia [41]. While damage to the striatum is indicated by reduced [^{18}F]-fluorodopa uptake in the putamen and caudate nucleus, usually no clinical signs of basal ganglia dysfunction are observed [11]. However, deficits of grip force coordination have been described in patients with cerebellar disease [26,34,37]. Thus the motor deficits in grip force coordination associated with the ataxia of patients with abetalipoproteinemia may also serve as a target for quantitative neurophysiological assessment using the methods described above.

CONCLUSION

Studies of motor function in neuroacanthocytosis syndromes are warranted to further investigate the nature of motor impairment in these patients. The paradigms reported here, or comparable techniques requiring a higher complexity of motor coordination [43,49] should be evaluated in these patients. In addition, these studies may contribute to a better understanding of the physiology of force coordination if the distinct neuropathology seen in these syndromes [41] is correlated to specific deficits. As in HD, the identification of reliable and objective biomarkers would be helpful for the assessment of disease severity and progression in clinical trials.

Acknowledgements

The work of the author reported in this chapter was supported by a grant of the "Deutsche Forschungsgemeinschaft" (German Research Foundation, DFG-Re-1330/1-1) and by the "Westfälische Stiftung für Neuromedizin" (Westphalian Neuromedical Foundation). The author is indebted to Andrew Gordon, Ph.D., and Karen Marder, M.D., M.P.H., both at Columbia University, New York, USA, for their support. The continuous support of all patients with HD and their relatives is especially appreciated.

REFERENCES

1. Aylward EH, Anderson NB, Bylsma FW, Wagster MV, Barta PE, Sherr M, Feeney J, Davis A, Rosenblatt A, Pearlson GD, Ross CA (1998) Frontal lobe volume in patients with Huntington's disease. *Neurology* **50:** 252-258.
2. Aylward EH, Brandt J, Codori AM, Mangus RS, Barta PE, Harris GJ (1994) Reduced basal ganglia volume associated with the gene for Huntington's disease in asymptomatic at-risk persons. *Neurology* **44:** 823-828.
3. Aylward EH, Codori AM, Barta PE, Pearlson GD, Harris GJ, Brandt J (1996) Basal ganglia volume and proximity to onset in presymptomatic Huntington disease. *Arch Neurol* **53:** 1293-1296.
4. Aylward EH, Codori AM, Rosenblatt A, Sherr M, Brandt J, Stine OC, Barta PE, Pearlson GD, Ross CA (2000) Rate of caudate atrophy in presymptomatic and symptomatic stages of Huntington's disease. *Mov Dis* **15:** 552-560.
5. Aylward EH, Li Q, Stine OC, Ranen N, Sherr M, Barta PE, Bylsma FW, Pearlson GD, Ross CA (1997) Longitudinal change in basal ganglia volume in patients with Huntington's disease. *Neurology* **48:** 394-399.
6. Beal MF, Hantraye P (2001) Novel therapies in the search for a cure for Huntington's disease. *Proc Natl Acad Sci USA* **98:** 3-4.
7. Berardelli A, Noth J, Thompson PD, Bollen EL, Curra A, Deuschl G, van Dijk JG, Topper R, Schwarz M, Roos RA (1999) Pathophysiology of chorea and bradykinesia in Huntington's disease. *Mov Dis* **14:** 398-403.

8. Bonfiglioli C, De Berti G, Nichelli P, Nicoletti R, Castiello U (1998) Kinematic analysis of the reach to grasp movement in Parkinson's and Huntington's disease subjects. *Neuropsychologia* **36:** 1203-1208.
9. Danek A, Rubio JP, Rampoldi L, Ho M, Dobson-Stone C, Tison F, Symmans WA, Oechsner M, Kalckreuth W, Watt JM, Corbett AJ, Hamdalla HH, Marshall AG, Sutton I, Dotti MT, Malandrini A, Walker RH, Daniels G, Monaco AP (2001) McLeod neuroacanthocytosis: genotype and phenotype. *Ann Neurol* **50:** 755-764.
10. Danek A, Tison F, Rubio J, Oechsner M, Kalckreuth W, Monaco AP (2001) The chorea of McLeod syndrome. *Mov Dis* **16:** 882-889.
11. Dexter DT, Brooks DJ, Harding AE, Burn DJ, Muller DP, Goss-Sampson MA, Jenner PG, Marsden CD (1994) Nigrostriatal function in vitamin E deficiency: clinical, experimental, and positron emission tomographic studies. *Ann Neurol* **35:** 298-303.
12. Dotti MT, Battisti C, Malandrini A, Federico A, Rubio JP, Circiarello G, Monaco AP (2000) McLeod syndrome and neuroacanthocytosis with a novel mutation in the *XK* gene. *Mov Dis* **15:** 1282-1284.
13. Feigin A, Zgaljardic D (2002) Recent advances in Huntington's disease: implications for experimental therapeutics. *Curr Opin Neurol* **15:** 483-489.
14. Fellows S, Schwarz M, Schaffrath C, Domges F, Noth J (1997) Disturbances of precision grip in Huntington's disease. *Neurosci Lett* **226:** 103-106.
15. Fellows SJ, Noth J, Schwarz M (1998) Precision grip and Parkinson's disease. *Brain* **121:** 1771-1784.
16. Flanagan JR, Jakobson LS, Munhall KG (1999) Anticipatory grip adjustments are observed in both goal-directed movements and movement tics in an individual with Tourette's syndrome. *Exp Brain Res* **128:** 69-75.
17. Forssberg H, Ingvarsson PE, Iwasaki N, Johansson RS, Gordon AM (2000) Action tremor during object manipulation in Parkinson's disease. *Mov Dis* **15:** 244-254.
18. Gordon AM (1997) Object release in patients with Parkinson's disease. *Neurosci Lett* **232:** 1-4.
19. Gordon AM (1998) Task-dependent deficits during object release in Parkinson's disease. *Exp Neurol* **153:** 287-298.
20. Gordon AM (2001) The development of hand motor control. In: *Brain and Behavior in Human Development* (Kalverboer AF, Gramsbergen A, eds), Dordrecht, Netherlands: Kluwer Academic Publishers
21. Gordon AM, Ingvarsson PE, Forssberg H (1997) Anticipatory control of manipulative forces in Parkinson's disease. *Exp Neurol* **145:** 477-488.
22. Gordon AM, Quinn L, Reilmann R, Marder K (2000) Coordination of prehensile forces during precision grip in Huntington's Disease. *Exp Neurol* **163:** 136-148.
23. Gordon AM, Reilmann R (1999) Getting a grasp on research: Does treatment taint testing of parkinsonian patients? *Brain* **122:** 1597-1598.
24. Hardie RJ, Pullon HW, Harding AE, Owen JS, Pires M, Daniels GL, Imai Y, Misra VP, King RH, Jacobs JM (1991) Neuroacanthocytosis. A clinical, haematological and pathological study of 19 cases. *Brain* **114:** 13-49.
25. Harper, P. S. *Huntington's Disease.* 1996. London, Saunders. (Major Problems in Neurology Vol. 22)
26. Hermsdörfer J, Wessel K, Mai N, Marquardt C (1994) Perturbation of precision grip in Friedreich's ataxia and late-onset cerebellar ataxia. *Mov Dis* **9:** 650-654.
27. Huntington Study Group (1996) Unified Huntington's Disease Rating Scale: reliability and consistency. *Mov Dis* **11:** 136-142.
28. Johansson RS (1998) Sensory input and control of grip. *Novartis Found Symp* **218:** 45-59; discussion 59-63.
29. Johansson RS, Cole KJ (1992) Sensory-motor coordination during grasping and manipulative actions. *Curr Opin Neurobiol* **2:** 815-823.

30. Kirkwood SC, Siemers E, Bond C, Conneally PM, Christian JC, Foroud T (2000) Confirmation of subtle motor changes among presymptomatic carriers of the Huntington disease gene. *Arch Neurol* **57:** 1040-1044.
31. Kirkwood SC, Siemers E, Stout JC, Hodes ME, Conneally PM, Christian JC, Foroud T (1999) Longitudinal cognitive and motor changes among presymptomatic Huntington disease gene carriers. *Arch Neurol* **56:** 563-568.
32. Lange H, Thorner G, Hopf A, Schröder KF (1976) Morphometric studies of the neuropathological changes in choreatic diseases. *J Neurol Sci* **28:** 401-425.
33. Louis ED, Lee P, Quinn L, Marder K (1999) Dystonia in Huntington's disease: Prevalence and clinical characteristics. *Mov Dis* **14:** 95-101.
34. Mai N, Bolsinger P, Avarello M, Diener HC, Dichgans J (1988) Control of isometric finger force in patients with cerebellar disease. *Brain* **111:** 973-998.
35. Marder K, Zhao H, Myers RH, Cudkowicz M, Kayson E, Kieburtz K, Orme C, Paulsen J, Penney JBJ, Siemers E, Shoulson I (2000) Rate of functional decline in Huntington's disease. Huntington Study Group. *Neurology* **54:** 452-458.
36. McMurray CT (2001) Huntington's disease: New hope for therapeutics. *Trends Neurosci* **24:** S32-S38.
37. Müller F, Dichgans J (1994) Dyscoordination of pinch and lift forces during grasp in patients with cerebellar lesions. *Exp Brain Res* **101:** 485-492.
38. Muratori L, Reilmann R, Gordon AM (2000) Coordination of isometric grip forces in multiple system atrophy (Shy-Drager type). *Mov Dis* **15:** 92.
39. Odergren T, Iwasaki N, Borg J, Forssberg H (1996) Impaired sensory-motor integration during grasping in writer's cramp. *Brain* **119:** 569-583.
40. Quinn L, Reilmann R, Marder K, Gordon AM (2001) Altered movement trajectories and force control during object transport in Huntington's disease. *Mov Dis* **16:** 469-480.
41. Rampoldi L, Danek A, Monaco AP (2002) Clinical features and molecular bases of neuroacanthocytosis. *J Mol Med* **80:** 475-491.
42. Rampoldi L, Dobson-Stone C, Rubio JP, Danek A, Chalmers RM, Wood NW, Verellen C, Ferrer X, Malandrini A, Fabrizi GM, Brown R, Vance J, Pericak-Vance M, Rudolf G, Carre S, Alonso E, Manfredi M, Nemeth AH, Monaco AP (2001) A conserved sorting-associated protein is mutant in chorea-acanthocytosis. *Nat Genet* **28:** 119-120.
43. Reilmann R, Gordon AM, Henningsen H (2001) Initiation and development of fingertip forces during whole-hand grasping. *Exp Brain Res* **140:** 443-452.
44. Reilmann R, Kirsten F, Quinn L, Henningsen H, Marder K, Gordon AM (2001) Objective assessment of progression in Huntington's disease: a 3-year follow-up study. *Neurology* **57:** 920-924.
45. Reilmann R, Kirsten F, Quinn L, Marder K, Henningsen H, Gordon AM (2002) Objective neurophysiological analysis of chorea in Huntington's Disease. *Neurology* **58:** 308.
46. Rosas HD, Liu AK, Hersch S, Glessner M, Ferrante RJ, Salat DH, van Der KA, Jenkins BG, Dale AM, Fischl B (2002) Regional and progressive thinning of the cortical ribbon in Huntington's disease. *Neurology* **58:** 695-701.
47. Serrien DJ, Burgunder JM, Wiesendanger M (2000) Disturbed sensorimotor processing during control of precision grip in patients with writer's cramp. *Mov Dis* **15:** 965-972.
48. Siesling S, van Vugt JP, Zwinderman KA, Kieburtz K, Roos RA (1998) Unified Huntington's disease rating scale: a follow up. *Mov Dis* **13:** 915-919.
49. Smith MA, Brandt J, Shadmehr R (2000) Motor disorder in Huntington's disease begins as a dysfunction in error feedback control. *Nature* **403:** 544-549.
50. Snowden JS, Craufurd D, Griffiths HL, Neary D (1998) Awareness of involuntary movements in Huntington disease. *Arch Neurol* **55:** 801-805.
51. Thieben MJ, Duggins AJ, Good CD, Gomes L, Mahant N, Richards F, McCusker E, Frackowiak RS (2002) The distribution of structural neuropathology in pre-clinical Huntington's disease. *Brain* **125:** 1815-1828.

CHAPTER 27

TREATMENT OPTIONS IN HUNTINGTON'S DISEASE

MATTHIAS DOSE
Bezirkskrankenhaus Taufkirchen, Taufkirchen, Germany

Abstract. Besides monogenetic inheritance, neuroacanthocytosis syndromes like McLeod- and Levine-Critchley syndrome share aspects of motor and mental symptomatology with Huntington's disease (HD). Since an expansion of a trinucleotide repeat on chromosome 4p16.3 had been identified as the mutation causing HD in 1993, much hope has been raised that it would take only a short time span to decode the molecular pathway "from gene to symptoms" and subsequently develop causal therapeutic strategies for this devastating progressive disease. However, though many fragments of the molecular "puzzle" of HD pathology have been revealed nearly ten years after the discovery of the HD gene, causal targeting of its molecular pathology is still missing. Besides conventional symptomatic treatment of psychiatric and neurologic symptoms of HD with antihyperkinetic (e.g. dopamine antagonists, tetrabenazine) and psychotropic (e.g. antipsychotics, antidepressants) drugs, substances which are supposed to exert "neuroprotective" effects (e.g. remacemide, coenzyme Q10, riluzole, creatine, memantine) have been or are being examined in controlled clinical studies in the US and Europe. Up to now, none of these substances has been found to exert statistically significant effects on slowing down the progression of HD symptomatology, but some of the studies are still under way. Data from *in vitro* experiments and HD mutant animal models suggest strategies to either prevent aggregation of expanded polyglutamines, activation of microglia, or inhibition of caspase cleavage. Spectacular results have been published one year after transplantation of fetal basal ganglia tissue in single HD cases. However, long-term observations are missing and from a methodological point it is quite questionable, whether local transplants into HD brains affected by systematic neurodegeneration which is not confined to the basal ganglia will be able to exert enduring positive effects. Overall, after nearly ten years of research in therapeutic options since the discovery of the HD gene we are still far from real therapeutic breakthroughs. With respect to HD sufferers and their families, scientists in the field should be reluctant to awaken illusionary hopes and expectations by publishing or interpreting preliminary results which are far from being able to be realized as promising therapeutic options.

A. Danek (ed.), Neuroacanthocytosis Syndromes, 243–250.

INTRODUCTION

Huntington's disease (HD) is an autosomal-dominant neurodegenerative disorder named after George Huntington (1850-1916). He had succeeded his father and grandfather in general practice and was able to give one of the first descriptions of a hereditary adult onset chronic progressive chorea with mental deterioration [11]. The term chorea originates from the Greek and Latin word "chorus" for "dance". Chorea until recently was used to characterize HD but should not be used any longer. The gait and hyperkinetic movements it describes are only a part of the complex symptomatology of HD. Rigidity, hyperkinesia and seizures with onset at 4 and 10 years of age were described by Hoffmann in 1888 in two daughters from a three-generation HD family [12]. Westphal (after whom early onset HD is often named "Westphal variant"), in contrast, did attribute the clinical features of a 18-year old patient to a cause separate from HD [22]. Psychiatric symptoms including psychoses, depression ("melancholia"), suicidality, irritability and "steady mental degeneration" were first systematically studied in 1892 by Phelps in Rochester, Minnesota [15].

Although Huntington had given a clear description of dominant inheritance, it was Punnett who in 1908 listed HD as likely to follow a Mendelian pattern [16] since Mendel's work (of 1865) had not been appreciated until 1900 [9]. Later on, geneticists were more confident in listing HD as an autosomal-dominant disorder [3]. This also made room for radical eugenic views which ended up in compulsory sterilization in several countries and in the murder of HD patients in Nazi Germany [13].

Since 1978, lay organizations from different countries have joined in the International Huntington's Association (IHA) which is closely linked to the World Federation of Neurology (WFN) Research Group on Huntington's chorea – thus trying to provide research with guidance according to patient needs and to prevent inhuman misuse as in the recent past. One of the results of this constructive collaboration are "Guidelines for predictive testing" which had been developed after the HD gene had been localized in 1983 and DNA-markers allowed for "indirect" testing in families with diagnosed patients and a sufficient number of relatives willing to undergo the test [23]. Since these "Guidelines" covered most of the relevant issues for people at risk for HD, they only had to be adapted when the HD mutation was discovered on the short arm of chromosome 4 in 1993 [20] and "direct" testing had become possible.

CLINICAL FEATURES OF HD

Chorea, defined by the WFN as "a state of excessive, spontaneous movements, irregularly timed, randomly distributed and abrupt", is the most frequent and conspicuous feature of HD. It varies from restlessness with mild, intermittent exaggeration of gesture and expression, fidgeting movements of the hands, unstable, dance-like gait to disabling, violent movements [19]. Its severity tends to increase during the initial years of illness and – in contrast to continuous progression of other motor symptoms – reaches a plateau in late stages. Other neurological signs comprise dystonia, rigidity, bradykinesia, akinesia, abnormalities of eye movements, myoclonus, epilepsy, dysarthria, dysphagia and cerebellar abnormalities. Schizophrenia-like symptoms, delusions, affective disorders (mostly depression), anxiety, obsessive-compulsive symptoms, irritability, aggression and cognitive decline occur in every stage of the disease. They are first symptoms in up to 39% of all HD patients. Another 39% show neurological symptoms first and 22% show a "mixed" symptomatology [21]. Weight loss and cachexia which may occur independently of hyperkinesias and, partially, of dysphagia characterize advanced stages. They possibly constitute a disease manifestation since they may persist in spite of high caloric intake through stomach tubes.

GENETICS AND PATHOLOGY

An expansion of a trinucleotide repeat (CAG) on chromosome 4p16.3 within the coding region of a gene termed *IT15* was identified as the mutation causing HD [20]. The HD gene encodes for a protein of unknown function named huntingtin which is expressed ubiquitously. Abnormal lengthening of the CAG repeat sequence in the gene results in expanded glutamine stretches within huntingtin which contribute to HD pathology.

Polyglutamines encoded by triplet repeats are present in a variety of proteins and are common and non-pathogenic up to a critical threshold. Larger expansions, however, cause a variety of neurodegenerative diseases such as HD, spinocerebellar ataxias (SCA), spinobulbar muscular atrophy (SMA) and dentato-rubro-pallido-Luysian atrophy (DRPLA) with the common feature of progressive neuronal loss and decline of motor and cognitive functions [10].

In HD brains, the N-terminus of mutant huntingtin has been localized in neuronal intranuclear inclusions (NII) and in dystrophic neurites in juvenile brains with a higher frequency than in adult brains [4]. This observation suggests a contribution of intraneuronal and intranuclear huntingtin aggregates that are dependent on CAG repeat length to HD symptomatology and neuronal death. It has, however, also been argued that NIIs merely reflect an epiphenomenon of neurotoxicity or that their formation may possibly be cell-protective, since huntingtin could be demonstrated to induce apoptosis without a correlation between cell death and formation of NII's [18].

Recently, it was claimed that in cultured cells apoptosis through caspase-8 and caspase-3 in the presence of mutant huntingtin is activated by a newly identified protein, Hippi. In addition, caspase-3 by cleavage of mutant huntingtin produces fragments that clump together and form inclusions in neurons and their nuclei. This leaves the question open as to whether the formation of inclusions is essential for cell death or not [8].

At post-mortem, HD brains show a reduction of weight by 200-300 g from a generalized atrophy which particularly affects the neostriatum (caudate nucleus and putamen). Although it has been claimed that the grade of pathology relates to CAG repeat length, several studies failed to establish a positive relationship [17]. Microscopically, a selective loss of spiny neurons (most pronounced in the caudate nucleus) is observed. This suggests an influence on various neurotransmitters. Increased thalamic excitatory activity resulting from excessive inhibition of the subthalamus by loss of GABA- and enkephalin-ergic neurons has been implicated in the etiology of chorea while direct thalamic inhibition by loss of GABA- and substance-P-neurons should be implicated in rigidity and bradykinesia. However, altered levels of acetylcholine and dopamine with altered receptor levels are also thought to be involved in symptom generation [14].

SYMPTOMATIC TREATMENT

Although HD at present is not curable, individual symptomatic treatment of disabling or disturbing motor and mental symptoms can at least relieve the burden of HD for sufferers and caregivers and maintain or improve quality of life. Chorea responds to substances which inhibit or decrease dopaminergic transmission like tetrabenazine (presynaptic dopamine depletion), tiapride (preferentially striatal dopamine D_2-receptor antagonist) or classical antipsychotic substances. In chorea, the latter should be used very cautiously in order to avoid extrapyramidal syndromes (like akathisia and tardive dyskinesia which may be undistinguishable from choreatic symptoms) as well as dysphoric and anergic syndromes which may mimic HD-related psychiatric symptoms (e.g. depression). Bradykinesia and dystonia may respond to anti-Parkinsonian drugs such as amantadine, while rigidity may require treatment with anti-spastic medications like baclofen, lisuride and amantadine. No special treatment options exist for dysarthria and dysphagia, although a well balanced combination of antihyperkinetics with an anti-Parkinsonian drug may provide improvement in some cases. For the treatment of seizures that are most prominent in juvenile cases valproic acid is recommended because of its putative GABA-ergic mechanism of action and its benign pharmacokinetic interactions with other drugs, as compared to carbamazepine (see the more extensive discussion in Chapter 13). Treatment of psychiatric syndromes in HD corresponds to usual recommendations only with the exception that substances with anticholinergic side effects (e.g. tricyclic antidepressants) should be avoided because of possible potentiation of choreatic hyperkinesias. Non-pharmacological therapies like physiotherapy, speech therapy, occupational and creative therapy, supportive psychotherapy, nutrition and social support contribute to individual symptom relief and improvement of quality of life [5].

NEUROPROTECTIVE TREATMENT OPTIONS

A supposed gain of toxic functions (or loss of physiologic functions) of proteins with polyglutamine expansions can theoretically be the result of different pathomechanisms providing specific options for pharmacological interventions. Hughes and Olson [10] have recently summarized these hypotheses and their possible therapeutic implications which are briefly reported here.

Protein Folding

One of the fundamental defects conferred by polyglutamine expansion is misfolding and altered solubility of proteins. If the capacity of the chaperone system (enzymes facilitating protein folding *in vivo*) in combination with the cell's ability to degrade mutant polyglutamine proteins by the ubiquitin/proteasome system were important to maintain cellular function, activation of chaperones or enhancement of proteasome clearance by small molecules may be of benefit.

Aberrant Protein Interactions and Effects on Transcription

Physical interactions between polyglutamine-containing proteins and transcription factors may either lead to primary cellular defects by perturbing histone acetylation or deacetylation or alter levels of mRNAs encoding proteins involved in neuronal signaling and ion homeostasis. Since it is far from being understood which of the observable gene expression changes are pathogenic and which may be compensatory or markers of physiological changes there are at present no means of interventions along these pathways.

Pathogenic Signaling between Neurons and Neuronal Environment

Neurodegenerative changes of HD can be mimicked in animal models by application of excitotoxic neurotransmitter agonists inducing a cascade of calcium influx, formation of free radicals, blockade of mitochondrial respiratory chain and – in some cases – cell death. Mitochondrial poisons can replicate a similar pattern of neurodegeneration. Neuroprotective effects of glutamate receptor antagonists and drugs which enhance mitochondrial function in different animal models of neurodegenerative diseases have encouraged clinical trials of NMDA antagonists (ketamine, remacemide, memantine), inhibitors of glutamate-release (baclofen, lamotrigine, riluzole) and mitochondrial support agents (coenzyme Q10, creatine).

None of these clinical trials which included well-designed and properly controlled studies (e.g. with remacemide or coenzyme Q10) has so far shown significant protective effects in HD. A placebo-controlled 3-years-study of riluzole ($n = 350$) at different European centers was scheduled for evaluation in 2003.

Activated microglia may be another pathogenic factor of neurodegeneration. Two substances in particular have been suggested for efficacy testing in neurodegenerative diseases: Cyclooxygenase inhibitors reduced microgliosis in animal models of Alzheimer's disease and cyclosporin eventually inhibited gliosis in a HD-patient who had received it for immunosuppression after fetal stem-cell transplantation and who died from heart failure.

Apoptosis and Caspase Cleavage

Besides activation of caspases (proteases involved in apoptosis), polyglutamine containing proteins have been shown to be substrates of caspase cleavage themselves. Since truncation of huntingtin increases its cellular toxicity some have speculated whether proteoloytic cleavage of huntingtin is involved in the initiation of HD and could be influenced by caspase inhibitors. In HD mice inhibition of caspases (in one case with minocycline) slows down disease progression, although – in the case of minocycline which would be suitable in humans – it remains unclear whether the modest clinical improvement is really due to caspase inhibition and not due to anti-inflammatory or calcium-chelating effects.

TRANSPLANTATION

Intrastriatal transplantation of fetal striatal neuroblasts has been reported to improve some motor and cognitive functions in three of five HD patients [1] and has been proposed as a future therapeutic option [6]. In forty patients with severe Parkinson's syndrome, however, a randomized double-blind placebo controlled study demonstrated significant improvement only for younger patients, but not for the total group of bilaterally transplanted patients. In addition, 15% of the transplanted patients developed severe dystonias and dyskinesias, not reported until now, which raised speculations concerning the preparation of the transplantation material [7]. Although more than 300 Parkinson's patients have been implanted with embryonic or fetal striatal tissues, numerous questions are still to be resolved: localization, choice of material, tissue cultivation, addition of neurotrophic factors, modification of progenitor cells, immunosuppression and treatment evaluation by unified scores [2]. Even more questions arise in the context of a systemic process like HD where neurodegeneration of striatal cells may only be an endpoint of pathology that probably starts in brain regions (e.g. cortex) which will not be influenced by striatal cell implants. Besides these more fundamental considerations, the present legal situation (at least in Germany) does not allow for implantation of fetal material because of ethical reasons with regard to "consuming" embryos for research: the utilization of fetal tissue is quite considerable since four embryos at least are needed for one transplantation.

CONCLUSIONS

When the expansion of a CAG repeat on chromosome 4 was identified as the cause of HD in 1993 there was great optimism that it would not take long to elucidate the pathways "from gene to symptoms" and to be able to offer – if not a cure of HD – a slow-down of disease progression with medication at least. Ten years later, the goal has not yet been reached. This does not mean that we must be pessimistic: the scientific community searching for a cure for HD is quite well-organized, in part generously funded by organizations like the Hereditary Disease Foundation (HDF) and closely cooperating with the International Huntington Association (IHA). Amazing parts of the puzzle of understanding HD, including models *in vitro* and in animals, have recently been elucidated and will shortly enable us to see and understand the whole picture. It is, however, our responsibility as scientists to point out again and again that – even if the puzzle of HD will be solved – there will still be a long way to go until curative and/or neuroprotective treatments will be available on the basis of controlled studies in a sufficient number of patients. Until this goal is reached we should modestly offer the present art of symptomatic treatment and support to our patients, including a comprehensible transmission of scientific information and the confidence that "we shall overcome" some day (according to a well-known song adapted to HD by the US branch of the IHA).

REFERENCES

1. Bachoud-Levi AC, Remy P, Nguyen JP et al (2000) Motor and cognitive improvements in patients with Huntington's disease after neural transplantation. *Lancet* **356:** 1975-1979.
2. Bauer P, Knoblich R, Mix E, Pahnke J, Rolfs A (2002) Aktueller Stand und Perspektive der Zelltransplantation bei neurodegenerativen Erkrankungen. *Nervenheilkunde* **21:** 88-93.
3. Davenport CB, Muncey EB (1916) Huntington's chorea in relation to heredity and eugenics. *Eugenics Record Office Bulletin* **17:** 195-222.
4. DiFiglia M, Sapp E, Chase KO, Davies SW, Bates GP, Vonsattel JP, Aronin N (1997) Aggregation of huntingtin in neuronal intranuclear inclusions and dystrophic neurites in brain. *Science* **277:** 1990-1993.
5. Dose M (1997) Klinische Diagnostik und Therapie bei Anlageträgern der Huntington-Krankheit. *Medizinische Genetik* **9:** 570-579.
6. Flint-Beal M et al (2001) Novel therapies in the search for a cure for Huntington's disease. *Proc Natl Acad Sci USA* **98:** 3-4.
7. Freed CR, Greene PE, Breeze RE et al (2001) Transplantation of embryonic dopamine neurons for severe Parkinson's disease. *New Engl J Med* **344:** 710-719.
8. Gervais FG et al (2002) Recruitment and activation of *caspase-8* by the Huntington-interacting protein Hip-1 and a novel partner Hippi. *Nat Cell Biol* **4:** 95-105.
9. Harper P (1991) *Huntington's Disease.* Saunders, London
10. Hughes RE, Olson JM (2001) Therapeutic opportunities in polyglutamine diseases. *Nature Med* **4:** 419-423.
11. Huntington G (1872) On chorea. *Medical and Surgical Reporter* **26:** 320-321.
12. Hoffmann J (1888) Über Chorea chronica progressiva (Huntington'sche Chorea, Chorea hereditaria). *Virchows Archiv* **111:** 513-548.

13. Klee E (1983) *"Euthanasie" im NS-Staat.* Fischer, Frankfurt/Main
14. Maat-Schieman ML, Dorsman JC, Smoor MA, Siesling S et al (1999) Distribution of inclusions in neuronal nuclei and dystrophic neurites in Huntington's disease brain. *J Neuropathol Exp Neurol* **58:** 129-137.
15. Phelps RM (1892) A new consideration of hereditary chorea. *J Nervous Mental Dis* **19:** 765-776.
16. Punnett RC (1908) Mendelian inheritance in man. *Proc Royal Soc Med* **1:** 135-168.
17. Rosenblatt A, Margolis RL, Becher MW et al (1998) Does CAG repeat number predict the rate of pathological changes in Huntington's disease? *Ann Neurol* **44:** 708-709.
18. Saudou F, Finkbeiner S, Devys D, Greenberg ME (1998) Huntingtin acts in the nucleus to induce apoptosis but death does not correlate with the formation of intranuclear inclusions. *Cell* **95:** 55-66.
19. Siesling S, van Vugt JP, Roos RA (1997) Hypokinesia: A prominent feature in Huntington's disease. *Kinesis* **2:** 17-22.
20. The Huntington's Disease Collaborative Study Group (1993) A novel gene containing a trinucleotide repeat that is expanded and unstable on Huntington's disease chromosomes. *Cell* **72:** 971-983.
21. Weindl A, Ceballos-Baumann AO (2001) Chorea. In: *Neurologische Therapie.* Lehmann-Horn F, Ludolph H (eds) Urban und Fischer, München, 126-149
22. Westphal C (1883) Über eine dem Bilde der cerebrospinalen grauen Degeneration ähnliche Erkrankung des zentralen Nervensystems ohne anatomischen Befund, nebst einigen Bemerkungen über paradoxe Kontraktion. *Arch Psychiatr Nervenkrankh* **14:** 87-96;767-773.
23. World Federation of Neurology Research Group on Huntington's Disease (1989) Ethical issues policy statement on Huntington's disease molecular predicitive test. *J Neurol Sci* **94:** 327-332.

CHAPTER 28

IS SURGICAL TREATMENT AN OPTION FOR CHOREA-ACANTHOCYTOSIS?

JENS VOLKMANN
Department of Neurology, Christian-Albrechts-Universität, Kiel, Germany

Abstract. Dystonic and choreatic hyperkinesias in neuroacanthocytosis syndromes may lead to severe motor disability in affected patients. Symptomatic medical treatment with neuroleptics or dopamine depleting agents can provide moderate improvement but hardly ever satisfactory long-term benefit. With the low morbidity rates of modern stereotactic techniques exploratory procedures in rare neurological conditions, which will never be addressed by controlled clinical trials, have become acceptable. This article discusses the pathophysiological rationale behind such surgical interventions and the treatment outcome in the few patients reported in the literature.

INTRODUCTION

The current model of basal ganglia pathophysiology suggests that hyperkinetic movement disorders such as chorea, dystonia or hemiballism result from abnormally decreased firing rates of neurons in subthalamic nucleus and internal globus pallidus (GPi) [5]. Inhibitory pallidal output to the thalamus is thus reduced and the resulting overactivity of the motor thalamus may account for involuntary movements generated in premotor cortical projection areas. Intraoperative microelectrode recordings in patients suffering from dystonia or hemiballism have confirmed such low discharge rates but also abnormal phasic discharge patterns of pallidal neurons when compared to patients suffering from Parkinson's disease or recordings in non-human primates [14].

A. Danek (ed.), Neuroacanthocytosis Syndromes, 251–257.

Pallidal and Thalamic Targets for Surgery

Interestingly, both pallidal lesioning and high-frequency stimulation are highly effective surgical procedures for the alleviation of levodopa-induced dyskinesias in Parkinson's disease [15] or of motor symptoms of dystonia [4]. The apparent paradox that reduced pallidal outflow leads to chorea, dystonia or hemiballism, whereas total abolition of pallidal activity may improve hyerkinetic movement disorders, has been a matter of intense debate [10]. A possible solution could be that the motor system is less likely to compensate for the input of abnormal ("noisy") neuronal signals than for the complete absence of basal ganglia input [10].

An alternative surgical approach for hyperkinetic movement disorders, which has been extensively used in the 1970s and 1980s, is lesioning of the nucleus ventrooralis anterior thalami (VOA). VOA thalamotomy has been effective in reducing dystonia, hemiballism and chorea [1,3,9,12] but bears a high risk of severe dysarthria and frontal dysexecutive syndromes when applied bilaterally. The VOA is the thalamic receiving area of pallidal input. VOA thalamotomy, therefore targets the same pallidothalamic circuit as pallidotomy. Unfortunately, no comparative studies are available demonstrating which target is superior in clinical efficacy or safety for the treatment of various hyperkinesias.

Possible Application to Chorea-Acanthocytosis

Chorea-acanthocytosis (ChAc) is a rare progressive neurological disease characterized by a wide variety of clinical features such as generalized chorea, orofaciolingual dyskinesias with dysphagia and dysarthria, muscle wasting, hyporeflexia, seizures, behavioral and psychiatric disturbances [8]. Despite the diversity of symptoms, hyperkinesias are most disabling especially in the first years of the disease and a successful symptomatic treatment of hyperkinesias would significantly improve quality of life and reduce dependence in these patients. Standard medical treatment with neuroleptics or dopamine depleting agents is very restricted and offers no satisfactory long-term benefit. There is as yet no animal model of ChAc available for studying the pathophysiology of motor symptoms. Any assumption on an underlying dysfunction of neuronal activity within internal pallidum and ventrolateral thalamus is simply based on the clinical similarity of hyperkinesias in ChAc with features of Huntington's disease (HD) or idiopathic dystonia. Nevertheless, given the low morbidity of modern functional stereotaxy, the progressive and disabling nature of motor symptoms in ChAc and the lack of symptomatic medical treatment it appears ethically justifiable to try surgical treatment in selected patients who fully understand the experimental nature of the procedure.

There is one case report in the literature describing successful treatment of hyperkinesias in ChAc with bilateral posteroventral pallidotomy [7]. Unlike pallidotomy, deep brain stimulation (DBS) is safe in bilateral procedures, reversible, adaptable to the clinical situation [15] and therefore especially suited for a therapeutic trial in disorders of unknown responsiveness to functional stereotactic procedures. We treated the first case of chorea-acanthocytosis by bilateral pallidal DBS and have reported on the outcome before [16].

CASE REPORT

A 38-year-old previously healthy male was referred to our unit with hyperkinesias of the lower limbs leading to progressive gait disturbance with frequent falls at the age of 35 years. In parallel he developed difficulties eating due to frequent involuntary protrusions of the tongue forcing food out of the mouth during mastication. A few months later a rapidly generalized chorea emerged with extensive oromandibular hyperkinesias leading to increasing self-mutilation of the tongue and the buccal mucosa. Within one year of onset of the symptoms the patient had lost 47 kg of weight due to severe dyskinesias and feeding difficulties. There was no history of seizures and no evidence for dementia or cognitive impairment. Involuntary vocalizations occurred comprising grunting and whistling which the patient described to have begun around the age of 17, but sharply increased after the emerge of chorea at the age of 35 years. Coprolalia was not reported.

The patient was the third child of healthy unrelated parents. Pregnancy, birth and developmental milestones were unremarkable. Both parents died of unrelated causes in their seventh decade with medical records unavailable. An older brother of the patient was born with bilateral hearing impairment and died at the age of 33 years from an unclear sudden cardiac arrest. However, the patient remembers that his brother had similar vocalization tics starting at young age, but never developed involuntary hyperkinesias before death. An older sister of the patient is said to be unaffected, although contact was lost many years ago.

Neurological examination of the cachectic patient revealed infrequent involuntary vocalizations such as grunting and showed marked generalized chorea predominantly of the trunk, with back arching and severe oromandibular hyperkinesias. The anterior parts of the tongue displayed signs of frequent biting and feeding dystonia was present. Dysphagia and dysarthria were absent. The gait was hyperkinetic with dystonic posturing of both feet and poor balance. Postural reflexes were markedly impaired with retropulsion. There was generalized wasting of muscles with equinovarus deformity of both feet. Examination of the limbs revealed generalized chorea and mild bradykinesia. Tendon reflexes of the lower limbs were absent and plantar responses were silent. The right arm was ataxic and dysmetric. Sensory examination was normal.

Neuropsychological assessment showed a marked reactive depression. The patient performed well in the cognitive tests and no mental deficits were found. A progressive personality disorder evolved with obsessive-compulsive features.

The diagnosis of ChAc was made on clinical grounds with fresh blood smears containing between 20 and 25% acanthocytes compared to less than 3% in non-affected controls. Blood testing for Kell antigen status confirmed that the patient was McLeod phenotype negative. Genetic testing for HD and dentato-rubro-pallido-Luysian Atrophy (DRPLA) were negative.

Subsequently, the patient was tested positive for a mutation in the *CHAC* gene on 9q21 [6].

Effects of Drug Treatment

Before admission to our hospital the patient had successively been treated with a variety of neuroleptic drugs, including haloperidol, risperidone, olanzapine, sulpiride (600 mg/d), tiapride and perphenazine, buspirone (up to 60 mg/d) and diazepam, without any significant symptomatic improvement. He was then started on clozapine (200 mg/d) which initially almost completely suppressed the hyperkinesias. The feeding dystonia and tongue biting ceased, with total restitution of tongue function. The patient was able to masticate food undisturbed and gained 40 kg of weight in the following weeks. After two months almost completely without symptoms the chorea gradually returned and showed no response to increasing the dosage of clozapine (300 mg/d). After another ten weeks, the symptoms and signs had returned to the patient's status before clozapine. He subsequently gave informed consent for a therapeutic trial of pallidal stimulation to alleviate the hyperkinesias.

Effects of Pallidal Stimulation

The patient underwent bilateral implantation of a quadrupolar electrode (model DBS 3387, Medtronic Inc., Minneapolis, USA) into the GPi under local anesthesia using the Leibinger stereotactic system. The regularly spaced (1.5 mm) quadrupolar electrode, which we also routinely use for pallidal stimulation in Parkinson‘s disease, was chosen to cover a larger tissue volume. Targeting and electrode implantation were performed as described elsewhere [15]. For physiological target verification we used electrical macrostimulation along the electrode trajectory, starting 6 mm above the intended target and extending up to 4 mm below. No microelectrode recordings were performed intraoperatively. As an immediate effect after placement of the electrodes at the target point (GPi right: 3.2 mm anterior to the midcommissural point (MCP), 5 mm inferior to the intercommissural line (ICL), 19.4 mm lateral to the midline of the third ventricle; GPi left: 3.2 mm anterior to MCP, 5 mm inferior to ICL, 19.7 mm lateral to the midline of the third ventricle; according to the atlas of Schaltenbrand and Wahren [11]) the generalized chorea almost completely disappeared. 1 to 2 mm below the target point stimulation with 0.1 ms pulse width, 130 Hz and up to 3.5 V started to elicit phosphene sensations in the contralateral visual field without producing capsular responses. Postoperatively, the benefit from microlesioning due to electrode placement faded within less than one week and bilateral stimulation via the provisional external lead was initiated.

In contrast to the excellent perioperative result with marked reduction of the hyperkinesias due to the microlesioning effect after electrode implantation, neither acute nor prolonged high-frequency stimulation of the GPi could reproduce the initial benefit. Each electrode contact was tested as active cathode in a bipolar setting. Raising the stimulation amplitude and pulse width up to a maximum of 10 V and 450 μsec (185 Hz), as far as tolerated with respect to side effects, did not reveal a reproducible reduction of chorea for any bipolar electrode combination during acute testing. Instead, increasing side effects such as pyramidal signs (facial hemispasms), optic sensations, scotoma and psychovegetative phenomena (nausea and epigastric sensations) were noted and confirmed the correct positioning of the stimulating electrodes. Increasing the frequency of stimulation up to 1000 Hz within the therapeutic amplitude range resulted in worsening of chorea without inducing other side-effects. An alternatively performed low frequency stimulation (10-50 Hz) did not show any reproducible positive effects on the hyperkinesia either, but led to reproducible deterioration of speech and gait. We then tried chronic stimulation with the bipolar electrode combination on each side eliciting phosphene sensations at lowest threshold and using stimulation parameters just below induction of visual side effects. The trial was terminated after 20 days of continuous stimulation without notable reduction of chorea.

Case Summary

In conclusion there was no benefit from high- or low-frequency stimulation of the GPi in our patient. On the opposite, stimulation at frequencies > 500 Hz led to a worsening of his symptoms. Consequently the leads were explanted after 3 weeks of extensive testing because no consistent clinical effect could be elicited.

Since microlesioning due to electrode implantation initially improved the hyperkinesias we discussed with the patient the possibility of a staged bilateral pallidotomy as ultimate treatment option. The patient at that time was reluctant with respect to a repeat surgical intervention and therapy with clozapine was re-instituted. Clozapine, however, proved insufficient and choreatiform hyperkinesias increased and lip biting and vocalizations recurred. Later, the patient was started on tetrabenazine 75 mg/d as add-on therapy, with a mild additional benefit. Currently the patient, due to his severe gait disorder and feeding problems, is at the brink of dependence on permanent medical and nursing care.

DISCUSSION

We described the first case of drug-resistant ChAc that was treated with bilateral high-frequency stimulation of the GPi. Pallidotomy and pallidal deep brain stimulation are highly effective treatments for levodopa-induced dyskinesias in Parkinson's disease, which may present with ballistic, choreatic, athetoid and dystonic features [15]. The beneficial effect of both procedures is possibly due to releasing the motor thalamus from abnormal neuronal discharge patterns in the GPi associated with the dyskinetic "on"-state in Parkinson's disease. A case report of successful treatment of hyperkinesias in ChAc by posteroventral pallidotomy suggested that symptomatic chorea might benefit from pallidal surgery, too [7]. Since the advantages of stimulation compared to pallidotomy lie in potential reversibility, adaptability to the clinical course and reduced likelihood of permanent side-effects we performed high-frequency stimulation of the GPi in a patient with severe ChAc. Unfortunately, our patient failed to respond to deep brain stimulation of the internal globus pallidus in the expected way. The benefit that the patient had from the lesioning effect of electrode placement suggests that a lesioning approach instead of high-frequency stimulation might have had a greater chance of success. Since the microlesioning effect extends over a certain volume along the electrode trajectory, suboptimal placement of the electrode contacts with respect to the hyperkinesia-generating subareas within the GPi possibly is an alternative explanation.

Recently, Burbaud and colleagues [2] reported a second case of ChAc treated by DBS. Their patient suffered from violent truncal spasms, muscle hypotonia, oromandibular dystonia and severe dysarthria. Electrodes were implanted bilaterally in the VOA. Thalamic DBS led to a marked reduction of truncal spasms during the entire 6 months of follow-up. Hypotonia was unchanged. There is no account on whether dysarthria or oromandibular dystonia changed. Given the special features of this case with violent, ballistic truncal hyperkinesias it is difficult to discern whether the VOA in general would be preferable to the GPi for treating hyperkinesias in choreatic syndromes. This case, however, clearly demonstrates that the pallidothalamic circuit must be involved in the pathophysiology of hyperkinesias in ChAc.

The present experience suggests that deep brain stimulation could be a treatment option for motor symptoms in severe cases of ChAc, but the optimal target remains uncertain. For future patients one may therefore consider implanting DBS leads simultaneously in VOA and GPi to verify which target has a greater impact on hyperkinesias and whether they act synergistically or not. This approach is technically feasible and has previously helped to identify the optimal stereotactic target for treating dystonia [13].

REFERENCES

1. Andrew J, Fowler CJ, Harrison MJ (1983) Stereotaxic thalamotomy in 55 cases of dystonia. *Brain* **106:** 981-1000.
2. Burbaud P, Rougier A, Ferrer X et al (2002) Improvement of severe trunk spasms by bilateral high-frequency stimulation of the motor thalamus in a patient with chorea-acanthocytosis. *Mov Dis* **17:** 204-207.
3. Cardoso F, Jankovic J, Grossman RG et al (1995) Outcome after stereotactic thalamotomy for dystonia and hemiballismus. *Neurosurgery* **36:** 501-7; discussion 507-508.
4. Coubes P, Roubertie A, Vayssiere N, et al (2000) Treatment of DYT1-generalised dystonia by stimulation of the internal globus pallidus. *Lancet* **355:** 2220-2221.
5. DeLong MR (1990) Primate models of movement disorders of basal ganglia origin. *Trends Neurosci* **13:** 281-285.
6. Dobson-Stone C, Danek A, Rampoldi L et al (2002) Mutational spectrum of the *CHAC* gene in patients with chorea-acanthocytosis. *Eur J Hum Genet* **10:** 773-781.
7. Fujimoto Y, Isozaki E, Yokochi F et al (1997) A case of chorea-acanthocytosis successfully treated with posteroventral pallidotomy. *Rinsho Shinkeigaku* **37:** 891-894.
8. Hardie RJ, Pullon HW, Harding AE et al (1991) Neuroacanthocytosis. A clinical, haematological and pathological study of 19 cases. *Brain* **114:** 13-49.
9. Kawashima Y, Takahashi A, Hirato M, et al (1991) Stereotactic Vim-Vo-thalamotomy for choreatic movement disorder. *Acta Neurochir Suppl* **52:** 103-106.
10. Marsden CD, Obeso JA (1994) The functions of the basal ganglia and the paradox of stereotaxic surgery in Parkinson's disease. *Brain* **117:** 877-897.
11. Schaltenbrand G, Wahren W (1977) *Atlas for stereotaxy of the human brain.* Georg Thieme, Stuttgart.
12. Tasker RR, Doorly T, Yamashiro K (1988) Thalamotomy in generalized dystonia. *Adv Neurol* **50:** 615-631.
13. Trottenberg T, Paul G, Meissner W et al (2001) Pallidal and thalamic neurostimulation in severe tardive dystonia. *J Neurol Neurosurg Psychiatry* **70:** 557-559.
14. Vitek JL, Chockkan V, Zhang JY et al (1999) Neuronal activity in the basal ganglia in patients with generalized dystonia and hemiballismus. *Ann Neurol* **46:** 22-35.

CHAPTER 29

THE POTENTIAL FOR GENE THERAPY OF NEURODEGENERATIVE DISORDERS

ALICE ANDREU AND ANDREW D. MILLER
Imperial College Genetic Therapies Centre,
Department of Chemistry, Flowers Building, Armstrong Road,
Imperial College of Science, Technology and Medicine, London
UK

Abstract. Gene therapy has been in the medical spotlight for over a decade now and its enormous potential offers hope to many sufferers of diseases involving a genetic component. Although progress has been made, clinical gene therapy is still quite a remote possibility especially in the relatively new field of targeting the central nervous system (CNS). The treatment of CNS disorders by gene therapy offers unique therapeutic challenges due to the fragility and complexity of the brain; delivery through the protective envelope of the brain is associated with many risks, and gene transfer into post-mitotic neurons is restrictive. Nevertheless, we believe that neurological gene therapy still has good future prospects. This short review will attempt to describe current gene therapeutic strategies, delivery methods and the most promising vectors for use in neurodegenerative disorders. We also provide a short description of our own system, currently under development.

INTRODUCTION

Recent advances in genetics are opening new doors for modern medical techniques, such as gene therapy. Although still primarily preclinical with most human clinical trials only in phase I, the latent potential of gene therapy is both unique and seemingly unlimited. Where gene therapies are in development for CNS disorders, the traditional approach has been to provide cells with a functional copy of a defective gene to produce locally missing proteins (these are mostly too large to go through the brain/blood barrier [27]). Other strategies for brain cancer are being developed, such as the introduction of genes to trigger cancer cells to commit cell suicide. Whilst the concept of delivering the

A. Danek (ed.), Neuroacanthocytosis Syndromes, 259–267.

normal gene in place of a mutated gene is in itself quite simple, the related practical issues have proven rather complex and unique difficulties associated with the CNS have discouraged early research in this field. Physically and biochemically protected by the skull and the brain/blood barrier respectively, the CNS is one of the most difficult organs to access. Moreover, the CNS is also the most complex and least understood. Genetically, the brain expresses more genes than any other organ to manage its crucial functions; loss of control over this gearbox of human integrity gives a tragic element to neurological diseases. Difficulties experienced in the transfection or transduction of post-mitotic target cells, immune rejection of gene carriers or transduced cells and high risks associated with CNS gene delivery are all major obstacles for the successful development of therapeutic gene delivery in the CNS. The improvement of gene transfer vectors is crucial not only for therapeutic purposes, but also for further studies of neurological mechanisms.

STRATEGIES AND LIMITATIONS OF GENE THERAPY FOR NEURODEGENERATIVE DISORDERS

Current treatment approaches to neurodegenerative diseases, whether surgical or pharmacological, are mainly addressing the symptoms and not the underlying causes. Correcting the problem at the genetic level would be a much more efficient strategy. Gene therapy aims at compensating for a missing or ineffective gene by locally producing desired proteins through precise gene delivery and expression. Unfortunately, getting cells to accept a carrier of genetic material, carry the genetic material to the nucleus and then express the encoded protein has proven far more complex than had been expected initially.

Delivery

The delivery aspect of gene therapy remains the main subject of current research and has proven quite challenging for the CNS. Administration of the therapeutic gene can be done either *in vivo*, where it is directly administered to the host, or *ex vivo*, where targeted cells are cultured *in vitro* before gene transfer, and are subsequently re-implanted. Either way, the procedure involves complex and risky neurosurgery.

Most *in vivo* deliveries are achieved through direct surgical injection, resulting in very localized gene transfer. Distribution is restricted by rapid and unlimited cellular uptake of vectors in the immediate vicinity of the injection site [17]. The vectors' large size and strong affinity for the cell membrane further prevents good spread, along with small injection volumes. Unfortunately, most hereditary neurodegenerative diseases require global transgene expression, requiring adequate strategies to be developed. The temporary rupture of the brain/blood barrier is achievable [17], however it is potentially very dangerous and alternative methods would be preferable.

Control of Gene Expression

It is important to aim at long term controlled gene expression to avoid repeating risky and complex administration procedures. Current techniques show diminution of transgene expression over time, mostly attributed to direct cytotoxicity of vector or transgenic product, immunological elimination, as well as shut down of the transgene promoters [17,27]. Whilst only 10% of normal levels of the missing enzyme/protein are often enough to restore normal function, their over-expression can be toxic [17]. This highlights the importance of controlled expression.

GENE DELIVERY SYSTEMS FOR NEURODEGENERATIVE DISORDERS

As transgenic DNA is recognised as foreign and therefore degraded by cellular nucleases, therapeutic genes need to be protected until they reach the site of expression. To date, none of the gene delivery vehicles available has proven to be entirely safe and efficient. *In vivo* approaches use viral vectors or synthetic ones, and *ex vivo* strategies graft cells that have been genetically engineered *in vitro*.

Viral Vectors [16,17]

The use of "domesticated" viruses has been widely thought to be the ideal solution to the CNS delivery challenge. Viruses have evolved to infect human cells and use host cell transcription/translation machinery to express their own proteins. However, wild type viruses usually kill host cells and therefore need to be genetically modified to become non replicative for therapeutic use.

Whilst viral systems have been generally found to be quite effective in laboratory and preclinical trials, they do introduce crucial problems [10]. Firstly, genetically modified viruses can regain pathogenic properties *in situ* through recombination [17]. Secondly, although the blood/brain barrier secludes the brain from the body's immune defences, viral antigens do trigger an immune response, leading to inflammation and immune rejection of transduced cells or direct cell death [17]. Finally, chronic high-level expression may overwhelm the cell's protein synthesis process [14].

A number of different viruses have been used in gene delivery experiments. Vectors derived from herpes simplex virus, adenovirus and more recently lentivirus are the most studied viral systems for treatment of neurodegenerative diseases.

Herpes Simplex Virus [20,36]. Herpes simplex virus (HSV) derived vectors were the first ones to be evaluated for use in the CNS [25]. HSV has strong neurotropic properties; it naturally infects human neuronal cells to produce indefinitely stable asymptomatic latent infections. HSV can be grown to high titers in cultures and has a large genome size (150 kbp), which in the case of recombinant HSV can accommodate up to 30-40 kbp of foreign genetic material.

However, wild type HSV is a natural human pathogen and its introduction into the brain would result in lethal encephalitis due to viral replication. Therefore, by means of genetic manipulations, various strategies have been developed to remove genes essential for viral replication. Long-term expression is also a problem as only latency-associated transcripts (LAT) are expressed [11] during the dormant infection. Their promoters can be used to express the transgene during latency [6,21,40]. Despite these tremendous efforts in making HSV safe for gene therapy use, significant cellular cytotoxicity and immunogenicity is still being found.

Adenovirus [27,36]. In 1993, the potential use of adenoviridae as gene delivery vectors into the CNS has been demonstrated [2,3,7,31]. This small virus (90 nm diameter, 36 kbp genome) gives highly effective gene transfer into *in vitro* post-mitotic neurons. Although adenoviridae are human pathogens and can cause respiratory infections, they are not life threatening to humans. Adenoviridae are not specific to neurons but they can infect a broad range of cells.

Recombinant viruses are made non-replicative by deletion of genes that activate expression. The resulting vectors can be grown in specialized cultures, yielding moderately high titers, although wild type virus may regenerate after multiple passages. Recombinant vectors' inability to spread well results in moderate *in vivo* transfection efficiency, which decreases as transduced cells are either rejected or die of direct cytotoxicity. Associated inflammatory and immunogenic properties have led to the development of gutless vectors; these contain a plasmid with a minimal viral genomic sequence and can be produced in the presence of a helper virus. Although their production is more problematic, there is an improvement in capacity from $\sim$ 5 kbp to $\sim$ 30 kbp, and they show good efficiency with low toxicity and immunogenicity.

Adeno-Associated Virus [16,17]. Adeno-associated viruses (AAV) are the smallest and structurally the simplest DNA animal viruses. They are capable of infecting a wide variety of hosts although mostly as co-infectors. Most AAVs are believed to be non-pathogenic, result in relatively long term transgene expression and their small size allows good spread *in vivo*. They can be rendered replication defective by deletion of genes responsible for expression of structural and regulatory proteins. AAV vectors have little or no associated toxicity or immunogenicity due to their lack of viral genes, thereby allowing multiple exposures. However, they give mediocre *in vivo* transduction in post-mitotic neurons, only have a very limited DNA insert capacity ($\sim$ 5 kbp) and their production is challenging.

Lentivirus [8]. Traditional retroviridae, which have been well developed for gene delivery, benefit from low pathogenicity and produce stable long-term expression. Unfortunately, retroviridae can only be prepared in low titers, may lead to insertional mutagenesis, and most importantly infect only mitotic cells. They are therefore restricted within the CNS area to *ex vivo* applications.

However, gutless lentiviridae, derived from HIV, were first considered for use as *in vivo* gene delivery agents in 1996 [24,28]. They are strongly neurotropic, result in efficient and stable long-term *in vivo* transduction of post-mitotic cells and have a moderate DNA insert capacity. Transfections *in vivo* show little apparent cytotoxicity or immune response. Although these have great potential for the treatment of CNS diseases, issues over safety and possible regeneration of the wild type virus (HIV) have hindered advances. As with all other retroviridae, they can only be produced in low titres and carry the risk of insertional mutagenesis.

Potentials of Viral Vectors. Viruses are naturally potent gene delivery systems. However, as illustrated by an unfortunate death in a viral gene therapy clinical trial [17,20], the potential risk to subjects remains relatively high. Cytotoxicity, immunogenicity and the risk of replication-competent virus regeneration *in situ* are obstacles that prove to be difficult to overcome, especially given that most humans have circulating antibodies to all the virus vectors currently in use. Safer new generation viral vectors of the gutless type are promising but are more difficult to produce in high titres and do not completely eliminate the risks of wild type regeneration and cytotoxicity. Other delivery methods such as biological mini-pumps and non-viral vectors may prove to be more suitable alternatives in the longer term.

Non-Viral Vectors [10]

Alternatives to viral systems are synthetic vectors, which aim to mimic the beneficial properties of viruses but have reduced immunogenicity. This concept is very attractive, but represents a potentially vast piece of work to try and imitate in tens of years that which viruses have achieved as a result of millions of years of evolution.

Naked DNA [33]. Naked DNA plasmids can be injected directly into the relevant part of the brain, although this exposes the DNA to potential damage and does not allow for any targeting facility. This strategy has been shown to give poor, short-term protein expression, comparable to results observed for naked DNA administration into muscles.

Liposomes [39]. There are a wide variety of non-viral delivery systems in development and many are based around cationic liposomes/lipids. Other non-viral vectors exist, such as a variety of polycationic polymers and cationic DEAE-dextrane [23]. Even artificially engineered human chromosomes are under development (i.e. non disruptive autonomic DNA holding large sized genes, inducing no immune response as they are made from naturally occurring human DNA) [5].

Formulation [10]. The traditional cationic liposome/lipid based non-viral vector not only protects complexed nucleic acids from degradation in the extracellular environment, but also allows non-specific endocytosis with the cell membrane, releasing complexed nucleic acids into the cytoplasm. Cationic liposomes/lipids interact with anionic nucleic acids spontaneously to yield cationic liposome/lipid-nucleic acid complexes (lipoplexes: LD) that retain sufficient positive charge to interact with anionic cell membranes and enter cells. The efficiency of lipofection (cationic liposome/lipid mediated transfection) has improved during the last few years specifically due to improvements in lipoplex formulation such as the use of cationic peptides or proteins to precondense nucleic acids prior to admixture with cationic liposomes, and the introduction of targeting molecules to the surface of the vector to increase uptake efficiency. The inclusions of nuclear targeting agents and protective factors have also been seen to promote further cationic liposome/lipid-based nucleic acid delivery.

Results [14,19]. Successful delivery and expression of cationic liposome/lipid based DNA vectors in the CNS have been reported in some neuronal cell cultures. *In vivo* studies are less promising, although direct injection into the brain has been shown to give some significant protein expression around the injection site. No inflammatory response has been observed, but low transfection efficiencies are primarily due to rapid degradation in the extracellular environment.

Potential of Cationic Liposomes/Lipids [19]. Cationic liposome/lipid based non-viral gene delivery systems have several significant advantages over viral carriers, these mostly being safety, easy and fast preparation, as well as versatility of formulation. However, current efficiency of gene delivery *in vivo* remains too low for current cationic liposome/lipid systems to have realistic applications *in vivo* as therapeutic agents. Nevertheless, recent advances aimed at combining viral core peptides with cationic liposome/lipid systems may well produce efficient yet safe non-viral vectors with real promise for applications in clinical CNS gene therapies.

Ex Vivo Approach [1,17,30]

An alternative approach to traditional gene therapy is to directly implant encapsulated genetically engineered cells to secrete missing enzymes. This methodology is mostly used for dopamine and growth factor production. Preliminary *in vivo* studies have proved successful. Non-neuronal dividing cells, preferably derived from the subject's own, can be cultured and transfected *in vitro* only to be re-implanted into the brain to produce the appropriate factors. The type of cells used is crucial, as mitotic cells may form tumors in the brain: however, skin fibroblasts and endothelial cells seem to integrate well. Polymer encapsulation can reduce immune rejection, tumor formation and virus spread.

Implanted cells can also be engineered to contain a drug-inducible apoptotic gene for easy elimination [13]. The development of universal cells is currently underway, where immune recognition factors are suppressed.

Neuronal stem cells are currently being investigated extensively for *ex vivo* gene therapy [35]. They are undifferentiated and can migrate extensively in the damaged adult brain where they generally lose their mitotic ability, thereby avoiding formation of tumors [4,22,29,37]. A small number are still naturally present in adult brains where they can be isolated, manipulated and re-implanted. This would be an ideal strategy for replacement of damaged neurons.

Although the *ex vivo* strategy offers more control over the *in vitro* gene transfer process itself, the major drawbacks of this technology include the surgical grafting procedures necessary, possible immune rejection of the graft, and potential tumorigenesis.

OUR CURRENT PROJECT

Global distribution has not yet been achieved for CNS gene delivery, due to poor spreading of vectors. We believe that a non-viral targeted delivery would not only be safe, sufficiently efficient, but also give long-term expression (upon transfection of post-mitotic neurons, expression should last as long as the cell lives to render DNA insertion unnecessary).

Our current project combines our non-viral vector Liposome-μ(mu)-DNA (LMD) system previously developed in the Miller group [38], with a targeting protein. A mixture of neutral and cationic lipids is formulated into a liposomal suspension that is combined with μ-DNA nanoparticles created by precise condensation of plasmid DNA by the mu peptide. The mu peptide is not only an impressive template for DNA condensation, but also harbors a nuclear localization signal with consequences for efficient intracellular trafficking of plasmid DNA post cell entry. The targeting protein moiety under investigation is the neurotropic non-toxic Tetanus Toxin fragment c (TTc, 52 kDa). A maleimide linker covalently attaches TTc to the body of the LMD system. TTc can be purified from recombinant expression in *E. coli*, although originally obtained by papain digestion of Tetanus toxin (TeTx, 150 kDa). TTc has all the properties of TeTx apart from the toxic effects (1 mg in guinea pig, no toxicity found) [9,15]. This protein is well known to diffuse in body fluids from wound sites and bind within 6 hours to the presynaptic membrane of the neuromuscular junction [34]. There, it is internalized by the motor neurons to undergo transsynaptic retrograde transport to the spinal cord within two days [12,26]. TTc's potential as a targeting moiety for neuronal gene delivery has already been shown in both viral and non-viral vector systems [18,32]. Therefore, we have significant hope for the future of the LMD-TTc system for the delivery of therapeutic genes to the CNS.

CONCLUSION

Whilst diseases affecting such an organ as the brain are particularly tragic, the potential treatments will need to be proportionally well devised, as risks involved are proportionately higher than in dealing with the therapy of other body organs. If gene therapy is to be realized for the treatment of neurodegenerative disorders, then a better understanding of physiological mechanisms and the genetic bases of CNS disorders is essential to develop clinically viable gene therapy treatments. Efficiency, safety, global delivery and expression control are all crucial areas to tackle.

REFERENCES

1. Aebischer P, Ridet J (2001) Recombinant proteins for neurodegenerative diseases: the delivery issue. *Trends Neurosci* **24:** 533-540.
2. Akli S, Caillaud C et al (1993) Transfer of a foreign gene into the brain using adenovirus vectors. *Nature Genet* **3:** 224-228.
3. Bajocchi G, Feldman S et al (1993) Direct in vivo gene transfer to ependymal cells in the central nervous system using recombinant adenovirus vectors. *Nature Genet* **3:** 229-234.
4. Gage F (2000) Mammalian neural stem cells. *Science* **287:** 1433-1438.
5. Choo K (2001) Engineering human chromosomes for gene therapy studies. *Trends Mol Med* **7:** 235-237.
6. Coffin R, Thomas S et al (1998) The herpes simplex virus 2kb latency associated transcript (LAT) leader sequence allows efficient expression of downstream proteins which is enhanced in neuronal cells: possible function of LAT ORFs. *J Gen Virol* **79:** 3019-3026.
7. Davidson B, Allen E et al (1993) A model system for in vivo gene transfer into the CNS using an adenoviral vector. *Nature Genet* **3:** 219-223.
8. Deglon N, Aebischer P (2002) Lentiviruses are vectors for CNS diseases. *Curr Top in Microbiol Immunol* **261:** 191-209.
9. Evinger C, Erichsen J (1986) Transsynaptic retrograde transport of fragment c of tetanus toxin demonstrated by immunohistochemical localisation. *Brain Res* **380:** 383-388.
10. Felgner P (1997) Nonviral strategies for gene therapy. *Sci Am* **276:** 102-106.
11. Fink D, Deluca N et al (1996) Gene transfer to neurons using herpes simplex virus-based vectors. *Annu Rev Neurosci* **19:** 265-287.
12. Fishman P, Carrigan D (1986) Retrograde transneuronal transfer of the c-fragment of tetanus toxin. *Brain Res* **406:** 275-279.
13. Garin M, Garrett R et al (2001) Molecular mechanisms for ganciclovir resistance in human T-lymphocytes transduced with retroviral vectors carrying the herpes simplex virus thymidine kinase gene. *Blood* **97:** 122-129.
14. Hecker J, Hall L et al (2001) Non viral gene delivery to the lateral ventricles in the rat brain: Initial evidence for widespread distribution and expression in the central nervous system. *Molecular Therapy* **3:** 375-384.
15. Helting T, Zwisler O (1977) Structure of tetanus toxin. *J Biol Chem* **252:** 187-193.
16. Hermens W, Verhaagen J (1998) Viral vectors, tools for gene transfer in the nervous system. *Progr Neurobiol* **55:** 399-432.
17. Hsich G, Sena-Esteves M et al (2002) Critical issues in gene therapy for neurologic disease. *Human Gene Therapy* **13:** 579-604.
18. Knight A, Cravajal J et al (1999) Non-viral neuronal gene delivery mediated by the Hc fragment of tetanus toxin. *Eur J Biochem* **259:** 762-769.

19. Kofler P, Wiesenhofer B et al (1998) Liposome-mediated gene transfer into established CNS cell lines, primary glial cells, and in vivo. *Cell Transplantation* **7:** 175-185.
20. Latchman D (2001) Gene delivery and gene therapy with herpes simplex virus-based vectors. *Gene* **264:** 1-9.
21. Lokensgard J, Bloom D et al (1994) Long term promoter activity during herpes simplex virus latency. *J Virol* **68:** 7148-7158.
22. McKay R (2000) Stem cells-hype and hope. *Nature* **406:** 361-364.
23. Miller AD (1999) Nonviral delivery systems for gene therapy. Chapter 4 In: *Understanding Gene Therapy,* Bios, Edited by NR Lemoine
24. Naldini L, Blomer U et al (1996) In vivo gene delivery and stable transduction of non dividing cells by a lentiviral vector. *Science* **272:** 263-267.
25. Palella T, Hidaka Y et al (1989) Expression of human HPRT mRNA in brains of mice infected with a recombinant herpes simplex virus-1 vector. *Gene* **80:** 137-44.
26. Pellizzari R, Rossetto O, Schiavo G, Montecucco C (1999) Tetanus and botulinum neurotoxins: Mechanism of action and therapeutic uses. *Philos Trans R Soc Lond B Biol Sci* **354:** 259-268.
27. Peltekian E, Parrish E et al (1997) Adenovirus mediated gene transfer to the brain: Methodological assessment. *J Neurosci Methods* **71:** 77-84.
28. Reiser J, Harmison G et al (1996) Transduction of nondividing cells using pseudotyped defective high titre HIV type 1 particles. *Proc Natl Acad Sci USA* **93:** 15266-15271.
29. Renfranz PJ, Cunningham MG, McKay RD (1991) Region-specific differentiation of the hippocampal stem cell line HiB5 upon implantation into the developing mammalian brain. *Cell* **66:** 713-729.
30. Ross C, Ralph M et al (2000) Somatic gene therapy for a neurodegenerative disease using microencapsulated recombinant cells. *Exp Neurol* **166:** 276-286.
31. Salle GLGL, Robert J et al (1993) An adenovirus vector for gene transfer into neurons and glia in the brain. *Science* **259:** 988-990.
32. Schneider H, Groves M et al (2000) Retargeting of adenoviral vectors to neurons using the Hc fragment of tetanus toxin. *Gene Therapy* **7:** 1584-1592.
33. Schwartz B, Benoist C et al (1996) Gene transfer by naked DNA into adult mouse brain. *Gene Therapy* **3:** 405-411.
34. Sinah K, Box M et al (2000) Analysis of mutants of tetanus toxin fragment c: ganglioside binding, cell binding and retrograde axonal transport properties. *Mol Biol* **37:** 1041-1051.
35. Shihabuddin L, Palmer T et al (1999) The search for neural progenitor cells: prospects for the therapy of neurodegenerative diseases. *Mol Med Today* **5:** 474-480.
36. Slack R, Miller F (1996) Viral vectors for modulating gene expression in neurons. *Curr Opin Neurobiol* **6:** 576-583.
37. Snyder E, Deitcher D et al (1992) Multipotent neural cell lines can engraft and participate in development of mouse cerebellum. *Cell* **68:** 33-51.
38. Tagawa T, Manvell M et al (2002) Characterisation of LMD virus-like nanoparticles self-assembled from cationic liposomes, adenovirus core peptide mu and plasmid DNA. *Gene Therapy* **9:** 564-576.
39. Yang K, Clifton G et al (1997) Gene therapy for central nervous system injury: the use of cationic liposomes: an invited review. *J Neurotrauma* **14:** 281-297.
40. Zhu J, Kang W et al (2000) Significantly increased expression of beta-glucuronidase in the central nervous system of mucopolysaccharidosis type VII mice from the latency associated transcript promoter in a nonpathogenic herpes simplex virus type 1 vector. *Mol Ther* **2:** 82-94.

CHAPTER 30

RESEARCH AGENDA IN NEUROACANTHOCYTOSIS

ADRIAN DANEK
Neurologische Klinik, Ludwig-Maximilians-Universität, München, Germany

Abstract. The Seeon meeting has reviewed what is presently known and has posed many questions to be solved before the outcome of patients affected by one of the neuroacanthocytosis syndromes can be improved. Insight into successful treatments might be found if we understood the normal roles of proteins such as XK, chorein, and junctophilin whose malfunction leads to basal ganglia degeneration. The long-term tasks are disentangling the pathogenetic cascade with its presumed final common pathway and of developing disease models by using isolated red cells or studying experimental animals. Short-term goals are to find better symptomatic relief (e.g. of orofacial dyskinesia), improved diagnostic techniques and more widespread knowledge about the neuroacanthocytosis syndromes. The network of experts and families built in Seeon could play an important role shaping the future of research in this field.

INTRODUCTION

The Seeon meeting and the developments since have clearly shown that "neuroacanthocytosis" is not a unitary entity. Nevertheless, there was agreement to retain the name until better-defined disease concepts develop and eventually replace the umbrella term. "Neuroacanthocytosis" is not a diagnosis but a syndrome that calls for further investigation. It is also a reminder that the shared properties of red cell membranes and neurons still need explanation.

TREATMENT NEEDS

Given the suffering of those affected by the relentless course of neuroacanthocytosis syndromes with their progressive motor disability and cognitive deterioration, cure must be the ultimate aim of neuroacanthocytosis research.

A. Danek (ed.), Neuroacanthocytosis Syndromes, 269–275.

The discoveries of XK and chorein protein malfunction as the causes of McLeod syndrome and chorea-acanthocytosis, respectively, have only provided first steps. Reconstitution of protein synthesis by directly fixing the gene defect is a logical next move but as the preceding chapter has shown, gene therapy still has a long way to go.

As an alternative strategy, the normal role of the proteins involved must be understood so that ways can be devised to replace their missing functions. So far, the roles neither of XK and its partner Kell in neuronal physiology nor of chorein have been clarified.

Nevertheless, the clinical observation of similarity of phenotypes in various diseases of the basal ganglia, Huntington's disease included, and of their late onset in adulthood may provide clues. First, they point to a final common pathogenetic pathway and secondly, there appear to be factors that delay disease onset in spite of the absence of one important protein. Thus, the possible interactions of XK, chorein, huntingtin and other proteins involved in basal ganglia degeneration are of major research interest.

Importantly, the *CHAC* gene has just been found to belong to a novel human gene family named after the yeast protein Vps13p. This new family that has accordingly been called VPS13 presently has four known members [33]. The *CHAC* gene on chromosome 9q21 has now been renamed *VPS13A*.

A gene on 8q22, *VPS13B*, had previously been reported as *COH1* and is responsible for Cohen syndrome which clinically is very unlike ChAc [6,17]. *VPS13C* (15q21) and *VPS13D* (1p36) need to be further characterized and tested for their possible involvement in other human diseases. It is quite possible that VPS13 proteins share some functions in the cell and that lack of chorein can be compensated by other VPS13 family members in most tissues. Therefore, only cells where chorein plays a non-redundant role would be vulnerable in ChAc patients.

In spite of the present lack of treatment approaches at the gene or protein level, the patient's everyday needs require systematic studies that must necessarily be based on a network of collaboration. What is the most efficient way to deal with the orofacial dyskinesia in ChAc? Do botulinum toxin injections play a role in the control of feeding dystonia? Is the effect of verapamil on red cell shape and on chorea [5] only anecdotal? Are sulpiride, tetrabenazine and other treatments used in Huntington's and Tourette's patients really helpful in neuroacanthocytosis syndromes? What is the most efficient cardiologic management in McLeod syndrome? Is there a place for heart transplantation? What can be done to delay the onset of MLS in boys suffering from chronic granulomatous disease due to large X-chromosomal deletions?

The Seeon meeting has perhaps provided a point of departure for an international network of collaboration. There was general agreement to install mechanisms for the exchange of information and of materials. Among the initiatives that resulted is the structured collection of clinical observations on patients with proven *VPS13A* mutations (A. Danek) and an attempt to collect the data from the few sleep studies performed in neuroacanthocytosis patients (L. Dolenc-Grošelj). It was also felt that banking of tissues could be of great help in the future: several participants agreed to collect samples centrally for use in upcoming studies (erythrocyte samples: G. Bosman; DNA: A.P. Monaco; muscle: H. Jung; nerve: A. Malandrini; brain specimens: J.G. de Yébenes). The network initiated at Seeon is, of course, open for additional researchers to join.

SHORT TERM GOALS

Little is still known about the tissue distribution of XK, Kell, chorein and the other VPS13 family members. Main foci of interest, of course, are the tissues and organs that are prominently affected by neuroacanthocytosis manifestations, such as skeletal muscle, heart muscle, peripheral nerve and the central nervous system. Of particular interest is the question whether the specific vulnerability of the basal ganglia in MLS and ChAc is mirrored by a similarly restricted expression of XK and chorein. Thus, specific antibodies and basic studies on unaffected control brains are wanted and appear feasible in the near future.

Perhaps because the research need has not yet been properly communicated to primary physicians, neuroacanthocytosis samples are still too few, but they are greatly needed to perform more advanced analyses than have been possible in the past. With respect to transmitter systems, only glutamic acid decarboxylase, choline acetyltransferase, substance P and dopamine metabolites have been measured in the brains of ChAc patients [3,10,12,15]. Additional studies are indispensable, e.g. to identify the cell types that are most vulnerable to the pathogenetic effects of neuroacanthocytosis syndromes, and to define the role, if any, of the endothelin neurotransmitter system.

Brain imaging studies have been performed in many patients world-wide but the rarity of neuroacanthocytosis at any one center might have prevented a complete evaluation of the information contained. To follow standardized neuroimaging protocols, e.g. for PET transmitter imaging or MRI volumetric studies, should prove feasible so that correlations with neuropsychiatric measures can be properly studied. More specific cognitive tasks should better define the phenotype of neuroacanthocytosis syndromes, similar to what has been done in Huntington's disease where e.g. recognition of face emotions and social cognition have been studied recently [27,29].

Sleep and autonomic features [9] as well as hypothalamic and other endocrine abnormalities [31] have been noted in some patients but deserve further investigation as do the diverse autoantibodies that have been described in a few cases [4,14,16,18].

Of great interest is the study of female carriers of McLeod gene mutations and of subjects who are heterozygous for mutations in the *VPS13A* gene. The associated possibility of seemingly autosomal-dominant transmission of chorea-acanthocytosis has not yet been fully explored [1,24] and the families described by Levine and by Critchley should be traced and re-investigated — not only because of historical interest.

Finally, there is a need to better define the natural course of neuroacanthocytosis syndromes and to find suitable markers of progression, such as the motor parameters proposed for Huntington's disease (see Chapter 26), that in future therapeutic studies could be used for monitoring.

LONG TERM APPROACHES

In the long run, two approaches appear promising in the attempt to identify the mechanisms leading from faulty genes to clinical manifestations in neuroacanthocytosis. The first approach uses the acanthocytic red cell as an easily accessible substitute for the whole organism; the second approach is similar but models the disease in laboratory animals.

So far, the basic defect of the acanthocytic membrane in ChAc has not yet been uncovered, in spite of a variety of physico-chemical studies [2,7,23,23,25, 30,32]. Interestingly, chorein is expressed in the red cell membrane, too, and is absent from ChAc erythrocytes [11]. Studies of red cell membranes will unquestionably be useful since erythrocytes probably share the same defect with neurons. Also, the idea that certain drugs might normalize red cell shape as well as decrease the movement disorder in a neuroacanthocytosis syndrome is thrilling, even if the claim was perhaps premature for verapamil [5]. Such an approach, however, needs improved communication between hematology and neuroscience research.

A mouse model for MLS is available, but unfortunately still unpublished. First steps to establish chorein knock-out mice are being taken after the gene homologue on mouse chromosome 19 has been analysed. These mice will represent an invaluable tool to advance our understanding of the molecular and physiological abnormalities triggered by ChAc and will even be more powerful if studied in parallel with HDL2 [22] and MLS mice. Eventually, treatment must be attempted in these disease models.

DIAGNOSIS

Improving the representation of the neuroacanthocytosis syndromes in medical education will increase awareness of these conditions but also calls for better availability of diagnostic means in order to provide the best available care for patients and to improve the research into this group of diseases.

An increasingly wide range of conditions is to be considered in the differential diagnosis of degenerative basal ganglia diseases (for a current tabulation see ref. 8) and full use should be made of the procedures that can be employed before costly molecular diagnostics are called for.

Reproducible elevation of creatine kinase (CK), even if slight, in a patient with a choreatic syndrome is an important diagnostic clue for a neuroacanthocytosis syndrome. In contrast to MLS and ChAc, CK is normal in Huntington's chorea [26] and the earlier explanation of CK elevation on the basis of increased muscle activity [3] is wrong.

Acanthocyte determination is more complex (see Chapter 9) and the deformed cells may escape detection for some time [20,28]. Actually, the paradoxical situation results that non-detection of acanthocytes does not, even after multiple attempts, disprove the assumption of a neuroacanthocytosis syndrome.

More helpful than acanthocyte determination can be immunohematological typing for the McLeod phenotype, but again caution is necessary: a mere "Kell negative" does not address the issue in question. The McLeod phenotype can only be excluded or confirmed if the proper Kell antibodies are employed. Unfortunately, these are not available everywhere. Such antibodies could also help determine the prevalence of the McLeod phenotype in the general population which has not yet been established. Anonymous testing, however, appears necessary to avoid the ethical problem of diagnosing an as yet incurable neurological disorder before onset.

Molecular genetic analysis for MLS is relatively straightforward as there are only three exons and is available in various research laboratories (for addresses see: http://www.geneclinics.org) but analysis of the *VPS13A* gene is a long and cumbersome process due to the large number of exons (see Chapter 19). Recently, the situation has improved since chorein was found missing in ChAc erythrocytes [11]. Absence of the protein from red cells thus might be highly indicative of a diagnosis of ChAc and molecular analysis of the chorein gene in the future may be reserved for questionable cases, for confirmation of the diagnosis and for genotype studies.

Increasing recognition of neuroacanthocytosis cases will further widen the spectrum of these syndromes and will lead to a better appreciation of the role of acanthocytes in mitochondrial disease and in PKAN [13,21]. This development will also help to differentiate among cases diagnosed as Huntington's disease but negative for *IT15*-mutations and will hopefully better define entities such as "FADAEP", familial autosomal-dominant acanthocytosis with exertion-induced paroxysmal dyskinesias [19].

SUMMARY

The first international symposium on neuroacanthocytosis held in Seeon in 2002 has initiated a network of research and initiative that has the potential of bringing about major steps towards a cure of a dreadful type of basal ganglia degeneration. There is also a developing patient interest group, initiated by the Irvine family. Together with their "Advocacy for Neuroacanthocytosis Patients" we are looking forward to the research developments that will be presented at the 2005 follow-up meeting of the Seeon symposium.

REFERENCES

1. Andermann E, Badhwar A, Al Asmi A et al (2002) Chorea-acanthocytosis in a large French-Canadian kindred with dominant or pseudo-dominant inheritance and variable expressivity. *Am J Hum Genet* **71:** 530.
2. Asano K, Osawa Y, Yanagisawa N, Takahashi Y, Oshima M (1985) Erythrocyte membrane abnormalities in patients with amyotrophic chorea with acanthocytosis. Part 2. Abnormal degradation of membrane proteins. *J Neurol Sci* **68:** 161-173.
3. Bird TD, Cederbaum S, Valpey RW, Stahl WL (1978) Familial degeneration of the basal ganglia with acanthocytosis: A clinical, neuropathological, and neurochemical study. *Ann Neurol* **3:** 253-258.
4. Bosman GJCGM, Bartholomeus IGP, De Grip WJ, Horstink MWIM (1994) Erythrocyte anion transporter and antibrain immunoreactivity in chorea-acanthocytosis. A contribution to etiology, genetics, and diagnosis. *Brain Res Bull* **33:** 523-528.
5. Brenes LG, Sanchez MI, Antillon A (1990) Verapamil induces complete remission of the clinical and laboratory findings in a patient with chorea-acanthocytosis. *Clin Res* **38:** 93A.
6. Chandler KE, Kidd A, Al Gazali L et al (2003) Diagnostic criteria, clinical characteristics, and natural history of Cohen syndrome. *J Med Genet* **40:** 233-241.
7. Copeland BR, Todd SA, Furlong CE (1982) High resolution two-dimensional gel electrophoresis of human erythrocyte membrane proteins. *Am J Hum Genet* **34:** 15-31.
8. Danek A, Jung HH, Melone MAB et al (2004) Neuroacanthocytosis: New developments in a neglected group of dementing disorders. *J Neurol Sci*: in press
9. Danek A, Rubio JP, Rampoldi L et al (2001) McLeod neuroacanthocytosis: Genotype and phenotype. *Ann Neurol* **50:** 755-764.
10. de Yébenes JG, Brin MF, Mena MA et al (1988) Neurochemical findings in neuroacanthocytosis. *Mov Dis* **3:** 300-312.
11. Dobson-Stone C, Velayos A, Monaco AP (2003) Molecular studies of chorein, the protein altered in chorea-acanthocytosis. *Am J Hum Genet* **73:** 2214.
12. Galatioto S, Serra S, Batolo D, Marafioti T (1993) Amyotrophic choreo-acanthocytosis: A neuropathological and immunocytochemical study. *Ital J Neurol Sci* **14:** 49-54.
13. Hayflick SJ, Westaway SK, Levinson B et al (2003) Genetic, clinical, and radiographic delineation of Hallervorden-Spatz Syndrome. *N Engl J Med* **348:** 33-40.
14. Hirayama M, Hamano T, Shiratori M et al (1997) Chorea-acanthocytosis with polyclonal antibodies to ganglioside GM1. *J Neurol Sci* **151:** 23-24.
15. Kanazawa I (1981) Regional distribution of markers for neurotransmitters in the basal ganglia of 'choreic' disorders. In: *Katsuki S, Tsubaki T and Toyokura Y (eds) Neurology: Proceedings of the 12th World Congress of Neurology, Kyoto, Japan, September 20-25, 1981,* pp 220-232. Excerpta Medica, Amsterdam.
16. Kay MM, Goodman J, Goodman S, Lawrence C (1990) Membrane protein band 3 alteration associated with neurologic disease and tissue-reactive antibodies. *Exp Clin Immunogenet* **7:** 181-199.
17. Kolehmainen J, Black GCM, Saarinen A et al (2003) Cohen syndrome is caused by mutations in a novel gene, *COH1*, encoding a transmembrane protein with a presumed role in vesicle-mediated sorting and intracellular protein transport. *Am J Hum Genet* **72:** 1359-1369.
18. Lagreze HL, Kornguth SE, Brooks BR, Levine RL (1988) Delayed immune response in chorea-amyotrophy with spherocytosis. *Neurology* **38:** 1642-1643.
19. Lerche H, Storch A, Pekrun A et al (2000) A novel form of autosomal dominant neuroacanthocytosis with exertion-induced paroxysmal dyskinesias. *J Neurol* **247 Suppl.3:** III/18.

20. Malandrini A, Fabrizi GM, Palmeri S et al (1993) Choreo-acanthocytosis like phenotype without acanthocytes: Clinicopathological case report. *Acta Neuropathol (Berl)* **86:** 651-658.
21. Mukoyama M, Kazui H, Sunohara N et al (1986) Mitochondrial myopathy, encephalopathy, lactic acidosis, and stroke-like episodes with acanthocytosis: A clinicopathological study of a unique case. *J Neurol* **233:** 228-232.
22. Nishi M, Hashimoto K, Kuriyama K et al (2002) Motor discoordination in mutant mice lacking junctophilin type 3. *Biochem Biophys Res Commun* **292:** 318-324.
23. Oshima M, Osawa Y, Asano K, Saito T (1985) Erythrocyte membrane abnormalities in patients with amyotrophic chorea with acanthocytosis. Part 1. Spin labeling studies and lipid analyses. *J Neurol Sci* **68:** 147-160.
24. Saiki S, Sakai K, Kitagawa Y et al (2003) Mutation in the *CHAC* gene in a family of autosomal dominant chorea-acanthocytosis. *Neurology* **61:** 1614-1616.
25. Sakai T, Antoku Y, Iwashita H et al (1991) Chorea-acanthocytosis: abnormal composition of covalently bound fatty acids of erythrocyte membrane proteins. *Ann Neurol* **29:** 664-669.
26. Sakai T, Mawatari S, Iwashita H, Goto I, Kuroiwa Y (1981) Choreoacanthocytosis. Clues to clinical diagnosis. *Arch Neurol* **38:** 335-338.
27. Snowden JS, Gibbons ZC, Blackshaw A et al (2003) Social cognition in frontotemporal dementia and Huntington's disease. *Neuropsych* **41:** 688-701.
28. Sorrentino G, De Renzo A, Miniello S, Nori O, Bonvita V (1999) Late appearance of acanthocytes during the course of chorea-acanthocytosis. *J Neurol Sci* **163:** 175-178.
29. Sprengelmeyer R, Young AW, Calder AJ et al (1996) Loss of disgust. Perception of faces and emotions in Huntington's disease. *Brain* **119:** 1647-1665.
30. Terada N, Fujii Y, Ueda H et al (1999) Ultrastructural changes of erythrocyte membrane skeletons in chorea-acanthocytosis and McLeod syndrome revealed by the quick-freezing and deep-etching method. *Acta Haematol* **101:** 25-31.
31. Terao S, Sobue G, Takahashi M et al (1995) Disturbances of hypothalamic-pituitary hormone secretion in familial chorea-acanthocytosis. *No To Shinkei* **47:** 57-61.
32. Ueno E, Oguchi K, Yanagisawa N (1982) Morphological abnormalities of erythrocyte membrane in the hereditary neurological disease with chorea, areflexia and acanthocytosis. A study with freeze fracture electron microscopy. *J Neurol Sci* **56:** 89-97.
33. Velayos A, Vettori A, Dobson-Stone C, Monaco AP (2003) Identification and analysis of the new, *Vps13*-like gene (*V13LC*) family that includes *CHAC*, the gene altered in chorea-acanthocytosis. *Am J Hum Genet* **73:** 1026.

INDIVIDUAL SPONSORS

The Advocacy for Neuroacanthocytosis Patients received large and small contributions from the families and friends of patients specifically for the publication and distribution of this first text on the disease.

Grateful thanks for generous support go to:

Carol Allen
Dr. L.K. Altman
Mary Ardant
Peter J. Ball
Rita and Martin Bennett
Robert M. and Vida Bleiweiss
Marcy Leavitt Bourne
Alisandra Lanto and James Bowers
Geoffrey Carter
Sue and Chuck Chaney
Ruby G. Clark
Sally Cutler
Robin Davies and Chowee Leow
Robert and Arleen Daugherty
Steven and Joyce Davis
Roger and Rita Day
David and Gillian Dillistone
Nancy Duncan
Anita and Stephen Esslinger
Roseanna and Gerry Farinella
Rev. Penny Fleming
Janet Grant
Paula and Schuyler Henderson
Joan and David Hill
Paula McLean Holcomb
John and Becky Irvine
Peter and Emma Letley
Rex and Karla Masten
John and Justine McGill
Helen and Hamilton Meserve
Etsuko and John Morris
Monty and Betsy Morris
Michael and Helga Mueller
Lynn and David Murby
Kimberly and Scribner Ochsenschlager
Joan and Robert Ogden
W.A. Page
Gill and Gordon Parry
Pamela and Jean-Marc Perraud
Marcel Perron
Axel and Inge Petersen
Dr. Betsy and Richard Rand
Friederich Reichert
Alistair and Annalisa Rellie
Cathy Robertson
Tony and Jan Robinson
Hans and Monika Rosenkranz
Patricia and Robert Ross
Dr. Richard and Lucy Sallick
Linda Senat
Joan and William Sheffield
Mark and Deirdre Simpson
Karen and Sam Smith
Dr. James and Ina Spicer
Chris and Hazel Stoddart
J. Foster and Marion Swetenham
Dr. Heino and Uschi Thiele
Jill and Michael Todd
Penny and Tom Wacker
Sir Peter and Meryl Walters
Nicholas and Veronica Ward
Madelyn and Philippe Willems
Christopher R. White

ABBREVIATIONS

A	alanine
aa	amino acids
AAV	adeno-associated viruses
ABC	avidin-biotin complex
ABL	abetalipoproteinemia
AD	autosomal-dominant
ADP	adenosine diphosphate
AE	anion exchanger
AED	antiepileptic drug
AF	atrial fibrillation
Ala	alanine
AMP	adenosine monophosphate
AMPK	cyclic AMP dependent kinase
ANA	antinuclear antibodies
APOB	apolipoprotein B gene
apoB	apolipoprotein B
AR	autosomal-recessive
Arg	arginine
ARVC	arrhythmogenic right ventricular cardiomyopathy
Asn	asparagine
Asp	aspartic acid
ATP	adenosine triphosphate
ATPase	adenosine triphosphatase
AZT	azidothymidine, zidovudine (Retrovir ©)
B	lip/tongue biting
BAC	bacterial artificial chromosome
big ET	big endothelin
B_{max}	maximal binding capacity
C	cysteine
C. elegans	*Caenorhabditis elegans*
C-A	chorea-acanthocytosis
CAG expansion	cytosine, adenine, and guanine nucleotide triplet repeats that are abnormally long and cause expression of expanded polyglutamine protein stretches (poly-Q)

CBP	CREB binding protein CREB: cyclic AMP-response element binding protein
CBZ	carbamazepine
CC	cardiac catheterization
cDNA	complementary DNA
ced	cell death abnormal
CHAC	chorea-acanthocytosis gene, now renamed *VPS13A*, coding for chorein
ChAc	chorea-acanthocytosis
CI	calcium ionophore
CK	creatine kinase
CNS	central nervous system
CoA	coenzyme A
COX	cytochrome oxidase
Cr	creatine
CRP	C-reactive protein
CSF	cerebrospinal fluid
Cys	cysteine
D	aspartic acid
D	dystonia
DAB	diaminobenzidine
DBS	deep brain stimulation
DCM	dilated cardiomyopathy
DNA	deoxyribonucleic acid
Dp	dysphonia
DRPLA	dentato-rubro-pallido-Luysian atrophy
Dt	dysarthria
E	glutamic acid
ECE	endothelin converting enzyme
ECG	electrocardiogram
Echo	echocardiogram
EDTA	ethylen-diamine-tetracetic acid
EEG	electroencephalography
ef	executive functions
EM	electron microscopy
EMG	electromyography
EOG	electro-oculography
EST	expressed sequence tag
ET	endothelin
F	phenylalanine
FADAEP	familial autosomal-dominant acanthocytosis with exertion-induced paroxysmal dyskinesias
FALS/FTD	familial amyotrophic lateral sclerosis with frontotemporal dementia

FBM	felbamate
FDC	familial dilated cardiomyopathy
FDG	fluorodeoxyglucose
fef	figural executive functions
FHBL	familial hypobetalipoproteinemia
FHC	familial hypertrophic cardiomyopathy
fm	figural memory
FSHMD	facioscapulohumeral muscular dystrophy
G	glycine
GABA	gamma amino butyric acid
GBP	gabapentin
GFAP	glial fibrillary acidic protein
Gln	glutamine
Glu	glutamic acid
Gly	glycine
GPi	internal globus pallidus
GTP	guanosine 5'-triphosphate
GVB	vigabatrin
H	histidine
HARP	hypoprebetalipoproteinemia, acanthocytosis, retinitis pigmentosa, and pallidal degeneration
HB	heart block
HD	Huntington's disease
HDF	Hereditary Disease Foundation
hdh	mouse HD gene homologue
HDL	Huntington's disease-like
His	histidine
HIV	human immunodeficiency virus
HMSN	hereditary motor and sensory neuropathy
HSS	Hallervorden-Spatz syndrome
HSV	herpes simplex virus
htt	huntingtin
I	isoleucine
ICL	intercommissural line
IgG	immunoglobulin G
IHA	International Huntington's Association
Ile	isoleucine
IQ	intelligence quotient
IT15	interesting transcript 15 (HD gene)
JPH3	junctophilin 3 gene
K	lysine
K_d	equilibrium dissociation constant (its reciprocal: a measure of affinity)
kDa	kilodalton

KEL	gene coding for the Kell protein
Kx	X-linked antigen of the Kell system (main antigen on the McLeod syndrome protein, XK)
L	leucine
la	language
LAFB	left anterior fasicular block
LAT	latency-associated transcripts
LBBB	left bundle branch block
LC	limb chorea
LD	cationic DEAE(diethylaminoethyl)-dextrane lipoplexes
LDH	lactate dehydrogenase
LDL	low density lipoprotein
Leu	leucine
LMD	liposome-μ(mu)-DNA
LTG	lamotrigine
LV	left ventricular
LVD	left ventricular dilatation
LVH	left ventricular hypertrophy
Lys	lysine
M	methionine
MCP	midcommissural point
MCTG	medium-chain triglycerides
md	movement disorder
MELAS	mitochondrial encephalopathy, lactic acidosis, and stroke-like episodes
mem	memory
MERRF	myoclonic epilepsy with ragged-red fibres
Met	methionine
MH	malignant hyperthermia
MIM	Mendelian Inheritance in Man (http://www.ncbi.nlm.nih.gov/entrez/query.fcgi?db=OMIM)
MLS	McLeod syndrome
MMSE	mini mental status examination
MRI	magnetic resonance imaging
mRNA	messenger RNA
mtDNA	mitochondrial DNA
MTP	microsomal triglyceride transfer protein
Mw	molecular weight
N	asparagine
NA	neuroacanthocytosis
n/a	not available
NADH	reduced form of nicotinamide adenine dinucleotide
NADH-TR	nicotinamide adenine dinucleotide (reduced) — tetrazolium reductase

NADPH	reduced form of NADP: nicotinamide adenine dinucleotide phosphate
NART	National Adult Reading Test
NBIA	Neurodegeneration with Brain Iron Accumulation
NCV	nerve conduction velocity
ND	not determined
NII	neuronal intranuclear inclusions
NMD	nonsense-mediated mRNA decay
NMDA	N-methyl-D-asparate
NO	nitric oxide
np	not performed
OCBZ	oxcarbazepine
OFD	orofacial dyskinesias
P	proline
PANK	pantothenate kinase
PAS	periodic acid Schiff
PB	phenobarbital
pc	personal communication
PCr	phosphocreatine
PDI	protein disulfide isomerase
PET	positron emission tomography
Phe	phenylalanine
PHT	phenytoin
PIQ	performance IQ
PKAN	pantothenate kinase associated neurodegeneration
poly-Q	polyglutamine
Pro	proline
PSG	polysomnographic recording
PTC	premature termination codon
PVC	prevacuolar compartment
Q	glutamine
R	arginine
RA	retinoic acid
RBBB	right bundle branch block
RBC	red blood cell
RCM	restrictive cardiomyopathy
REM	rapid eye movement
RPR	rapid plasma reagent
RT-PCR	reverse transcription-polymerase chain reaction
S	serine
SCA	spinocerebellar ataxia
SD	standard deviation
SD	sudden death
SDS-PAGE	sodium dodecyl sulphate polyacrylamide gel

	electrophoresis
Ser	serine
SMA	spinobulbar muscular atrophy
SWS	slow wave sleep
T	threonine
TEM	transmission electron microscopy
TeTx	tetanus toxin
TGase	transglutaminase
TGB	tiagabine
TGN	trans-Golgi network
Thr	threonine
TPM	topiramate
TPR	tetratrico peptide repeat
Trp	tryptophan
TST	total sleep time
TTc	tetanus toxin fragment c
tTGase	tissue transglutaminase
TUNEL	TdT-mediated dUTP nick-end labelling TdT: terminal deoxynucleotidyltransferase dUTP: 2'-deoxyuridine 5'-triphosphate
Tyr	tyrosine
U/l	enzymatic units per liter
UHDRS	Unified Huntington's Disease Rating Scale
UK	United Kingdom
V, Val	valine
vef	verbal executive functions
VIQ	verbal IQ
VLDL	very low density lipoprotein
vm	verbal memory
VOA	nucleus ventrooralis anterior thalami (anterior ventrooral nucleus of the thalamus)
vp	visuo-perceptual functions
VPA	valproate
VPS13	gene family, named after yeast protein Vps13p
W	tryptophan
WAIS	Wechsler Adult Intelligence Scale
WAIS-III	Wechsler Adult Intelligence Scale, 3rd revision
WCST	Wisconsin Card Sorting Test
WFN	World Federation of Neurology
X	position of an amino acid that was changed to a stop codon
XK	McLeod syndrome protein
XK	McLeod syndrome gene
Y	tyrosine

INDEX

INDEX OF CONTRIBUTORS